Progress in Mathematics
Volume 203

David E. Blair

Riemannian Geometry of Contact and Symplectic Manifolds

Birkhäuser
Boston • Basel • Berlin

David E. Blair
Department of Mathematics
Michigan State University
East Lansing, MI 48824-1027
U.S.A.

Library of Congress Cataloging-in-Publication Data

Blair, David E., 1940
   Riemannian geometry of contact and symplectic manifolds / David E. Blair.
     p. cm. – (Progress in mathematics (Boston, Mass.))
   Includes bibliographical references and index.
   ISBN 0-8176-4261-7 (alk. paper) – ISBN 3-7643-4261-7 (alk. paper)
    1. Contact manifolds. 2. Symplectic manifolds. 3. Geometry, Riemannian. I. Title. II.
Series.

   QA614.3.B53 2001
   516.3'73–dc21
                                    2001052707
                                        CIP

AMS Subject Classifications: 53B35, 53C15, 53C25, 53C26, 53C40, 53C42, 53C55, 53C56, 53D05, 53D10, 53D12, 53D15, 53D22, 53D25, 53D35, 58E11

Printed on acid-free paper
©2002 Birkhäuser Boston   *Birkhäuser*

ISBN 0-8176-4261-7   SPIN 10845070
ISBN 3-7643-4261-7

Reformatted from author's files by TEXniques, Inc., Cambridge, MA
Printed and bound by Hamilton Printing Company, Rensselaer, NY
Printed in the United States of America

9 8 7 6 5 4 3 2 1

To Rebecca
in appreciation of all her
love and support

Contents

Preface

The author's lectures, "Contact Manifolds in Riemannian Geometry," volume 509 (1976), in the Springer-Verlag *Lecture Notes in Mathematics* series have been out of print for some time and it seems appropriate that an expanded version of this material should become available. The present text deals with the Riemannian geometry of both symplectic and contact manifolds, although the book is more contact than symplectic. This work is based on the recent research of the author, his students, colleagues, and other scholars, the author's graduate courses at Michigan State University and the earlier lecture notes.

Chapter 1 presents the general theory of symplectic manifolds. Principal circle bundles are then discussed in Chapter 2 as a prelude to the Boothby–Wang fibration of a compact regular contact manifold in Chapter 3, which deals with the general theory of contact manifolds. Chapter 4 focuses on Riemannian metrics associated to symplectic and contact structures. Chapter 5 is devoted to integral submanifolds of the contact subbundle. In Chapter 6 we discuss the normality of almost contact structures, Sasakian manifolds, K-contact manifolds, the relation of contact metric structures and CR-structures, and cosymplectic structures. Chapter 7 deals with the important study of the curvature of a contact metric manifold. In Chapter 8 we give a selection of results on submanifolds of Kähler and Sasakian manifolds, including an illustration of the technique of A. Ros in a theorem of F. Urbano on compact minimal Lagrangian submanifolds in $\mathbb{C}P^n$. Chapter 9 discusses the symplectic structure of tangent bundles, contact structure of tangent sphere bundles, general vector bundles and normal bundles of Lagrangian and integrals submanifolds giving rise to new examples of symplectic and contact manifolds.

In Chapter 10 we study a number of curvature functionals on spaces of associated metrics and their critical point conditions; we show also that in the symplectic case, the "total scalar curvature" is a symplectic invariant and in the contact case is a natural functional whose critical points are the metrics for which the characteristic vector field generates isometries. In the presence of a certain amount of negative curvature, special directions appear in the contact subbundle; we discuss these and their relations to Anosov and conformally Anosov flows in Chapter 11. Chapter 12 deals with the subject of complex contact manifolds. We conclude with a brief treatment of 3-Sasakian manifolds in Chapter 13.

The text attempts to strike a balance between giving detailed proofs of basic properties, which will be instructive to the reader, and stating many results whose proofs would take us too far afield. It has been impossible, however, to be encyclopedic and include everything, so that unfortunately some important topics have been omitted or covered only briefly. An extensive bibliography is given.

It is the author's hope that the reader will find this both a good introduction to the Riemannian geometry of contact and symplectic manifolds and a useful reference to recent research in the area.

The author expresses his appreciation to C. Baikoussis, B.-Y. Chen, D. Chinea, T. Draghici, B. Foreman, Th. Koufogiorgos, Y.-H. P. Pang and D. Perrone for reading parts of the manuscript and offering valuable suggestions. The author also expresses his appreciation to Ann Kostant of Birkhäuser and to Elizabeth Loew of TeXniques for her kind assistance in the production of this book.

David E. Blair   October, 2001

1
Symplectic Manifolds

1.1 Definitions and examples

To set the stage for our development, we begin this book with a treatment of the basic features of symplectic geometry. By a *symplectic manifold* we mean an even-dimensional differentiable (C^∞) manifold M^{2n} together with a global 2-form Ω which is closed and of maximal rank, i.e., $d\Omega = 0$, $\Omega^n \neq 0$. By a *symplectomorphism* $f : (M_1, \Omega_1) \longrightarrow (M_2, \Omega_2)$ we mean a diffeomorphism $f : M_1 \longrightarrow M_2$ such that $f^*\Omega_2 = \Omega_1$.

Before continuing with symplectic manifolds, we present some basic linear algebra. On a vector space V^{2n}, if $\Omega \in \bigwedge^2 V$ with $\mathrm{rk}\Omega = 2n$, then there exist $\theta^1, ..., \theta^{2n} \in V^*$, linearly independent and such that

$$\Omega = \theta^1 \wedge \theta^2 + \theta^3 \wedge \theta^4 + \cdots + \theta^{2n-1} \wedge \theta^{2n}.$$

To see this, for a basis $\{\omega^i\}$ of V^* write

$$\Omega = \sum_{i<j} a_{ij}\omega^i \wedge \omega^j = \omega^1 \wedge \sum_{1<j} a_{1j}\omega^j + \omega^2 \wedge \sum_{2<j} a_{2j}\omega^j + \text{terms in } \omega^3, ..., \omega^{2n}$$

$$= \omega^1 \wedge \beta^1 + \omega^2 \wedge \beta^2 + \text{terms in } \omega^3, ..., \omega^{2n}$$

where β^2 involves only $\omega^3, ..., \omega^{2n}$ and $\beta^1 = a\omega^2 + \beta^3$ and β^3 involves only $\omega^3, ..., \omega^{2n}$. Therefore

$$\Omega = \omega^1 \wedge \beta^1 + \frac{1}{a}\beta^1 \wedge \beta^2 - \frac{1}{a}\beta^3 \wedge \beta^2 + \text{terms in } \omega^3, ..., \omega^{2n}$$

$$= (\omega^1 - \frac{1}{a}\beta^2) \wedge \beta^1 + \text{terms in } \omega^3, ..., \omega^{2n}$$

which is of the form $\theta^1 \wedge \theta^2 + \Omega_1$ where Ω_1 involves only $\omega^3, ..., \omega^{2n}$. Now repeat the process for Ω_1.

We shall often choose the labeling such that

$$\Omega = \theta^1 \wedge \theta^{n+1} + \cdots + \theta^n \wedge \theta^{2n}.$$

As a corollary we see that there exists a basis $\{e_i, e_{n+i}\}$ of V^{2n} such that $\Omega(e_i, e_{n+j}) = \delta_{ij}$, $i, j = 1, ..., n$. A change of basis which leaves invariant the normal form $\Omega = \sum_{i=1}^{n} \theta^i \wedge \theta^{n+i}$ is given by a symplectic matrix, i.e.,

$$\begin{pmatrix} A & B \\ C & D \end{pmatrix}^{-1} = \begin{pmatrix} D & -C \\ -B & A \end{pmatrix}^T \quad \text{if and only if}$$

$$\begin{pmatrix} A & B \\ C & D \end{pmatrix}^T \begin{pmatrix} 0 & I \\ -I & 0 \end{pmatrix} \begin{pmatrix} A & B \\ C & D \end{pmatrix} = \begin{pmatrix} 0 & I \\ -I & 0 \end{pmatrix}.$$

In particular the structural group of the tangent bundle of a symplectic manifold is reducible to $Sp(2n, \mathbb{R})$. Further, using the fact that M^{2n} may be given a Riemannian metric and is orientable, the structural group is reducible to $SO(2n)$ and hence in turn to $U(n)$. Thus in particular M^{2n} carries an almost complex structure; this will be discussed in greater detail below and in Chapter 4. The name *symplectic* is due to H. Weyl [1939, p. 165] changing the Latin *com/plex* to the Greek *sym/plectic*.

Two canonical examples of symplectic manifolds are the following:

1. $\mathbb{R}^{2n}$ with coordinates $(x^1, ..., x^n, y^1, ..., y^n)$ admits the symplectic form $\Omega = \sum dx^i \wedge dy^i$. The classical theorem of Darboux states that on any symplectic manifold there exist local coordinates with respect to which the symplectic form can be written in this way. We will give a modern proof of this result in Section 1.3.

2. Let M be a differentiable manifold; then its cotangent bundle T^*M has a natural symplectic structure. For $z \in T^*M$ and $V \in T_z T^*M$ define a 1-form β, often called the *Liouville form*, by $\beta(V)_z = z(\pi_* V)$ where $\pi : T^*M \longrightarrow M$ is the projection map. If $x^1, ..., x^n$ are local coordinates on M, $q^i = x^i \circ \pi$ and fibre coordinates $p^1, .., p^n$ form local coordinates on T^*M; β then has the local expression $\sum_{i=1}^{n} p^i dq^i$. The natural symplectic structure on T^*M is given by $\Omega = -d\beta$.

The reader may recognize the second example from classical mechanics; indeed the cotangent bundle of the configuration space may be thought of as the phase space of a dynamical system and we may obtain Hamilton's equations of motion as follows. Let H be a real-valued function on a symplectic manifold (M, Ω) and define a vector field X_H by $\Omega(X_H, Y) = YH$. X_H is called the *Hamiltonian vector field* generated by H. Two basic properties of X_H are $\mathcal{L}_{X_H}\Omega = 0$, $\mathcal{L}$ being Lie differentiation, and $X_H H = 0$. In fact the classical Poisson bracket is $\{f_1, f_2\} = \Omega(X_{f_2}, X_{f_1}) = X_{f_1} f_2$. In local coordinates $(q^1, ..., q^n, p^1, ..., p^n)$ given by the Darboux Theorem, $\Omega = \sum dq^i \wedge dp^i$ and $X_H = \sum \left(-\frac{\partial H}{\partial q^i}\frac{\partial}{\partial p^i} + \frac{\partial H}{\partial p^i}\frac{\partial}{\partial q^i} \right)$. Thus the differential equations for the integral curves of X_H are

$$\dot{p}^i = -\frac{\partial H}{\partial q^i}, \quad \dot{q}^i = \frac{\partial H}{\partial p^i},$$

Hamilton's equations of motion.

Before giving further examples, we mention the relationship with Riemannian geometry which will become central for our study, viz. "associated metrics". Given a symplectic manifold (M, Ω), there exist a Riemannian metric g and an almost complex structure J such that $\Omega(X, Y) = g(X, JY)$. In fact we shall see in Chapter 4 that there are many such metrics, g and J being created simultaneously by polarization.

On the other hand, given an almost complex structure J, i.e., a tensor field J of type (1,1) such that $J^2 = -I$, on an almost complex manifold, a Riemannian metric is said to be *Hermitian* if $g(JX, JY) = g(X, Y)$ and the pair (J, g) is called an *almost Hermitian structure*. Now defining a 2-form Ω by $\Omega(X, Y) = g(X, JY)$, Ω is called the *fundamental 2-form* of the almost Hermitian structure. If $d\Omega = 0$, the almost Hermitian structure is said to be *almost Kähler*, whereas the structure is *Kähler* if J or equivalently Ω is parallel with respect to the Levi-Città connection of g. Thus associated metrics can be thought of as almost Kähler structures whose fundamental 2-form is the given symplectic form.

Since $d\Omega = 0$, $[\Omega] \in H^2(M, \mathbb{R})$, $[\Omega]$ denoting the deRham cohomology class determined by Ω. Using an associated metric, $\delta\Omega = 0$ and hence Ω is harmonic. To see this, use the fact that an almost Kähler manifold is *quasi-Kähler*, i.e., $(\nabla_k J_{ip})J_j{}^p = (\nabla_p J_{ij})J_k{}^p$, sum on the indices i and k and use $J^2 = -I$. Also $[\Omega]^n = [\Omega^n] \in H^{2n}(M, \mathbb{R})$. In particular for M compact the following are two necessary conditions for the existence of a symplectic structure.

i) M carries an almost complex structure.

ii) There exists an element $w \in H^2(M, \mathbb{R})$ such that $w^n \neq 0$.

Thus, for example, from i) S^4 is not symplectic and from ii) S^6 is not symplectic.

Gromov in his thesis proved that if M is an open manifold, then i) implies the existence of a 1-form ω such that $d\omega$ is symplectic (see A. Haefliger [1971, p. 133]). Also Kähler manifolds are symplectic so there are plenty of compact ones, e.g., complex projective space, $S^2 \times S^2$, algebraic varieties, etc. Note that the even-dimensional Betti numbers of a compact almost Kähler manifold are non-zero. It is also well known that the odd-dimensional Betti numbers of a compact Kähler manifold are even, but this is not true in the almost Kähler case. We now give two descriptions of an example of Thurston [1976] of a compact symplectic manifold with no Kähler structure; this manifold is known as the *Thurston manifold* or as the *Kodaira–Thurston manifold* (Kodaira [1964]).

Briefly, first take the product of a torus T^2, as a unit square with opposite sides identified, and an interval and glue the ends together by the diffeomorphism of T^2 given by the matrix $\begin{pmatrix} 1 & 1 \\ 0 & 1 \end{pmatrix}$. This gives a compact 3-manifold whose first Betti number is 2. Now taking the product with S^1 we have a 4-manifold M with first Betti number 3 and hence M cannot have a Kähler structure. Let θ_1, θ_2 be coordinates on T^2; then $d\theta_1 \wedge d\theta_2$ exists after the twisting on the 3-manifold. Thus if ϕ_1 is the coordinate on the interval and ϕ_2 the coordinate on the final circle, $d\theta_1 \wedge d\theta_2 + d\phi_1 \wedge d\phi_2$ is a symplectic form.

A second version of this example was given by E. Abbena [1984] who also gave a natural associated metric for this symplectic structure, computed its curvature and showed by using harmonic forms that the first Betti number was 3.

Let G be the closed connected subgroup of $GL(4, \mathbb{C})$ defined by

$$\left\{ \begin{pmatrix} 1 & a_{12} & a_{13} & 0 \\ 0 & 1 & a_{23} & 0 \\ 0 & 0 & 1 & 0 \\ 0 & 0 & 0 & e^{2\pi i a} \end{pmatrix} \,\middle|\, a_{12}, a_{13}, a_{23}, a \in \mathbb{R} \right\},$$

i.e., G is the product of the Heisenberg group and S^1. Let Γ be the discrete subgroup of G with integer entries and $M = G/\Gamma$. Denote by x, y, z, t coordinates on G, say for $A \in G$, $x(A) = a_{12}$, $y(A) = a_{23}$, $z(A) = a_{13}$, $t(A) = a$. If L_B is left translation by $B \in G$, $L_B^* dx = dx$, $L_B^* dy = dy$, $L_B^*(dz - xdy) = dz - xdy$, $L_B^* dt = dt$. In particular these forms are invariant under the action of Γ; let $\pi : G \longrightarrow M$, then there exist 1-forms $\alpha_1, \alpha_2, \alpha_3, \alpha_4$ on M such that $dx = \pi^*\alpha_1$, $dy = \pi^*\alpha_2$, $dz - xdy = \pi^*\alpha_3$, $dt = \pi^*\alpha_4$. Setting $\Omega = \alpha_4 \wedge \alpha_1 + \alpha_2 \wedge \alpha_3$ we see that $\Omega \wedge \Omega \neq 0$ and $d\Omega = 0$ on M giving M a symplectic structure.

The vector fields

$$\mathbf{e}_1 = \frac{\partial}{\partial x}, \quad \mathbf{e}_2 = \frac{\partial}{\partial y} + x\frac{\partial}{\partial z}, \quad \mathbf{e}_3 = \frac{\partial}{\partial z}, \quad \mathbf{e}_4 = \frac{\partial}{\partial t}$$

are dual to dx, dy, $dz - xdy$, dt and are left invariant. Moreover $\{\mathbf{e}_i\}$ is orthonormal with respect to the left invariant metric on G given by

$$ds^2 = dx^2 + dy^2 + (dz - xdy)^2 + dt^2.$$

On M the corresponding metric is $g = \sum \alpha_i \otimes \alpha_i$ and is called the *Abbena metric*.

Moreover M carries an almost complex structure defined by

$$J\mathbf{e}_1 = \mathbf{e}_4, \ \ J\mathbf{e}_2 = -\mathbf{e}_3, \ \ J\mathbf{e}_3 = \mathbf{e}_2, \ \ J\mathbf{e}_4 = -\mathbf{e}_1.$$

Then noting that $\Omega(X, Y) = g(X, JY)$, we see that g is an associated metric.

At the time of Thurston's example there was no known example of a compact symplectic manifold with no Kähler structure. Since the appearance of this example there have been many others, e.g., Cordero, Fernández and Gray [1986], Cordero, Fernández and de Leon [1985], Gompf [1994], Holubowicz and Mozgawa [1998], Jelonek [1996], McCarthy and Wolfson [1994], McDuff [1984], Watson [1983].

1.2   Lagrangian submanifolds

Let $\iota : L \longrightarrow M^{2n}$ be an immersion into a symplectic manifold (M^{2n}, Ω). We say that L is a *Lagrangian submanifold* if the dimension of L is n and $\iota^*\Omega = 0$. Two simple examples are the following.

1. The fibres of the cotangent bundle T^*M as discussed in the previous section are Lagrangian submanifolds with respect to the symplectic structure $d\beta$. Also suppose that ϕ is a section of T^*M. Then $\phi^*\beta = \phi$ as a 1-form on M; in particular

   $$(\phi^*\beta)(X)_m = \beta(\phi_*X)_{\phi(m)} = \phi(m)(\pi_*\phi_*X) = \phi(X)_m.$$

   Therefore $\phi^*\Omega = \phi^*(-d\beta) = -d\phi$ so that a section $\phi : M \longrightarrow T^*M$ is a Lagrangian submanifold if and only if ϕ is closed. When ϕ is exact, say dS, S is said to be a *generating function* for the submanifold.

2. Let (M_1, Ω_1) and (M_2, Ω_2) be symplectic manifolds and $f : M_1 \longrightarrow M_2$ a diffeomorphism. Then $(M_1, \Omega_1) \times (M_2, -\Omega_2)$ is symplectic, say $M = M_1 \times M_2$ with projections π_1 and π_2 and $\Omega = \pi_1^*\Omega_1 - \pi_2^*\Omega_2$. Let Γ_f denote the graph of f. Then f is a symplectomorphism if and only if Γ_f is a Lagrangian submanifold of (M, Ω).

There are several difficulties in studying Lagrangian submanifolds. First of all, since $\iota^*\Omega = 0$, there is no induced structure, so in a sense the geometry is transverse to the submanifold. From the standpoint of submanifold theory the codimension is high so that theory is more complicated. Lagrangian submanifolds are very abundant. For example, given any vector X at a point $m \in M$, there exists a Lagrangian submanifold through the point tangent to X. We shall see this as a corollary to the Darboux Theorem in the next section. Also Lagrangian submanifolds tend to get in the way of each other; loosely speaking two Lagrangian submanifolds that are C^1 close tend to intersect more than one would expect of two arbitrary C^1 close submanifolds. Going into this point in some detail, let M be a compact manifold and, identifying M with the zero section of T^*M, view M as a Lagrangian submanifold of T^*M. Now let L be a Lagrangian submanifold of T^*M near M. Regarding L as the image of a closed 1-form ϕ, the question of when L and M intersect reduces to the question of when ϕ has a zero. So to perturb M to a disjoint Lagrangian submanifold L, M must admit a closed 1-form without zeros. An obstruction to this was given by Tischler [1970] in the following Theorem.

Theorem 1.1 *If a compact manifold admits a closed 1-form without zeros, then the manifold fibres over the circle and conversely.*

In contrast the problem of perturbing M to an arbitrary disjoint submanifold is equivalent to finding a non-vanishing 1-form which is equivalent to finding a non-vanishing vector field and the obstruction to this is the Euler characteristic.

If M is simply connected, the situation is "worse". For now ϕ is exact and hence given by a function on M, but since M is compact, such a function must have at least two critical points. So a perturbation of S^2 to an arbitrary submanifold in T^*S^2 may intersect in only one point, but a perturbation to a Lagrangian submanifold must have at least two intersection points.

For L_1 and L_2, C^1 close Lagrangian submanifolds of a symplectic manifold (M, Ω), the same situation holds by virtue of the following theorem of Weinstein [1971] which we will prove in the next section.

Theorem (Weinstein) *If L is a Lagrangian submanifold of a symplectic manifold (M, Ω), then there exists a neighborhood of L in M that is symplectomorphic to a neighborhood of the zero section in T^*L.*

More general than the notion of a Lagrangian submanifold are the notions of isotropic and coisotropic submanifolds. A submanifold $\iota : N \longrightarrow M^{2n}$ is *isotropic* if $\iota^*\Omega = 0$, so in particular the dimension of N is $\leq n$. In the subject of almost Hermitian manifolds (M^{2n}, J, g) these submanifolds are called *totally real* submanifolds, Yau [1974], Chen and Ogiue [1974a]. The key point is that since $\Omega(X, Y) = g(X, JY)$, J maps the tangent space into the normal space. We remark that one sometimes sees another notion of totally real submanifold in the literature, namely a submanifold for which no tangent space contains a non-zero complex subspace; however we will use only the stronger notion in this text.

The isotropic or totally real condition at a point $m \in N$ can be written as $\iota_* T_m N \subseteq \{V | \Omega(V, \iota_* T_m N) = 0\}$. A submanifold $\iota : N \longrightarrow M^{2n}$ is *coisotropic* if $\iota_* T_m N \supseteq \{V | \Omega(V, \iota_* T_m N) = 0\}$; in terms of (J, g), J maps the normal space into the tangent space, so the dimension of N is $\geq n$.

In particular for a Lagrangian submanifold N^n in $\mathbb{C}^n$, J maps the tangent spaces onto the normal spaces; therefore $T\mathbb{C}^n|_N = TN^n \oplus iTN^n = TN \otimes \mathbb{C}$ and hence the complexified tangent bundle is trivial. Gromov [1971] (see also Weinstein [1977]) proved that if N^n is compact, then N^n admits a Lagrangian immersion into $\mathbb{C}^n$ if and only if the complexified tangent bundle is trivial.

The question of imbeddings is a different matter. It is known that the sphere S^n cannot be imbedded in $\mathbb{C}^n$ as a Lagrangian submanifold. This is a consequence of a more general result of Gromov [1985] that a compact imbedded Lagrangian submanifold in $\mathbb{C}^n$ cannot be simply connected (see also Sikorav [1986]). For an immersed sphere as a Lagrangian submanifold with only one double point, see Example 5.3.3, Weinstein [1977, p. 26] or Morvan [1983].

Our discussion also has the following application to the problem of fixed points of symplectomorphisms, Weinstein [1977, p. 29]. Let (M, Ω) be a compact simply connected symplectic manifold. Then a symplectomorphsim f sufficiently C^1 close to the identity has at least two fixed points. To see this let Δ be the diagonal of $(M, \Omega) \times (M, -\Omega)$ and Γ_f the graph of f. Then Δ and Γ_f intersect at least twice and hence f has at least two fixed points. For example let M be complex projective space $\mathbb{C}P^n$ and f an automorphism of the Kähler structure that is C^1 close to the identity. Any function on $\mathbb{C}P^n$ has at least $n + 1$ critical points, so that f must have at least $n + 1$ fixed points.

1.3   The Darboux–Weinstein theorems

We have mentioned the theorems of Darboux and Weinstein already; in this section we present a modern proof of both theorems using a technique of Moser [1965]. The use of this idea to prove the classical Darboux Theorem is due to Weinstein [1971, 1977] and independently to J. Martinet [1970]. In addition to the papers mentioned, a general reference is the book by P. Libermann and C.-M. Marle [1987, Chp. III, Section 15] which we follow here; for the Darboux Theorem see also N. Woodhouse [1980, pp. 7–9]. We begin with the following theorem of Weinstein. As a matter of notation, for a submanifold $\iota : N \longrightarrow M$ and a differential form Φ on M, $\Phi|_N$ denotes the form acting on $T_N M$, the restriction of TM to N, and not the pullback, $\iota^* \Phi$, of Φ to M (see e.g., Libermann and Marle [1987, p. 360]).

Theorem 1.2 *Let Ω_0 and Ω_1 be symplectic forms on a symplectic manifold M and N a submanifold (possibly a point) on which $\Omega_0|_N = \Omega_1|_N$. Then there* *exist tubular neighborhoods U and V and a symplectomorphism $\rho : U \longrightarrow V$ such that $\rho|_N$ is the identity.*

Proof. Since $d(\Omega_1 - \Omega_0) = 0$, by the Generalized Poincaré Lemma (see e.g., Libermann and Marle [1987, p. 361]) there exists a tubular neighborhood W of N and a 1-form α on W such that $\Omega_1 - \Omega_0 = d\alpha$ and $\alpha|_N = 0$. Now for $t \in \mathbb{R}$ set $\Omega_t = \Omega_0 + t(\Omega_1 - \Omega_0)$. Ω_t is non-degenerate on an open subset W_1 of $W \times \mathbb{R}$ containing $N \times \mathbb{R}$. Let X be the vector field on W_1 defined by

$$X(m,t)\lrcorner\,\Omega_t(m) = -\alpha(m)$$

where $\lrcorner$ denotes the left interior product. For a point m, consider the integral curve $t \longrightarrow \phi_m(t)$ of X through $(m,0)$ and regard the domain of $\phi : (m,t) \longrightarrow \phi_m(t)$ as an open subset W_2 of $W \times \mathbb{R}$ with $W_2 \subset W_1$. Since $\alpha|_N = 0$, X restricted to $N \times \mathbb{R}$ vanishes and hence $N \times \mathbb{R} \subset W_2$. Now since $[0, 1]$ is compact, any point $m \in N$ has a neighborhood $U_m \subset M$ such that $U_m \times [0, 1] \subset W_2$. Let $U = \cup_{m \in N} U_m$. For $(m, t_0) \in U \times [0, 1]$ we compute

$$\frac{d}{dt}(\phi_m(t)^*\Omega_t)(m)\Big|_{t=t_0} = \phi_m(t_0)^*(\pounds_X\Omega_{t_0} + \frac{d}{dt}\Omega_t\Big|_{t=t_0})(m) = 0$$

since

$$\pounds_X\Omega_t = X\lrcorner\,d\Omega_t + d(X\lrcorner\,\Omega_t) = -d\alpha = \Omega_0 - \Omega_1, \text{ and } \frac{d}{dt}\Omega_t = \Omega_1 - \Omega_0.$$

Thus if $\rho : U \longrightarrow V = \rho(U) \subset W$ is the diffeomorphism determined by $(\rho(m), 1) = \phi_m(1)$, then $\rho^*\Omega_1 = \Omega_0$ and $\rho(m) = m$ for $m \in N$. ∎

As a corollary we now have the classical theorem of Darboux.

Theorem 1.3 *Given (M^{2n}, Ω) symplectic and $m \in M$, there exist a neighborhood U of m and local coordinates $(x^1, ..., x^n, y^1, ..., y^n)$ on U such that $\Omega = \sum dx^i \wedge dy^i$.*

Proof. Let $(u^1, ..., u^n, v^1, ..., v^n)$ be local coordinates on a neighborhood of m such that $\frac{\partial}{\partial u^i}(m)$ and $\frac{\partial}{\partial v^i}(m)$ form a symplectic frame at the point m. Set $\Omega_0 = \Omega$ and $\Omega_1 = \sum du^i \wedge dv^i$. Then Ω_0 and Ω_1 agree at m. Now constructing ρ as in the previous theorem, $x^i = u^i \circ \rho$ and $y^i = v^i \circ \rho$ form the desired coordinates. ∎

We remark that one can easily choose the "Darboux" coordinates such that $\frac{\partial}{\partial y^1}(m)$ is any preassigned vector X at m and that $x^i = const., z = const.$ defines a Lagrangian submanifold. Thus we see that given a point m and a tangent vector X at m there exists a Lagrangian submanifold through the point and tangent to X, as we remarked in the last section.

We now prove the theorem of Weinstein that locally a symplectic manifold is the cotangent bundle of a Lagrangian submanifold.

Theorem 1.4 *If L is a Lagrangian submanifold of a symplectic manifold (M, Ω), then there exists a neighborhood of L in M that is symplectomorphic to a neighborhood of the zero section in T^*L.*

Proof. Let $T_L M$ be the restriction of the tangent bundle TM to L and $\mathbf{E}$ a Lagrangian complement of TL in $T_L M$. Such a vector bundle $\mathbf{E}$ exists but is by no means unique; e.g., relative to an associated metric as described above, $\mathbf{E}$ could be taken as the normal bundle of L. Define $j : \mathbf{E} \longrightarrow T^*L$ by

$$j(\zeta)(X) = \Omega(\zeta, X)$$

where $\zeta \in \mathbf{E}_m$ and $X \in T_m L$, $m \in L$. Moreover there exist a tubular neighborhood $\mathcal{U}$ of L in M and a diffeomorphism ϕ of $\mathcal{U}$ onto $\phi(\mathcal{U}) \subset \mathbf{E}$ such that $\phi|_L$ is the zero section and, identifying $T_{\phi(m)} \mathbf{E}_m$ with $\mathbf{E}_m$,

$$\phi_*(m)|_{\mathbf{E}_m} = id|_{\mathbf{E}_m}, \quad m \in L$$

(see e.g., Libermann and Marle [1987, p. 358] or in the Riemannian case use the inverse of the exponential map). Then $j \circ \phi$ is a diffeomorphism of $\mathcal{U}$ onto the open subset $j(\phi(\mathcal{U}))$ of T^*L whose restriction to L is the zero section, $s_0 : L \longrightarrow L' \subset T^*L$. Moreover $(j \circ \phi)_*(m)$ maps the complementary Lagrangian subspaces $T_m L$ and $\mathbf{E}_m$ onto $T_{s_0(m)} L'$ and $T_{s_0(m)}(T_m^* L)$ respectively. But $T_{s_0(m)} L'$ and $T_{s_0(m)}(T_m^* L)$ are complementary Lagrangian subspaces with respect to the symplectic form $d\beta$ on T^*L. Now identifying L and L', the restriction of $(j \circ \phi)_*(m)$ to $T_m L$ is just the identity and since $\phi_*(m)|_{\mathbf{E}_m} = id|_{\mathbf{E}_m}$, $(j \circ \phi)_*(m)$ restricted to $\mathbf{E}_m$ is j. In particular $(j \circ \phi)_*(m)\zeta$ is vertical and $X \in T_m L$, so using the local expression $\sum dp^i \wedge dq^i$ of $d\beta$ on T^*L,

$$d\beta((j \circ \phi)_*(m)\zeta, (j \circ \phi)_*(m)X) = j(\zeta)(X) = \Omega(\zeta, X).$$

The result now follows from Theorem 1.2. ∎

1.4   Symplectomorphisms

Recall that a diffeomorphism $f : M \longrightarrow M$ is a symplectomorphism if $f^*\Omega = \Omega$. A vector field X which generates a 1-parameter group of symplectomorphisms is called a *symplectic vector field*. Clearly $\pounds_X \Omega = 0$.

Theorem 1.5 *Let X be a symplectic vector field on (M, Ω), g an associated metric and J the corresponding almost complex structure. Then $X^i = J^{ik}\theta_k$ for some closed 1-form θ. Conversely, given a closed 1-form θ, $X^i = J^{ik}\theta_k$ defines a symplectic vector field.*

Proof. $0 = \mathcal{L}_X \Omega = d(X \lrcorner \Omega)$ implies $\theta = \frac{1}{2}(X \lrcorner \Omega)$ is a closed 1-form. Let T be the vector field given by $g(T, Y) = \theta(Y)$. Then $g(JT, Y) = -\theta(JY) = -\Omega(X, JY) = g(X, Y)$. Therefore $X^i = J^i{}_k T^k$ or $X^i = J^{ik}\theta_k$ as desired. Conversely given θ closed, define X by $X^i = J^{ik}\theta_k$ from which $-\theta_l = J_{li}X^i = -J_{il}X^i$, i.e., $\theta = \frac{1}{2}(X \lrcorner \Omega)$. Therefore $\mathcal{L}_X \Omega = d(X \lrcorner \Omega) = 2d\theta = 0$. ∎

Corollary 1.1 *For $f \in C^\infty(M)$, $J\nabla f$ is symplectic where ∇f is the gradient of f. Conversely, given X symplectic, X is locally $J\nabla f$.*

In particular X is locally the Hamiltonian vector field X_f. Compare this with the following classical treatment. Suppose that N is a level hypersurface of the function H on (M, Ω), on which $dH \neq 0$. Then X_H is a non-zero tangent vector field which is in the direction of J of the normal direction: $g(Y, \nabla H) = YH = \Omega(X_H, Y) = g(X_H, JY) = -g(JX_H, Y)$ giving $X_H = J\nabla H$.

Finally we prove a result of Hatakeyama [1966] that the group of symplectomorphisms acts transitively on a compact symplectic manifold.

Theorem 1.6 *The group of symplectomorphisms acts transitively on a compact symplectic manifold (M, Ω).*

Proof. We first prove the result for a Darboux neighborhood $\mathcal{U}$ about $p \in M$, i.e., we have local coordinates $(x^1, .., x^n, y^1, ..., y^n)$ such that $\Omega = \sum dx^i \wedge dy^i$ and $x^i(p) = y^i(p) = 0$. Let $q(\neq p) \in \mathcal{U}$ with coordinates (a^i, b^i) and define a function f on $\mathcal{U}$ by $f = \frac{1}{2}\sum(a^i y^i - b^i x^i)$. Then the vector field X defined by $X \lrcorner \Omega = 2df$ generates a 1-parameter group ϕ_t such that $\phi_1(p) = q$. For, writing $X = X^i \frac{\partial}{\partial x^i} + X^{i*} \frac{\partial}{\partial y^i}$, $X \lrcorner \Omega = X^i dy^i - X^{i*} dx^i = a^i dy^i - b^i dx^i$. Thus $X = a^i \frac{\partial}{\partial x^i} + b^i \frac{\partial}{\partial y^i}$ and its integral curves have the form $x^i = a^i t$, $y^i = b^i t$. Strictly speaking X is determined by $f \in C^\infty(M)$, where $f = \frac{1}{2}\sum(a^i y^i - b^i x^i)$ on $\mathcal{U}$ and vanishes outside some larger neighborhood; M compact then implies that ϕ_t is a diffeomorphism of M. Thus any two points in $\mathcal{U}$ may be joined by a symplectomorphism. Now for $p, q \in M$ join them by a curve and cover it by a finite number of Darboux neighborhoods $\mathcal{U}_\alpha$, $\alpha = 1, ..., k$. Choose a sequence of points p_α such that $p_0 = p, p_k = q$ and $p_\alpha \in \mathcal{U}_\alpha \cap \mathcal{U}_{\alpha+1}$ and apply the above result. ∎

For a generalization to symplectomorphisms mapping k points to k points, see Boothby [1969] or Kriegl–Michor [1997, p. 472].

2
Principal S^1-bundles

2.1 The set of principal S^1-bundles as a group

Let P and M be C^∞ manifolds, $\pi : P \longrightarrow M$ a C^∞ map of P onto M and G a Lie group acting on P to the right. Then (P, G, M) is called a *principal G-bundle* if

1. G acts freely on P,

2. $\pi(p_1) = \pi(p_2)$ if and only if there exists $g \in G$ such that $p_1 g = p_2$,

3. P is locally trivial over M, i.e., for every $m \in M$ there exists a neighborhood $\mathcal{U}$ of m and a map $F_{\mathcal{U}} : \pi^{-1}(\mathcal{U}) \longrightarrow G$ such that $F_{\mathcal{U}}(pg) = (F_{\mathcal{U}}(p))g$ and such that the map $\psi : \pi^{-1}(\mathcal{U}) \longrightarrow \mathcal{U} \times G$ taking p to $(\pi(p), F_{\mathcal{U}}(p))$ is a diffeomorphism.

For a general reference to the theory of principal fibre bundles see Bishop and Crittenden [1964, Chapters 3 and 5], Kobayashi and Nomizu [1963-69, Chapter II].

We now turn to the case where $G = S^1$, in which case we say that P is a *principal circle bundle* over M and we study the group structure of the set $\mathcal{P}(M, S^1)$ of all principal circle bundles over M. Our treatment is based on Kobayashi [1956].

Given $P, P' \in \mathcal{P}(M, S^1)$ with projections π, π', let $\Delta(P \times P') = \{(u, u') \in P \times P' | \pi(u) = \pi'(u)\}$. We say $(u_1, u_1') \sim (u_2, u_2')$ if there exists $s \in S^1$ such that $u_1 s = u_2$ and $u_1' s^{-1} = u_2'$. Note that since S^1 is abelian, $u_3 = u_2 t = u_1 st$, $u_3' = u_2' t^{-1} = u_1' s^{-1} t^{-1} = u_1' (st)^{-1}$.

Let $P + P' = \Delta(P \times P')/\sim$ and $\pi'' : P + P' \longrightarrow M$ the induced projection. S^1 acts on $\Delta(P \times P')$ by $(u, u')s = (us, u')$. Now if $(u_1, u_1') \sim (u_2, u_2')$, $u_1 t = u_2$ and $u_1' t^{-1} = u_2'$ and so $u_2 s = u_1 ts = (u_1 s)t$. Therefore $(u_1 s, u_1') \sim (u_2 s, u_2')$ and hence S^1 acts on $P + P'$.

S^1 acts freely: Suppose $u''s = u''$, $u'' \in P + P'$ and suppose (u, u') represents u''. Then $(u, u') \sim (us, u')$ so that $u's^{-1} = u'$ and hence $s = 1 \in S^1$.

S^1 acts transitively on fibres: Suppose $u_1'', u_2'' \in \pi''^{-1}(m)$ and (u_1, u_1'), (u_2, u_2') are representatives. Then $u_2 = u_1 s$, $u_2' = u_1' s'$, $s, s' \in S^1$. Now $(u_2, u_2') \sim (u_2 s', u_1') = (u_1 ss', u_1') = (u_1, u_1')ss'$ and hence $u_2'' = u_1'' ss'$.

$P + P'$ is locally trivial: If $F_\mathcal{U}(u) = g$, $F_\mathcal{U}'(u') = g'$, set $F_\mathcal{U}''(u, u') = gg'$. Then $F_\mathcal{U}''(us, u') = gsg' = gg's$.

Theorem 2.1 *Under the operation* $+$, $\mathcal{P}(M, S^1)$ *is an abelian group.*

Proof. Let P_0 be the trivial bundle and $\alpha : P \longrightarrow P + P_0$ defined by $\alpha(u) = [(u, (\pi(u), 1))]$. Then α is a bundle isomorphism:

$$\alpha(us) = [(us, (\pi(u), 1))] = [(u, (\pi(u), 1))s] = [(u, (\pi(u), 1))]s = \alpha(u)s;$$

$$\alpha^{-1}([(u, (\pi(u), g))]) = \alpha^{-1}([(ug^{-1}, (\pi(u), 1))]) = ug^{-1}.$$

Let $-P$ be a manifold diffeomorphic to P and $-u$ the point corresponding to u. Define $-\pi : -P \longrightarrow M$ by $-\pi(-u) = \pi(u)$. S^1 acts on $-P$ by $(-u)s = -(us^{-1})$. Then $-P \in \mathcal{P}(M, S^1)$. Now let $(u_1, -u_2) \in \Delta(P \times -P)$; then there exists a unique $s \in S^1$ such that $u_1 = u_2 s$. Let $\alpha : P + (-P) \longrightarrow P_0$ be defined by $\alpha([(u_1, -u_2)]) = (\pi(u_1), s)$. Then α is a bundle isomorphism.

Let $\Delta(P \times P' \times P'') = \{(u, u', u'') | \pi(u) = \pi'(u') = \pi''(u'')\}$ and define the equivalence $\sim$ by $(u, u', u'') \sim (us, u's^{-1}s', u''s'^{-1})$. Then $\Delta(P \times P' \times P'')/\sim$ is naturally isomorphic to $(P + P') + P''$, $((u's^{-1}, us)s', u''s'^{-1})$, and to $P + (P' + P'')$, $(us, (u's', u''s'^{-1})s^{-1})$. S^1 acts on $\Delta(P \times P' \times P'')$ by $(u, u', u'')s = (us, u', u'')$. Now if $(u_1, u_1', u_1'') \sim (u_2, u_2', u_2'')$, then $u_2 = u_1 t$, $u_2' = u_1' t^{-1} t'$, $u_2'' = u_1'' t'^{-1}$. Then $u_2 s = u_1 ts = (u_1 s)t$ so that the right action preserves $\sim$.

Finally $P + P'$ is isomorphic to $P' + P$ by $[(u, u')] \longleftrightarrow [(u', u)]$, $(us, u') \sim (u, u's)$. $\blacksquare$

Let G_m be the cyclic subgroup of S^1 of order m and $P \in \mathcal{P}(M, S^1)$. Since S^1 acts on P on the right, so does G_m. Then P/G_m is a principal bundle over M with group S^1/G_m. But $S^1/G_m \cong S^1$ and hence we can consider $P/G_m \in \mathcal{P}(M, S^1)$. More precisely: Let $[u]$ be an element of P/G_m which is represented by $u \in P$. Define the action of S^1 on P/G_m by setting $[u]s = [us']$ where $s = s'^m$. This definition is independent of the choice of u and s'. For if $g \in G_m$, $[ug]s = [ugs'] = [us'g] = [us'] = [u]s$ and if $s''^m = s$, then $(s'^{-1}s'')^m = 1$ so that $s'^{-1}s'' \in G_m$ and hence $[us''] = [us's'^{-1}s''] = [us']$.

Theorem 2.2 *Let P, G_m and P/G_m be as above. Then $P/G_m \cong m \cdot P$.*

Proof. From the definition above it follows by induction that $m \cdot P$ can be defined directly by

$$\Delta(P \times \cdots \times P) = \{(u_1, ..., u_m) \in P \times \cdots \times P | \pi(u_1) = \cdots = \pi(u_m)\}$$

two elements of which, say $(u_1, ..., u_m)$ and $(u_1s_1, ..., u_ms_m)$ are equivalent if and only if $s_1 \cdots s_m = 1$. The quotient space of $\Delta(P \times \cdots \times P)$ by this relation is $m \cdot P$. The action of S^1 on $m \cdot P$ is given by $[(u_1, ..., u_m)]s = [(u_1s, u_2, ..., u_m)]$. Define $\phi : P/G_m \longrightarrow m \cdot P$ by $\phi([u]) = [(u, ..., u)]$ which is independent of the choice of u, for if $g \in G_m$, $g^m = 1$, then $\phi([ug]) = [(ug, ..., ug)] = [(u, ..., u)]$. Now take $s \in S^1$ and s' such that $s'^m = s$, then $\phi([u]s) = \phi([us']) = [(us', ..., us')] = [(us, u, .., u)] = [(u, ..., u)]s = (\phi([u]))s$. Therefore ϕ is a bundle isomorphism of P/G_m onto $m \cdot P$ ∎

Corollary 2.1 *If P is simply connected and $m > 1$, then there is no bundle $P' \in \mathcal{P}(M, S^1)$ such that $P = m \cdot P'$.*

Proof. Suppose that P' exists, then $P \cong P'/G_m$ and so P' is a covering space of P. Since P is simply connected, this can happen only if $m = 1$. ∎

A principal bundle may also be thought of as an equivalence class of principal coordinate bundles which are given by their transition functions. Let $\{\mathcal{U}_i\}$ be a differentiably simple open cover of M (i.e., $\{\mathcal{U}_i\}$ is locally finite, each $\mathcal{U}_i$ has compact closure and any nonempty finite intersection is diffeomorphic to an open cell of $\mathbb{R}^n$). With respect to this cover let $f_{ij} : \mathcal{U}_i \cap \mathcal{U}_j \longrightarrow S^1$ be the transition functions of a bundle $P \in \mathcal{P}(M, S^1)$. The f_{ij} are defined by $f_{ij}(\pi(p)) = F_{U_i}(p)(F_{U_j}(p))^{-1}$. Then $f_{ij} \in \Gamma(\mathcal{U}_i \cap \mathcal{U}_j, \mathcal{S}^1)$, the set of all sections over $\mathcal{U}_i \cap \mathcal{U}_j$ with coefficients in $\mathcal{S}^1$, the sheaf of germs of local C^∞ maps from M into S^1. Thus $f = \{f_{ij}\}$ is a cochain of M. Now $f_{ik} = f_{ij}f_{jk}$. Thus $f_{i_0i_1i_2} = \delta f_{i_0i_1} = f_{i_0i_1}f_{i_0i_2}^{-1}f_{i_1i_2} = f_{i_0i_2}f_{i_0i_2}^{-1} = 1$ and f is a cocycle. Now P and P' are equivalent if and only if $P - P'$ is the trivial bundle, so P and P' are the same here if and only if ff'^{-1} is a coboundary where f' is the cocycle of P'. Therefore $\mathcal{P}(M, S^1) \cong H^1(M, \mathcal{S}^1)$.

The natural short exact sequence, $0 \longrightarrow \mathbb{Z} \longrightarrow \mathbb{R} \longrightarrow S^1 \longrightarrow 0$ induces a short exact sequence of the corresponding sheaves $0 \longrightarrow \mathcal{Z} \longrightarrow \mathcal{R} \longrightarrow \mathcal{S}^1 \longrightarrow 0$. From this we get the cohomology sequence

$$\cdots \longrightarrow H^1(M, \mathcal{R}) \rightarrow H^1(M, \mathcal{S}^1) \rightarrow H^2(M, \mathcal{Z}) \rightarrow H^2(M, \mathcal{R}) \rightarrow \cdots .$$

Now let $\{\phi_i\}$ be a partition of unity subordinate to $\{\mathcal{U}_i\}$ with $\phi_i|_{M-\mathcal{V}_i} = 0, \bar{\mathcal{V}}_i \subset \mathcal{U}_i$. Let $\{\beta_{ijk}\} \in Z^2(M, \mathcal{R})$. Consider $\alpha_{ij} = \sum_k \phi_k \beta_{ijk}$. If in a neighborhood of $m \in \mathcal{U}_i \cap \mathcal{U}_j$, β_{ijk} is not defined, we have $\phi_k(m') = 0$ for every m' in the neighborhood. Hence we see that α_{ij} is defined on $\mathcal{U}_i \cap \mathcal{U}_j$. Now

$$\delta\{\alpha_{ij}\} = \{\alpha_{ij} - \alpha_{ik} + \alpha_{jk}\} = \left\{\sum_l \phi_l(\beta_{ijl} - \beta_{ikl} + \beta_{jkl})\right\}$$

$$= \left\{\sum_l \phi_l \beta_{ijk}\right\} = \{\beta_{ijk}\}.$$

This can be done for other-dimensional cocycles as well. Hence $Z^i(M, \mathcal{R}) = B^i(M, \mathcal{R}), i > 0$. Thus in particular we have that

$$H^1(M, \mathcal{R}) = H^2(M, \mathcal{R}) = 0.$$

On the other hand the C^∞ maps of M into $\mathbb{Z}$ are just the constant integer functions and hence the corresponding cohomology groups must be isomorphic, in particular $H^2(M, \mathcal{Z}) = H^2(M, \mathbb{Z})$. Thus we finally have the isomorphisms

$$H^1(M, \mathcal{S}^1) \cong H^2(M, \mathbb{Z}), \quad \mathcal{P}(M, S^1) \cong H^2(M, \mathbb{Z}).$$

For example it is well known that for complex projective space $\mathbb{C}P^n$, $H^2(\mathbb{C}P^n, \mathbb{Z}) \cong \mathbb{Z}$. Thus by the above isomorphism, $\mathcal{P}(\mathbb{C}P^n, S^1) \cong \mathbb{Z}$. The most famous example of a principal circle bundle over $\mathbb{C}P^n$ is the Hopf fibration of an odd-dimensional sphere S^{2n+1}. Suppose that $S^{2n+1} \in \mathcal{P}(\mathbb{C}P^n, S^1)$ corresponds to $k \in \mathbb{Z}$, then since S^{2n+1} is simply connected, $k = \pm 1$ by the Corollary. Therefore we finally have the following result.

Theorem 2.3 $\mathcal{P}(\mathbb{C}P^n, S^1) \cong \mathbb{Z}$ with S^{2n+1} corresponding to 1 for a proper orientation of $\mathbb{C}P^n$.

2.2 Connections on a principal bundle

A *connection* on a principal G-bundle (P, G, M) is a C^∞ distribution H on P such that

1. $T_pP = H_p \oplus V_p$, $V_p = \ker \pi_*$,

2. $R_{g*}(H_p) = H_{pg}$ where R_g is right translation.

Vectors in H_p are said to be *horizontal* and for $t \in T_pP$ we denote its horizontal part by Ht. The map $\pi_*|_{H_p}$ is one-to-one and hence $\pi_*(H_p) = T_{\pi(p)}M$. Thus given a vector field X on M there exists a unique vector field $\tilde{X}$ on P such that $\tilde{X}(p) \in H_p$ and $\pi_*\tilde{X}(p) = X(\pi(p))$, i.e., $\tilde{X}$ is π-related to X. $\tilde{X}$ is called the *horizontal lift* of X. The following two properties of horizontal lifts follow easily from the definitions:

$$\widetilde{[X,Y]} = H[\tilde{X},\tilde{Y}], \quad R_{g*}\tilde{X} = \tilde{X}.$$

A p-form ω on P is *vertical* (resp. *horizontal*) if $\omega(t_1, ..., t_p) = 0$ when one or more of the t_i's is horizontal (resp. vertical).

Now regard $p \in P$ as a map of $G \longrightarrow P$ by $p(g) = pg$. Let $\mathfrak{g}$ denote the Lie algebra of G and define a Lie algebra homomorphism of $\mathfrak{g}$ into a Lie algebra $\bar{\mathfrak{g}}$ of vector fields on P by $\bar{X} = (p_*X)(e)$. $p : G \longrightarrow \pi^{-1}(\pi(p))$ is a diffeomorphism, so given $t \in V_p$, let $X(e) = p_*^{-1}t$; then there exists $\bar{X} \in \bar{\mathfrak{g}}$ such that $\bar{X}(p) = t$. On the other hand, given $\bar{X} \in \bar{\mathfrak{g}}$, $\pi_*\bar{X}(p) = \pi_*(p_*X)(e) = (\pi \circ p)_*X(e) = 0$, since $\pi \circ p$ is a constant map. Thus $\bar{X}$ is vertical.

Given a connection H on (P, G, M) define a $\mathfrak{g}$-valued 1-form ϕ on P by $\phi(t) = X \in \mathfrak{g}$ where $\bar{X}(p)$ is the vertical part of t. ϕ is called the *connection form* of H. The following lemmas are well known and their proofs can be found in the references (e.g., Bishop and Crittenden [1964, pp. 76-77], Kobayashi and Nomizu [1963-69, p. 64])

Lemma 2.1 $\phi \circ R_{g*} = Ad\, g^{-1} \circ \phi$, *i.e.* ϕ *is equivariant.*

Lemma 2.2 *If ϕ is a $\mathfrak{g}$-valued C^∞ equivariant 1-form such that $\phi(\bar{X}(p)) = X$, then there exists a unique connection H whose 1-form is ϕ.*

Given a $\mathfrak{g}$-valued p-form σ and a $\mathfrak{g}$-valued q-form ω, their bracket is defined by

$$[\sigma, \omega](X_1, ..., X_{p+q})$$

$$= \frac{1}{(p+q)!} \sum_{(i_1,...,i_{p+q})} sgn(i_1, ..., i_{p+q})[\sigma(X_{i_1}, ..., X_{i_p}), \omega(X_{i_{p+1}}, ..., X_{i_{p+q}})].$$

Let ω be a p-form on P and define a $(p+1)$-form $D\omega$ by

$$D\omega(t_1, ..., t_{p+1}) = d\omega(Ht_1, ..., Ht_p).$$

Clearly $D\omega$ is horizontal. If ϕ is the connection form of H, $\Phi = D\phi$ is called the *curvature form* of H. Φ is equivariant as can be seen as follows.

$$\Phi(R_{g*}t_1, R_{g*}t_2) = d\phi(HR_{g*}t_1, HR_{g*}t_2) = d\phi(R_{g*}Ht_1, R_{g*}Ht_2)$$

$$= -\frac{1}{2}\phi(R_{g*}[Ht_1, Ht_2]) = -\frac{1}{2}Ad\, g^{-1}\phi([Ht_1, Ht_2]) = Ad\, g^{-1}\Phi(t_1, t_2).$$

We now have the structural equation; again for the proof see the references (e.g., Bishop and Crittenden [1964, p. 81], Kobayashi and Nomizu [1963-69, Chapter II, p. 77]).

Theorem 2.4 $d\phi = -\frac{1}{2}[\phi, \phi] + \Phi$.

Let $P \in \mathcal{P}(M, S^1)$ and note that the Lie algebra of S^1 is $\mathbb{R}$ with the trivial bracket operation. Thus if η is a connection form on P, $[\eta, \eta] = 0$ and so if Φ is the curvature, the structural equation is simply $d\eta = \Phi$.

Again since S^1 is abelian, for $X, Y \in T_u P$ and $s \in S^1$ we have

$$\Phi(R_{s*}X, R_{s*}Y) = Ad\, s^{-1}\Phi(X, Y) = \Phi(X, Y).$$

Therefore there exists a unique 2-form Ω on M such that $\Phi = \pi^*\Omega$. Now $\pi^*(d\Omega) = d\Phi = 0$ and hence Ω is a closed 2-form M.

If now η' is another connection, then as before $(\eta - \eta') \circ R_{s*} = \eta - \eta'$. Hence there exists a unique 1-form β on M such that $\pi^*\beta = \eta - \eta'$. Now $\pi^*d\beta = d(\eta - \eta') = \Phi - \Phi' = \pi^*\Omega - \pi^*\Omega'$ and hence $d\beta = \Omega - \Omega'$. Thus the cohomology class of Ω is independent of the choice of the connection form and is called the *characteristic class* of P; again see Kobayashi [1956]. Since the transition functions are mappings from $\mathcal{U}_i \cap \mathcal{U}_j$ into S^1, they can be considered as real-valued functions (mod 1) and it can then be shown that Ω is integral, i.e., $\int_c \Omega$ = integer for any finite singular cocycle c with integer coefficients. This gives a homomorphism of $\mathcal{P}(M, S^1)$ onto the integral classes of the second deRham cohomology.

We end this chapter with the following theorem of Kobayashi [1963] which should again be compared with the isomorphism $\mathcal{P}(M, S^1) \cong H^2(M, \mathbb{Z})$.

Theorem 2.5 *Let Ω be a 2-form on M representing an element of $H^2(M, \mathbb{Z})$; then there exist a principal circle bundle P and a connection form η on P such that $d\eta = \pi^*\Omega$.*

Proof. Let P be the principal bundle corresponding to Ω and η' a connection form on P so that the closed 2-form Ω' defined by $d\eta' = \pi^*\Omega'$ is cohomologous to Ω. Let β be a 1-form on M such that $\Omega - \Omega' = d\beta$. Now set $\eta = \eta' + \pi^*\beta$. $\pi^*\beta$ is horizontal and equivariant and hence η is a connection form on P and $d\eta = \pi^*\Omega$ as desired. ∎

3
Contact Manifolds

3.1 Definitions

By a *contact manifold* we mean a C^∞ manifold M^{2n+1} together with a 1-form η such that $\eta \wedge (d\eta)^n \neq 0$. In particular $\eta \wedge (d\eta)^n \neq 0$ is a volume element on M so that a contact manifold is orientable. Also $d\eta$ has rank $2n$ on the Grassmann algebra $\bigwedge T_m^* M$ at each point $m \in M$ and thus we have a 1-dimensional subspace, $\{X \in T_m M | d\eta(X, T_m M) = 0\}$, on which $\eta \neq 0$ and which is complementary to the subspace on which $\eta = 0$. Therefore choosing ξ_m in this subspace normalized by $\eta(\xi_m) = 1$ we have a global vector field ξ satisfying

$$d\eta(\xi, X) = 0, \quad \eta(\xi) = 1.$$

ξ is called the *characteristic vector field* or *Reeb vector field* (Reeb [1952]) of the contact structure η. Computing Lie derivatives by the formula $\mathcal{L}_\xi = d \circ (\xi \lrcorner) + (\xi \lrcorner) \circ d$ we have immediately that

$$\mathcal{L}_\xi \eta = 0, \quad \mathcal{L}_\xi d\eta = 0.$$

We denote by $\mathcal{D}$ the *contact distribution* or *subbundle* defined by the subspaces $\mathcal{D}_m = \{X \in T_m M : \eta(X) = 0\}$. Roughly speaking the meaning of the contact condition, $\eta \wedge (d\eta)^n \neq 0$, is that the contact subbundle is as far from being integrable as possible; in particular $\mathcal{D}$ rotates as one moves around on the manifold. For a subbundle defined by a 1-form η to be integrable it is necessary and sufficient that $\eta \wedge (d\eta) \equiv 0$. In contrast we shall see in Chapter 5 that, for a contact manifold M^{2n+1}, the maximum dimension of an integral submanifold of $\mathcal{D}$ is only n. A one-dimensional integral submanifold of $\mathcal{D}$ will be called a *Legendre curve*, especially to avoid confusion with an integral curve of the vector field ξ.

A contact structure is *regular* if ξ is regular as a vector field, that is, every point of the manifold has a neighborhood such that any integral curve of the vector field passing through the neighborhood passes through only once (cf. Palais [1957]). Two well-known examples of non-regular vector fields on surfaces are the irrational flow on a torus and the flow around a Möbius band.

We now prove the classical theorem of Darboux.

Theorem 3.1 *About each point of a contact manifold* (M^{2n+1}, η) *there exist local coordinates* $(x^1, \dots, x^n, y^1, \dots, y^n, z)$ *with respect to which*

$$\eta = dz - \sum_{i=1}^{n} y^i dx^i.$$

Proof. In some coordinate neighborhood choose a $2n$-ball transverse to ξ; $d\eta$ is symplectic on this ball and hence there exist local coordinates $(x^1, \dots, x^n, y^1, \dots, y^n, u)$ such that $d\eta = \sum dx^i \wedge dy^i$. Now $d(\eta + \sum y^i dx^i) = 0$ so that $\eta + \sum y^i dx^i = df$ for some function f, i.e., $\eta = df - \sum y^i dx^i$. Now $\eta \wedge (d\eta)^n = df \wedge dx^1 \wedge \cdots \wedge dx^n \wedge dy^1 \wedge \cdots \wedge dy^n \neq 0$. Therefore df is independent of $dx^1, \dots, dx^n, dy^1, \dots, dy^n$ and hence we can regard x^i, y^i and $z = f$ as a coordinate system. ∎

In a Darboux coordinate system $\xi = \frac{\partial}{\partial z}$. For if $\xi = a\frac{\partial}{\partial z} + b^i\frac{\partial}{\partial x^i} + c^i\frac{\partial}{\partial y^i}$, then $1 = \eta(\xi) = a - b^i y^i$. Also $0 = d\eta(\xi, \frac{\partial}{\partial x^i}) = \sum dx^i \wedge dy^i(\xi, \frac{\partial}{\partial x^i})$ gives $c^i = 0$. Similarly $0 = d\eta(\xi, \frac{\partial}{\partial y^i})$ gives $b^i = 0$. Thus $a = 1$ and so $\xi = \frac{\partial}{\partial z}$.

A diffeomorphism f of M^{2n+1} or between open subsets of $\mathbb{R}^{2n+1}$ with the contact structure of the Darboux form of Theorem 3.1 is called a *contact transformation* if $f^*\eta = \tau\eta$ for some non-vanishing function τ on the domain of f. If $\tau \equiv 1$, f is called a *strict contact transformation*.

There is also the notion of a *contact structure in the wider sense*, often called simply a "contact structure" by many authors, which can be defined in a number of ways. For example, a contact manifold in the wider sense is a manifold with a differentiable structure modeled on the pseudogroup of contact transformations on $\mathbb{R}^{2n+1}$ (J. Gray [1959]), i.e., in the overlap of coordinate neighborhoods the transition functions preserve the Darboux form to within a non-vanishing function multiple. An alternate approach is to put the emphasis on the field of 2n-planes $\mathcal{D}$ and to define the structure as a hyperplane field defined locally by a contact form. In the overlap of coordinate neighborhoods $\mathcal{U} \cap \mathcal{U}'$, $\eta' = f^*\eta = \tau\eta$ and hence $d\eta' = f^*d\eta = d\tau \wedge \eta + \tau d\eta$ from which

$$\eta' \wedge (d\eta')^n = \tau^{n+1}\eta \wedge (d\eta)^n \neq 0.$$

Let M^{2n+1} be a contact manifold in the wider sense. On a coordinate neighborhood $\mathcal{U}_\alpha$ choose coordinates $(x^1, \ldots, x^{2n+1})$ such that $\eta_\alpha \wedge (d\eta_\alpha)^n = \lambda_\alpha dx^1 \wedge \cdots \wedge dx^{2n+1}$ with $\lambda_\alpha > 0$. Similarly on a neighborhood $\mathcal{U}_\beta$ choose coordinates $(y^1, \ldots, y^{2n+1})$ such that $\eta_\beta \wedge (d\eta_\beta)^n = \lambda_\beta dy^1 \wedge \cdots \wedge dy^{2n+1}$ with $\lambda_\beta > 0$. Now on $\mathcal{U}_\alpha \cap \mathcal{U}_\beta$, $\eta_\alpha = \tau_{\alpha\beta}\eta_\beta$ and hence $\eta_\alpha \wedge (d\eta_\alpha)^n = \tau_{\alpha\beta}^{n+1}\eta_\beta \wedge (d\eta_\beta)^n$. Therefore $\tau_{\alpha\beta}^{n+1}\lambda_\beta \left|\frac{\partial y^i}{\partial x^j}\right| = \lambda_\alpha$. From this one may easily obtain the following results (see also J. Gray [1959], Sasaki [1965], Stong [1974]).

Theorem 3.2 *Let M^{2n+1} be a contact manifold in the wider sense. If n is odd, then M^{2n+1} is orientable.*

Theorem 3.3 *Let M^{2n+1} be a contact manifold in the wider sense. If n is even and M^{2n+1} is orientable, then M^{2n+1} is a contact manifold.*

We also have the following theorem of Sasaki [1965].

Theorem 3.4 *Let M^{2n+1} be a contact manifold in the wider sense which is not a contact manifold in the restricted sense; then its 2-fold covering manifold is a contact manifold in the restricted sense.*

Proof. Let $\{\mathcal{U}_\alpha\}$ be an open cover of M^{2n+1} by coordinate charts. Recall the local contact forms η_α above and their transition $\eta_\alpha = \tau_{\alpha\beta}\eta_\beta$. Consider the 2-fold covering $\pi : \tilde{M}^{2n+1} \longrightarrow M^{2n+1}$ given as follows. $\tilde{M}^{2n+1}$ is the union of the sets $\{\mathcal{U}_\alpha \times \mathbb{Z}_2\}$ with the following equivalence relation, where $\mathbb{Z}_2$ is the set of integers ± 1. Let ϵ_α denote $+1$ or -1. Two elements $(p_\alpha, \epsilon_\alpha) \in \mathcal{U}_\alpha \times \mathbb{Z}_2$ and $(p_\beta, \epsilon_\beta) \in \mathcal{U}_\beta \times \mathbb{Z}_2$ are equivalent if $p_\alpha = p_\beta$ and $\epsilon_\alpha = sgn(\tau_{\alpha\beta}(p_\beta))\epsilon_\beta$. Now define local contact forms on $\mathcal{U}_\alpha \times \pm 1$ by $\eta_{(\alpha,\epsilon_\alpha)} = \epsilon_\alpha \pi^* \eta_\alpha$. Then

$$\eta_{(\alpha,\epsilon_\alpha)} = \epsilon_\alpha \pi^* \eta_\alpha = \epsilon_\alpha \pi^* (\tau_{\alpha\beta}\eta_\beta) = \epsilon_\alpha (\tau_{\alpha\beta} \circ \pi) \pi^* \eta_\beta = \epsilon_\alpha \epsilon_\beta (\tau_{\alpha\beta} \circ \pi) \eta_{(\beta,\epsilon_\beta)}$$

and $\epsilon_\alpha \epsilon_\beta (\tau_{\alpha\beta} \circ \pi) > 0$. The local forms $\eta_{(\alpha,\epsilon_\alpha)}$ can now be used to construct a global contact form on $\tilde{M}^{2n+1}$. ∎

In particular a connected and simply connected contact manifold in the wider sense is a contact manifold in the restricted sense; for an alternate approach to this idea see Monna [1983].

The name contact (Berührungstransformation) seems to be due to Sophus Lie [1890] and is natural in view of the simple example of Huygens' principle (Huygens [1690]; see also MacLane [1968, part II, p. 83]). Consider $\mathbb{R}^2$ with coordinates (x, y). The classical notion of a "line element" of $\mathbb{R}^2$ is a point together with a non-vertical line through the point. Thus a line element may be regarded as a point in $\mathbb{R}^3$ determined by the point and the slope p of the line. Given a smooth curve C in the plane without vertical tangents, say $y = f(x)$, its tangent lines determine a curve in $\mathbb{R}^3$ with coordinates (x, y, p)

which is a Legendre curve of the contact form $\eta = dy - pdx$. If now C is a wave front, by Huygens' principle the new wave front C_t at time t is the envelope of the circular waves centered at all the points of C, say of radius t taking the velocity of propagation to be 1. Corresponding to a point (x, y) on C, the point $(\bar{x}, \bar{y})$ on C_t lies on both the normal line and the circle of radius t centered at (x, y), i.e., $\bar{y} - y = -\frac{1}{p}(\bar{x} - x)$ and $(\bar{x} - x)^2 + (\bar{y} - y)^2 = t^2$. Thus $(\bar{x} - x)^2 = \frac{p^2 t^2}{p^2+1}$, so depending on the direction of propagation, e.g., choosing the negative root, the transformation of $\mathbb{R}^3$ mapping (x, y, p) to

$$\bar{x} = x - \frac{pt}{\sqrt{p^2 + 1}}, \quad \bar{y} = y + \frac{t}{\sqrt{p^2 + 1}}, \quad \bar{p} = p,$$

maps C to C_t. A simple calculation shows that $d\bar{y} - \bar{p}d\bar{x} = dy - pdx$ and so the transformation is a contact transformation. Moreover tangent wave fronts (curves) are mapped to tangent wave fronts (curves) and hence the name "contact".

3.2   Examples

In [1971] J. Martinet proved that every compact orientable 3-manifold carries a contact structure. However, we now have the following theorem of J. Gonzalo [1987] showing that there are three independent contact structures.

Theorem 3.5 *Every closed orientable 3-manifold has a parallelization by three contact forms.*

Before turning to some detailed examples we should mention that, in contrast to Martinet's result, there exist $(2n+1)$-dimensional manifolds, $n \geq 2$, with no contact structure even in the wider sense. In particular, Stong [1974] showed that for every $n \geq 2$ there is a closed oriented connected manifold of dimension $2n + 1$ with no contact structure in the wider sense. One such manifold is $SU(3)/SO(3)$ for $n = 2$ and $(SU(3)/SO(3)) \times S^{2n-4}$ for $n > 2$.

3.2.1 $\mathbb{R}^{2n+1}$

In effect we have already seen that $\mathbb{R}^{2n+1}(x^1, \ldots, x^n, y^1, \ldots, y^n, z)$ with the Darboux form $\eta = dz - \sum_{i=1}^{n} y^i dx^i$ is a contact manifold. The characteristic vector field ξ is $\frac{\partial}{\partial z}$ and the contact subbundle $\mathcal{D}$ is spanned by

$$X_i = \frac{\partial}{\partial x^i} + y^i \frac{\partial}{\partial z}, \quad X_{n+i} = \frac{\partial}{\partial y^i},$$

$i = 1, \ldots, n.$

3.2.2 $\mathbb{R}^{n+1} \times P\mathbb{R}^n$

We now give an example of a contact manifold in the wider sense which is not a contact manifold in our sense (J. Gray [1959]). Consider $\mathbb{R}^{n+1}$ with coordinates $(x^1, \ldots, x^{n+1})$ and real projective space $P\mathbb{R}^n$ with homogeneous coordinates, $(t_1, \ldots, t_{n+1})$ and let $M^{2n+1} = \mathbb{R}^{n+1} \times P\mathbb{R}^n$. The subsets $\mathcal{U}_i, i = 1, \ldots, n+1$ defined by $t_i \neq 0$, form an open cover of M^{2n+1} by coordinate neighborhoods. On $\mathcal{U}_i$ define a 1-form η_i by $\eta_i = \dfrac{1}{t_i} \sum_{j=1}^{n+1} t_j dx^j$; we then have $\eta_i \wedge (d\eta_i)^n \neq 0$ and $\eta_i = \dfrac{t_j}{t_i} \eta_j$. Thus, M^{2n+1} has a contact structure in the wider sense, but for n even, M^{2n+1} is non-orientable and hence cannot carry a global contact form.

3.2.3 $M^{2n+1} \subset \mathbb{R}^{2n+2}$ *with* $T_m M^{2n+1} \cap \{0\} = \emptyset$

Turning to more standard examples, we prove the following theorem (J. Gray [1959]).

Theorem 3.6 *Let* $\iota : M^{2n+1} \longrightarrow \mathbb{R}^{2n+2}$ *be a smooth hypersurface immersed in* $\mathbb{R}^{2n+2}$ *and suppose that no tangent space of* M^{2n+1} *contains the origin of* $\mathbb{R}^{2n+2}$; *then* M^{2n+1} *has a contact structure.*

Proof. Let x^A, $A = 1, \ldots, 2n+2$ be cartesian coordinates on $\mathbb{R}^{2n+2}$ and consider the 1-form

$$\alpha = x^1 dx^2 - x^2 dx^1 + \cdots + x^{2n+1} dx^{2n+2} - x^{2n+2} dx^{2n+1}.$$

Let $V_1, \ldots, V_{2n+1}$ be $2n+1$ linearly independent vectors at a point $x_0 = (x_0^1, \ldots, x_0^{2n+2})$ and define a vector W at x_0 with components

$$W^A = *dx^A(V_1, \ldots, V_{2n+1})$$

where $*$ is the Hodge star operator of the Euclidean metric on $\mathbb{R}^{2n+2}$. Then W is normal to the hyperplane spanned by $V_1, \ldots, V_{2n+1}$. Now regard x_0 as a vector with components x_0^A; then

$$(\alpha \wedge (d\alpha)^n)(V_1, \ldots, V_{2n+1}) = \sum x_0^A W^A.$$

Thus if no tangent space of M^{2n+1} regarded as a hyperplane in $\mathbb{R}^{2n+2}$ contains the origin, then $\eta = \iota^* \alpha$ is a contact form on M^{2n+1}. ∎

As a special case we see that an odd-dimensional sphere S^{2n+1} carries a contact structure. Moreover α is invariant under reflection through the origin, $(x^1, \ldots, x^{2n+2}) \to (-x^1, \ldots, -x^{2n+2})$ and hence the real projective space

$P\mathbb{R}^{2n+1}$ is also a contact manifold. J. A. Wolf [1968] then considered more general quotients of S^{2n+1} and proved that a complete connected odd-dimensional Riemannian manifold of positive constant curvature inherits a contact structure from the form α.

Similarly consider the 1-form $\beta = \sum_{i=1}^{n+1} x^i dx^{n+1+i}$ and denote by $\mathbb{R}_1^{n+1}$ and $\mathbb{R}_2^{n+1}$ the subspaces defined by $x^i = 0$ and $x^{n+1+i} = 0$ respectively, $i = 1, \ldots, n+1$. Then β induces a contact form on M^{2n+1} if and only if $M^{2n+1} \cap \mathbb{R}_1^{n+1} = \emptyset$ and $M^{2n+1} \cap \mathbb{R}_2^{n+1}$ is an n-dimensional submanifold and no tangent space of $M^{2n+1} \cap \mathbb{R}_2^{n+1}$ in $\mathbb{R}_2^{n+1}$ contains the origin of $\mathbb{R}_2^{n+1}$.

More generally, given a symplectic manifold (M, Ω), a hypersurface $\iota : S \longrightarrow M$ is said to be of *contact type* if there exists a contact form η on S such that $d\eta = \iota^*\Omega$.

3.2.4 T_1^*M, T_1M

We shall show that the cotangent sphere bundle and the tangent sphere bundle of a Riemannian manifold are contact manifolds (see e.g., Reeb [1952], Sasaki [1962]). Let M be an $n+1$-dimensional Riemannian manifold and T^*M its cotangent bundle . Also let $(x^1, \ldots, x^{n+1})$ be local coordinates on a neighborhood $\mathcal{U}$ of M and $(p^1, \ldots, p^{n+1})$ coordinates on the fibres over $\mathcal{U}$. If $\pi : T^*M \longrightarrow M$ is the projection map, then as in Chapter 1, $q^i = x^i \circ \pi$ and p^i are local coordinates on T^*M. Consider the Liouville form β which is given locally by $\beta = \sum_{i=1}^{n+1} p^i dq^i$. The bundle T_1^*M of unit cotangent vectors has empty intersection with the zero section of T^*M, its intersection with any fibre of T^*M is an n-dimensional sphere and no tangent space to this intersection contains the origin of the fibre. Thus, as in the discussion at the end of the last example, β induces a contact structure on the hypersurface T_1^*M of T^*M.

Similarly one obtains a contact structure on the bundle T_1M of unit tangent vectors. In fact if G_{ij} denotes the components of the metric on M with respect to the coordinates $(x^1, \ldots, x^{n+1})$ and if $(v^1, \ldots, v^{n+1})$ are the fibre coordinates on TM, define β locally by $\beta = \sum_{i,j} G_{ij} v^j dq^i$ where $q^i = x^i \circ \pi$ and $\pi : TM \longrightarrow M$ is the projection (see Section 9.2 for details).

3.2.5 $T^*M \times \mathbb{R}$

Let M be an n-dimensional manifold and T^*M its cotangent bundle. As in the previous example we can define a 1-form β by the local expression $\beta = \sum_{i=1}^{n} p^i dq^i$. Let $M^{2n+1} = T^*M \times \mathbb{R}$, t the coordinate on $\mathbb{R}$ and $\mu : M^{2n+1} \longrightarrow T^*M$ the projection to the first factor. Then $\eta = dt - \mu^*\beta$ is a contact form on M^{2n+1}.

3.2.6 T^3

We have mentioned that Martinet proved that every compact orientable 3-manifold carries a contact structure. Here we will give explicitly a contact structure on the 3-dimensional torus T^3. First consider $\mathbb{R}^3$ with the contact form

$$\eta = \sin y\, dx + \cos y\, dz; \quad \eta \wedge d\eta = -dx \wedge dy \wedge dz.$$

$\xi = \sin y \frac{\partial}{\partial x} + \cos y \frac{\partial}{\partial z}$ and $\mathcal{D}$ is spanned by $\{\frac{\partial}{\partial y}, \cos y \frac{\partial}{\partial x} - \sin y \frac{\partial}{\partial z}\}$. The rotation of $\mathcal{D}$ in the direction of the y-axis is dramatically clear in this example. Thus one sees the non-integrability of $\mathcal{D}$ as one moves around on the manifold even though $\mathcal{D}$ does not rotate when one moves along the line of intersection of $\mathcal{D}$ with a plane $y = const$.

Now η is invariant under translation by 2π in each coordinate and hence the 3-dimensional torus also carries this structure. For each value of y, ξ induces a flow on the 2-torus defined by $y = const$. Depending on the value of y the flow is a rational or irrational flow on T^2. Thus the contact structure on T^3 is not regular. We will see in Theorem 4.14, that no torus carries a regular contact structure. Note here in particular though, that some of the integral curves of ξ are closed and some are not.

3.2.7 T^5

In [1979] R. Lutz proved the existence of contact structures on principal T^2-bundles over 3-manifolds. In particular the 5-dimensional torus admits a contact structure given by the form

$$\eta = (\sin \theta_1 \cos \theta_3 - \sin \theta_2 \sin \theta_3)d\theta_4 + (\sin \theta_1 \sin \theta_3 + \sin \theta_2 \cos \theta_3)d\theta_5$$

$$+ \sin \theta_2 \cos \theta_2 d\theta_1 - \sin \theta_1 \cos \theta_1 d\theta_2 + \cos \theta_1 \cos \theta_2 d\theta_3.$$

Hadjar [1998] obtains other explicit contact forms on the 5-torus, especially ones for which the contact subbundle is transverse to the trivial fibration of T^5 over T^4 by circles.

There has been also been recent interest in constructing 5-dimensional contact manifolds by using other surfaces as fibre. Both Geiges [1997a] and Altschuler and Wu [2000] have proved the following theorem using very different techniques.

Theorem 3.7 *Let M be a compact orientable 3-manifold and Σ a compact orientable surface. Then $M \times \Sigma$ carries a contact form.*

Altschuler and Wu also show the existence of contact forms on the products $S^{2p+k+3} \times M^k \times \Sigma$ where M^k is compact, orientable and parallelizable and Σ a compact orientable surface as before.

3.2.8 Overtwisted contact structures

Examples 3.2.1 and 3.2.6 on $\mathbb{R}^3$, namely $\eta = (dz - ydx)$ and $\eta = (\sin y dx + \cos y dz)$, have cylindrical coordinate versions (see also Douady [1982/83, p. 131], Bennequin [1983, p. 93]). Let (r, θ, z) be the usual cylindrical coordinates on $\mathbb{R}^3 \setminus \{x = y = 0\}$. Making the naive substitutions $y \longrightarrow r$, $dx \longrightarrow r d\theta$, $dz \longrightarrow dz$ these examples become

$$\eta = dz - r^2 d\theta, \quad \eta \wedge d\eta = -2r dr \wedge d\theta \wedge dz$$

$$\xi = \frac{\partial}{\partial z}, \quad \mathcal{D} = \left\{ \frac{\partial}{\partial r}, \frac{\partial}{\partial \theta} + r^2 \frac{\partial}{\partial z} \right\},$$

and

$$\eta = \cos r\, dz + r \sin r\, d\theta, \quad \eta \wedge d\eta = (r + \sin r \cos r) dr \wedge d\theta \wedge dz,$$

$$\xi = \frac{\sin r}{r + \sin r \cos r} \frac{\partial}{\partial \theta} + \frac{r \cos r + \sin r}{r + \sin r \cos r} \frac{\partial}{\partial z},$$

$$\mathcal{D} = \left\{ \frac{\partial}{\partial r}, \cos r \frac{\partial}{\partial \theta} - r \sin r \frac{\partial}{\partial z} \right\}.$$

Note that in both examples the integral curves of $\frac{\partial}{\partial r}$ are Legendre curves and in the second example that the curve $r = \pi, z = constant$ is also a Legendre curve. Thus in the second example $\mathcal{D}$ is tangent to the disk $\Delta = \{z = 0, r \leq \pi\} \subset \mathbb{R}^3$ along the boundary. Now consider the topological disk $\Delta^\epsilon = \{z = \epsilon r^2, r \leq \pi\}$. $\mathcal{D}$ is tangent to Δ^ϵ only at the origin. On $\Delta^\epsilon \setminus \{(0,0,0)\}$ define a line field by the intersection of the tangent plane to the paraboloid Δ^ϵ and $\mathcal{D}$ at each point. These fields can be expressed by the vector fields

$$r \frac{\partial}{\partial r} + 2\epsilon \frac{\partial}{\partial \theta} + 2\epsilon r^2 \frac{\partial}{\partial z}$$

in the first case and

$$r \sin r \frac{\partial}{\partial r} - 2\epsilon r \cos r \frac{\partial}{\partial \theta} + 2\epsilon r^2 \sin r \frac{\partial}{\partial z}$$

in the second. For simplicity take the projection of these vector fields to the xy-plane. The integral curves in the first case are the logarithmic spirals $r = A e^{\frac{\theta}{2\epsilon}}$ and in the second case are the curves $\theta = -2\epsilon \ln \sin r + C$ which near the origin spiral indefinitely and approach a limit cycle on the boundary of the disk. Thus we have the following diagrams; a diagram of this type for the second example was introduced by Douady [1982/83] (see also Bennequin [1983, p. 94]) and we will call such a diagram the *Douady portrait* of the contact structure.

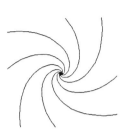 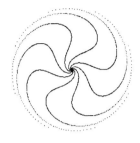

A 3-dimensional contact manifold is said to be *overtwisted* (Eliashberg [1989]) if there exist a contact embedding of a neighborhood of the disk Δ with the contact structure $\eta = \cos r\, dz + r\sin r\, d\theta$.

Roughly speaking the meaning of overtwisted is that as one moves radially from the image of origin, the plane $\mathcal{D}$ turns over in a finite distance. For the radial Legendre curves in the disk Δ, $\mathcal{D}$ turns over as one goes from $r = 0$ to $r = \pi$. This means that for some r between 0 and π the vector field ξ is tangent to the disk Δ, and in particular for r the solution of $r\cos r + \sin r = 0$ in $(0, \pi)$, $r \approx 2.02876$, the circle of this radius in Δ is an integral curve of ξ.

3.2.9 Contact circles

At the beginning of this Section we mentioned the result of Gonzalo [1987] that a compact orientable 3-manifold has three independent contact structures. Geiges and Gonzalo [1995] introduce the notion of a contact circle: A 3-manifold admits a *contact circle* if it admits a pair of contact forms (η_1, η_2) such that for any $(\lambda_1, \lambda_2) \in S^1$, $\lambda_1\eta_1 + \lambda_2\eta_2$ is also a contact form. This circle is a *taut contact circle* if the contact forms $\lambda_1\eta_1 + \lambda_2\eta_2$ define the same volume form for all $(\lambda_1, \lambda_2) \in S^1$; this is equivalent to the two conditions

$$\eta_1 \wedge d\eta_1 = \eta_2 \wedge d\eta_2, \quad \eta_1 \wedge d\eta_2 = -\eta_2 \wedge d\eta_1.$$

Geiges and Gonzalo then prove the following classification theorem.

Theorem 3.8 *A compact orientable 3-manifold admits a taut contact circle if and only if it is diffeomorphic to the quotient of a Lie group G by a discrete subgroup, acting by left multiplication, where G is either $SU(2)$, the universal cover of $PSL(2, \mathbb{R})$ or the universal cover of the group of Euclidean motions $E(2)$.*

In contrast to the overtwisted contact structures in Example 3.2.8 (and in contrast to taut contact circles), a contact structure is said to be *tight* if it is not overtwisted (see e.g. Eliashberg and Thurston [1998]). Geiges and Gonzalo [1995] also prove that the connected sum of any number of copies of the manifolds listed in Theorem 3.8, T^2-bundles over S^1 or $S^2 \times S^1$ admit a contact circle consisting of tight contact structures.

3.3   The Boothby–Wang fibration

We now give an important class of examples, namely principal circle bundles over symplectic manifolds of integral class and conversely we will prove the celebrated theorem of Boothby and Wang [1958] that a compact regular contact manifold is of this type. Examples of this type are often simply referred to as *Boothby–Wang fibrations*.

In Chapter 2 we saw that the set of principal circle bundles over a manifold M has a group structure isomorphic to the cohomology $H^2(M, \mathbb{Z})$. Now let (M^{2n}, Ω) be a symplectic manifold such that $[\Omega] \in H^2(M^{2n}, \mathbb{Z})$ and $\pi : M^{2n+1} \longrightarrow M^{2n}$ the corresponding circle bundle. By Theorem 2.5 there exists a connection form η on M^{2n+1} such that $d\eta = \pi^*\Omega$. Now if ξ is a vertical vector field, say with $\eta(\xi) = 1$, and $X_1, \ldots, X_{2n}$ linearly independent horizontal vector fields, then $(\eta \wedge (d\eta)^n)(\xi, X_1, \ldots, X_{2n})$ is non-zero. Thus regarding the Lie algebra valued form η as a real-valued form, we see that η is a contact structure on M^{2n+1}.

The most well-known special case of the Boothby–Wang fibration is the Hopf fibration of an odd dimensional unit sphere S^{2n+1} over complex projective space $\mathbb{C}P^n$ of constant holomorphic curvature equal to 4. The standard contact structure on S^{2n+1} obtained in Example 3.2.3 can also be obtained by the above construction. Additional details of the geometry of Boothby–Wang fibrations will be given from time to time, particularly in Examples 4.5.4 and 6.7.2.

Theorem 3.9 *Let (M^{2n+1}, η') be a compact regular contact manifold; then there exists a contact form $\eta = \tau\eta'$ for some non-vanishing function τ whose characteristic vector field ξ generates a free effective S^1 on M^{2n+1}. Moreover M^{2n+1} is the bundle space of a principal circle bundle $\pi : M^{2n+1} \longrightarrow M^{2n}$ over a symplectic manifold M^{2n} whose symplectic form Ω determines an integral cocycle on M^{2n} and η is a connection form on the bundle with curvature form $d\eta = \pi^*\Omega$.*

Proof. Since η' is regular, its characteristic vector field ξ' is a regular vector field and hence its maximal integral curves or orbits are closed subsets of M^{2n+1}; but M^{2n+1} is compact, so these integral curves are homeomorphic to circles. Moreover, as ξ' is regular, M^{2n+1} is a fibre bundle over a manifold M^{2n} (the set of maximal integral curves with the quotient topology, see e.g., Palais [1957]) and we denote the projection by π.

Now let $f'_t : M^{2n+1} \longrightarrow M^{2n+1}$ denote the 1-parameter group of diffeo-morphisms generated by ξ' and define the period λ' of ξ' at $m \in M^{2n+1}$ by $\lambda'(m) = \inf\{t|t > 0, f'_t(m) = m\}$. λ' is constant on each orbit and since there are no fixed points, λ' is never zero. Also as the orbits are circles, λ' is not infinite. We will show that λ' is constant on all of M^{2n+1}. Our argument is due to Tanno [1965]. Let k be a Riemannian metric on M^{2n} and let $g = \pi^*k + \eta' \otimes \eta'$. Then g is a Riemannian metric on M^{2n+1} and ξ' is a unit Killing vector field with respect to g since $\eta'(\xi') = 1$ and $\mathcal{L}_{\xi'}\eta' = 0$. If ∇ denotes the Levi-Cività connection of g, $g(\nabla_{\xi'}\xi', X) = -g(\nabla_X\xi', \xi') = 0$ and hence the orbits of ξ are geodesics. If γ is an orbit through m, let γ' be an orbit sufficiently near to γ that there exists a unique minimal geodesic from m to γ' meeting γ' orthogonally at m'. Then since f'_t is an isometry for all t, the image of the geodesic arc $\widehat{mm'}$ is orthogonal to γ and γ' for all t. Thus, as a point m on γ moves through one period along γ, the corresponding point on γ' moves through one period and hence λ' is constant on M^{2n+1}.

Now define η and ξ by $\eta = \frac{1}{\lambda'}\eta'$ and $\xi = \lambda'\xi'$. Since λ' is constant, ξ is the characteristic vector field of the contact form η. Moreover ξ has the same orbits as ξ' and its period function $\lambda = 1$. Thus the one-parameter group f_t of ξ depends only on the equivalence class mod 1 of t and the action so induced of S^1 is effective and free.

Since ξ is regular we may cover M^{2n+1} by coordinate neighborhoods with coordinates $(x^1, \ldots, x^{2n+1})$ such that the integral curves of ξ are given by $x^1 = const., \ldots, x^{2n} = const.$ Projecting such neighborhoods we obtain an open cover $\{\mathcal{U}_i\}$ of M^{2n}, and on each $\mathcal{U}_i$ we define a local cross section s_i by setting $x^{2n+1} = const.$ Then define $F_i : \mathcal{U}_i \times S^1 \longrightarrow S^1$ by $F_i(p, t) = s_i(p)t$. The transition functions for the bundle structure are then given by $f_{ij}(p) = F_i(p,t)F_j(p,t)^{-1}$.

We have already seen that $\mathcal{L}_\xi\eta = 0$ and $\mathcal{L}_\xi d\eta = 0$, so that η and $d\eta$ are invariant under the action of S^1. Now take $A = \frac{d}{dt}$ as a basis of $\mathfrak{S}^1 = \mathbb{R}$, the Lie algebra of S^1 and set $\tilde{\eta} = \eta A$ so that η may be regarded as a Lie algebra-valued 1-form. For an element $B \in \mathfrak{S}^1$ denote by B^* the induced vector field on M^{2n+1}. In particular $A^* = \xi$, so that $\tilde{\eta}(A^*) = A$. Moreover right translation by $t \in S^1$ is just f_t so that $R_t^*\tilde{\eta} = \tilde{\eta}$ by the invariance of η under the S^1 action. Thus, η (precisely $\tilde{\eta}$) is a connection form on M^{2n+1}.

If $\tilde{\Omega}$ is the curvature form of η, then the structural equation is $d\eta = -\frac{1}{2}[\eta, \eta] + \tilde{\Omega} = \tilde{\Omega}$ since S^1 is abelian. On the other hand $d\eta$ is horizontal and invariant, so there exists a 2-form Ω on M^{2n} such that $d\eta = \pi^*\Omega$. Now $\pi^*d\Omega = d\pi^*\Omega = d^2\eta = 0$ so that $d\Omega = 0$ and $\pi^*(\Omega)^n = (\pi^*\Omega)^n = (d\eta)^n \neq 0$ giving $\Omega^n \neq 0$. Therefore M^{2n} is symplectic. Finally, since the transition functions f_{ij} are real(mod 1)-valued, $[\Omega] \in H^2(M^{2n}, \mathbb{Z})$ (see e.g., Kobayashi [1956]). ■

3.4   The Weinstein conjecture

In Example 3.2.6 we studied a contact structure on the 3-dimensional torus and observed that some of the orbits of ξ are closed and some are not. It is a well-known conjecture of Weinstein [1979] that on a compact contact manifold M satisfying $H^1(M, \mathbb{R}) = 0$, ξ must have a closed orbit. The present author knows of no example of a compact contact manifold not satisfying $H^1(M, \mathbb{R}) = 0$ for which ξ does not have a closed orbit and believes the conjecture is true without the assumption of $H^1(M, \mathbb{R}) = 0$. In view of the example on the torus the following result of Petkov and Popov [1995] is interesting. Let (M^{2n+1}, η) be an analytic, connected contact manifold with complete characteristic vector field ξ. Since $\eta \wedge (d\eta)^n$ is invariant under the action of ξ, it induces an invariant Lebesgue measure on M^{2n+1}. A point $m \in M^{2n+1}$ is a *periodic point* if the integral curve of ξ through m is periodic. Petkov and Popov prove that either the set of periodic points has measure 0 or there exists $T > 0$ such that $exp(T\xi)(m) = m$ for every point m. Note that for some m, T could be a multiple of the period for that m (cf. the notion of an almost regular contact structure below).

Interest in the Weinstein conjecture has often been phrased in terms of the question of the existence of periodic orbits of Hamiltonian systems. In Section 1.1 we considered briefly a real-valued function H on a symplectic manifold (M, Ω) and its Hamiltonian vector field X_H. Since $X_H H = 0$, X_H is tangent to the level (energy) surfaces of H. Thus if a level surface S is of contact type, the Hamiltonian vector field X_H is collinear with the characteristic vector field ξ.

For hypersurfaces in $\mathbf{R}^{2n+2}$ with the standard symplectic structure, the Weinstein conjecture is known to be true if the hypersurface is convex (Weinstein [1978]), star-shaped (Rabinowitz [1978]) and, more recently, if the hypersurface is of contact type and without the assumption $H^1(M, \mathbb{R}) = 0$ (Viterbo [1987], see also Hofer and Zehnder [1987]). As an aside it is interesting to note that Ginzburg [1995] showed that if one gives up on the hypersurface being of contact type, then there exist imbeddings of S^{2n+1}, $n \geq 3$, in $\mathbf{R}^{2n+2}$ whose Hamiltonian vector field has no closed orbits.

In cotangent bundles T^*M, first note that if S is a compact connected hypersurface, then $T^*M \setminus S$ has exactly two components, one of which is bounded; Hofer and Viterbo [1988] prove that if the bounded component of $T^*M \setminus S$ contains the zero section and if S is of contact type, then ξ has a closed orbit.

Turning to compact contact manifolds in general (not necessarily hypersurfaces of contact type), Hofer [1993] showed that the Weinstein conjecture is true on a compact overtwisted contact manifold, on the 3-sphere for any contact form and on a closed orientable 3-manifold with $\pi_2 \neq 0$.

Banyaga [1990] showed that the Weinstein conjecture is true if the contact form is C^2 close to a regular contact form (again without assuming $H^1(M, \mathbb{R}) = 0$). C. B. Thomas [1976] introduced the notion of an almost regular contact structure. A contact structure is said to be *almost regular* if there exists a positive integer N such that every point has a neighborhood such that any integral curve of ξ passing through the neighborhood, passes through at most N times. With this idea in mind Banyaga and Rukimbira [1994] showed that the Weinstein conjecture is true if the contact form is C^1 close to an almost regular contact form.

A contact manifold is called an R-*contact manifold* (Rukimbira [1993]) if ξ is Killing with respect to some (not necessarily associated (Chapter 4)) Riemannian metric g for which $\eta(X) = g(X, \xi)$. For compact contact manifolds admitting such a metric the Weinstein conjecture is true (Rukimbira [1993]). An auxiliary result of Rukimbira [1994] is that for a compact K-contact manifold M^{2n+1} (see Subsection 4.5.4 or Section 6.2), the dimension of any leaf closure is at most the smaller of $n+1$ and $2n+1$ minus the rank of the vector space of harmonic vector fields. Also on a compact K-contact manifold, ξ has at least $n + 1$ closed orbits (Rukimbira [1995a]) and a compact simply connected K-contact M^{2n+1} with exactly $n + 1$ closed orbits is homeomorphic to a sphere (Rukimbira [1999]).

4
Associated Metrics

4.1 Almost complex and almost contact structures

We will generally regard the theory of almost Hermitian structures as well known and give here only definitions and a few properties that will be important for our study; many of these were already mentioned in Chapter 1. For more detail the reader is referred to Gray and Hervella [1980], Kobayashi–Nomizu [1963-69, Chapter IX] and Kobayashi–Wu [1983]; also, despite its classical nature, the book of Yano [1965] contains helpful information on many of these structures.

An *almost complex structure* is a tensor field J of type $(1,1)$ such that $J^2 = -I$. A *Hermitian metric* on an almost complex manifold (M, J) is a Riemannian metric which is invariant by J, i.e.,

$$g(JX, JY) = g(X, Y).$$

Noting that J is negative-self-adjoint with respect to g, i.e., $g(X, JY) = -g(JX, Y)$, $\Omega(X, Y) = g(X, JY)$ defines a 2-form called the *fundamental 2-form* of the *almost Hermitian structure* (M, J, g). If M is a complex manifold and J the corresponding almost complex structure, we say that (M, J, g) is a *Hermitian manifold*. If $d\Omega = 0$, the structure is *almost Kähler*. For geometers working strictly over the complex domain, a Hermitian metric is a Hermitian quadratic form and hence complex-valued; it takes its non-zero values as appropriate when one argument is holomorphic and the other anti-holomorphic. Our metric g becomes half the real part of $g(X - iJX, Y + iJY)$. For geometers concerned with a variety of structures presented for the most part in terms of real tensor fields, as we will be, it has become quite standard to use the word

Hermitian as we have done. A reader interested in this point may want to see Kobayashi–Wu [1983, pp. 80-81] for commentary.

Note also that every almost complex manifold admits a Hermitian metric, for if k is any Riemannian metric, g defined by

$$g(X,Y) = k(X,Y) + k(JX, JY)$$

is Hermitian.

Given (M, J, g) we can construct a particular local orthonormal basis as follows. Let $\mathcal{U}$ be a coordinate neighborhood on M and X_1 any unit vector field on $\mathcal{U}$. Let $X_{1*} = JX_1$. Now choose a unit vector field X_2 orthogonal to both X_1 and X_{1*}. Then JX_2 is also orthogonal to X_1 and X_{1*}. Continuing in this manner we have a local orthonormal basis of the form $\{X_1, \dots, X_n, JX_1, \dots, JX_n\}$. Such a basis is called a *J-basis*. Note in particular that an almost complex manifold is even-dimensional.

Again given (M, J) choose g Hermitian. Let $\{\mathcal{U}_\alpha\}$ be an open cover with *J*-bases $\{X_i, X_{i*}\}$, $\{\bar{X}_i, \bar{X}_{i*}\}$ on $\mathcal{U}_\alpha$ and $\mathcal{U}_\beta$ respectively. With respect to these bases J is given by

$$\begin{pmatrix} 0 & -I \\ I & 0 \end{pmatrix}.$$

If now $X \in T_m M$, $m \in \mathcal{U}_\alpha \cap \mathcal{U}_\beta$, then for the column vectors of components (X) and $(\bar{X})$ with respect to these bases

$$(\bar{X}) = \begin{pmatrix} A & B \\ C & D \end{pmatrix} (X)$$

where A, B, C, D are $n \times n$ matrices and $\begin{pmatrix} A & B \\ C & D \end{pmatrix} \in O(2n)$. Now

$$\begin{pmatrix} 0 & -I \\ I & 0 \end{pmatrix} \begin{pmatrix} A & B \\ C & D \end{pmatrix} (X) = (\overline{JX}) = \begin{pmatrix} A & B \\ C & D \end{pmatrix} (JX)$$

$$= \begin{pmatrix} A & B \\ C & D \end{pmatrix} \begin{pmatrix} 0 & -I \\ I & 0 \end{pmatrix} (X),$$

i.e., $\begin{pmatrix} A & B \\ C & D \end{pmatrix}$ and $\begin{pmatrix} 0 & -I \\ I & 0 \end{pmatrix}$ commute. Therefore $D = A$ and $C = -B$ and hence $\begin{pmatrix} A & B \\ C & D \end{pmatrix} \in U(n)$. In particular the structural group of the tangent bundle of an almost complex manifold is reducible to $U(n)$. Recall also that $\det \begin{pmatrix} A & B \\ -B & A \end{pmatrix} = |\det(A + iB)|^2 > 0$ and therefore an almost complex manifold is orientable.

Conversely suppose that we are given M such that the structural group of TM can be reduced to $U(n)$. Let $\{\mathcal{U}_\alpha\}$ be an open cover such that we can choose local orthonormal bases which transform in the overlaps of neighborhoods by the action of $U(n)$. In each $\{\mathcal{U}_\alpha\}$ define J_α by $\begin{pmatrix} 0 & -I \\ I & 0 \end{pmatrix}$; this matrix commutes with $U(n)$ and hence the set $\{J_\alpha\}$ determines a global tensor field J such that $J^2 = -I$. Thus an almost complex (Hermitian) structure on M can be thought of as a reduction of the structural group to $U(n)$.

As we will see in our discussion of associated metrics below, the structural group of a symplectic manifold is reducible to $U(n)$ and that of a contact manifold to $U(n) \times 1$ (Chern [1953]). For an odd-dimensional manifold M^{2n+1} J. Gray [1959] defined an *almost contact structure* as a reduction of the structural group to $U(n) \times 1$. In terms of structure tensors we say M^{2n+1} has an *almost contact structure* or sometimes (ϕ, ξ, η)-*structure* if it admits a tensor field ϕ of type (1,1), a vector field ξ and a 1-form η satisfying

$$\phi^2 = -I + \eta \otimes \xi, \quad \eta(\xi) = 1.$$

Many authors include also that $\phi\xi = 0$ and $\eta \circ \phi = 0$; however these are deducible from the other conditions as we now show.

Theorem 4.1 *Suppose M^{2n+1} has a (ϕ, ξ, η)-structure. Then $\phi\xi = 0$ and $\eta \circ \phi = 0$. Moreover the endomorphism ϕ has rank $2n$.*

Proof. First note that $\phi^2 = -I + \eta \otimes \xi$ and $\eta(\xi) = 1$ give $\phi^2\xi = -\xi + \eta(\xi)\xi = 0$ and hence either $\phi\xi = 0$ or $\phi\xi$ is a non-trivial eigenvector of ϕ corresponding to the eigenvalue 0. Using $\phi^2 = -I + \eta \otimes \xi$ again we have $0 = \phi^2\phi\xi = -\phi\xi + +\eta(\phi\xi)\xi$ or $\phi\xi = \eta(\phi\xi)\xi$. Now if $\phi\xi$ is a non-trivial eigenvector of the eigenvalue 0, $\eta(\phi\xi) \neq 0$ and therefore $0 = \phi^2\xi = \eta(\phi\xi)\phi\xi = (\eta(\phi\xi))^2\xi \neq 0$, a contradiction. Thus, $\phi\xi = 0$.

Now since $\phi\xi = 0$, we also have that $\eta(\phi X)\xi = \phi^3 X + \phi X = -\phi X + \phi(\eta(X)\xi) + \phi X = 0$ for any vector field X and hence $\eta \circ \phi = 0$.

Finally since $\phi\xi = 0$, $\xi \neq 0$ everywhere, rank$\phi < 2n + 1$. If a vector field $\bar{\xi}$ satisfies $\phi\bar{\xi} = 0$, $\phi^2 = -I + \eta \otimes \xi$ gives $0 = -\bar{\xi} + \eta(\bar{\xi})\xi$; thus $\bar{\xi}$ is collinear with ξ and so rank$\phi = 2n$. ∎

If a manifold M^{2n+1} with a (ϕ, ξ, η)-structure admits a Riemannian metric g such that

$$g(\phi X, \phi Y) = g(X, Y) - \eta(X)\eta(Y),$$

we say M^{2n+1} has an *almost contact metric structure* and g is called a *compatible* metric. Setting $Y = \xi$ we have immediately that

$$\eta(X) = g(X, \xi).$$

As in the case of an almost complex structure the existence of the compatible metric is easy. For if k' is any metric, first set $k(X, Y) = k'(\phi^2 X, \phi^2 Y) + \eta(X)\eta(Y)$; then $\eta(X) = g(X, \xi)$. Now define g by

$$g(X, Y) = \frac{1}{2}\left(k(X, Y) + k(\phi X, \phi Y) + \eta(X)\eta(Y)\right)$$

and check the details.

For a manifold M^{2n+1} with an almost contact metric structure (ϕ, ξ, η, g) we can also construct a useful local orthonormal basis. Let $\mathcal{U}$ be a coordinate neighborhood on M and X_1 any unit vector field on $\mathcal{U}$ orthogonal to ξ. Then $X_{1^*} = \phi X_1$ is a unit vector field orthogonal to both X_1 and ξ. Now choose a unit vector field X_2 orthogonal to ξ, X_1 and X_{1^*}. Then ϕX_2 is also a unit vector field orthogonal to ξ, X_1, X_{1^*} and X_2. Proceeding in this way we obtain a local orthonormal basis $\{X_i, X_{i^*} = \phi X_i, \xi\}$ called a ϕ-*basis*.

Now given a manifold M^{2n+1} with a (ϕ, ξ, η)-structure, let g be a compatible metric and $\{\mathcal{U}_\alpha\}$ an open cover with ϕ-bases $\{X_i, X_{i^*}, \xi\}$. With respect to such a basis ϕ is given by the matrix

$$\begin{pmatrix} 0 & -I & 0 \\ I & 0 & 0 \\ 0 & 0 & 0 \end{pmatrix}.$$

Proceeding as in the almost complex case, we see that the structural group of M^{2n+1} is reducible to $U(n) \times 1$. Conversely given an almost contact structure as defined by this reduction of the structural group and an open cover $\{\mathcal{U}_\alpha\}$ respecting the action of $U(n) \times 1$, define ϕ_α on $\mathcal{U}_\alpha$ by the above matrix and define η_α and ξ_α by row and column vectors of zeros with last entry 1. Then, again as in the almost complex case, this defines global structure tensors (ϕ, ξ, η) satisfying $\phi^2 = -I + \eta \otimes \xi$ and $\eta(\xi) = 1$. We shall subsequently speak of an almost contact structure (ϕ, ξ, η) and suppress the terminology "(ϕ, ξ, η)-structure".

4.2   Polarization and associated metrics

We begin with a discussion of the well-known decomposition, called "polarization", of a non-singular matrix A into the product of an orthogonal matrix F and a positive definite symmetric matrix G. Let $H(n)$ denote the set of positive definite symmetric $n \times n$ matrices and as usual $O(n)$ the orthogonal group. In treating the subject of constructing Riemannian metrics associated to 2-forms of rank $2r$, Y. Hatakeyama [1962] proved the analyticity of the polar decomposition; that the decomposition is continuous can be found in

Chevalley [1946, p. 14-16]. We prove Hatakeyama's result by a sequence of Lemmas.

Lemma 4.1 *For $G \in H(n)$, the map $\sigma_G : \mathfrak{gl}(n, \mathbb{R}) \longrightarrow \mathfrak{gl}(n, \mathbb{R})$ given by $\sigma_G(A) = GAG^{-1}$ has positive eigenvalues.*

Proof. Let $\lambda_i > 0$ be the eigenvalues of G. There exists $P \in O(n)$ such that PGP^{-1} is diagonal, say Δ. Then $\sigma_P^{-1}\sigma_\Delta\sigma_P = \sigma_G$ and hence σ_Δ and σ_G have the same eigenvalues. Now $\sigma_\Delta(A)_{il} = (\Delta A \Delta^{-1})_{il} = \sum_{jk} \lambda_i \delta_{ij} a_{jk} \delta_{kl} \frac{1}{\lambda_l} = \frac{\lambda_i}{\lambda_l} a_{il}$. Thus the n^2 eigenvalues of σ_G are the positive numbers $\frac{\lambda_i}{\lambda_l}$. ■

Lemma 4.2 *For $G \in H(n)$ and A skew-symmetric, AG symmetric implies that $A = 0$.*

Proof. $AG = G^T A^T = -GA$. Therefore $\sigma_G(A) = -A$ and so by the previous lemma $A = 0$. ■

Now $O(n)$ and $H(n)$ are analytic submanifolds of $GL(n, \mathbb{R})$; thus $\varphi : O(n) \times H(n) \longrightarrow GL(n, \mathbb{R})$ defined by $\varphi(F, G) = FG$ is analytic.

Lemma 4.3 *$d\varphi$ is one-to-one and hence φ^{-1} given by polarization is analytic.*

Proof. Let $X \in T_{(F,G)}O(n) \times H(n)$ and consider the curve $(Fe^{tA}, G + tB)$ where A is skew-symmetric and B is symmetric and which has tangent X at (F, G).

$$d\varphi(X) = \lim_{t \to 0} \frac{Fe^{tA}(G + tB) - FG}{t} = FAG + FB.$$

If now $d\varphi(X) = 0$, $F(AG + B) = 0$. Therefore $AG = -B$ which is symmetric. Thus by Lemma 4.2, $A = 0$ and hence also $B = 0$. ■

Theorem 4.2 *Polarization as a map from $GL(n, \mathbb{R}) \longrightarrow O(n) \times H(n)$ gives an analytic diffeomorphism between these manifolds with respect to the usual analytic structures.*

We now prove the existence of associated metrics.

Theorem 4.3 *Let (M^{2n}, Ω) be a symplectic manifold. Then there exists a Riemannian metric g and an almost complex structure J such that*

$$g(X, JY) = \Omega(X, Y).$$

Proof. Let k be any Riemannian metric on M and let $\{X_1, \ldots, X_{2n}\}$ be a local k-orthonormal basis. Let $A_{ij} = \Omega(X_i, X_j)$. A is a $2n \times 2n$ non-singular skew-symmetric matrix. By polarization we have $A = FG$ for some orthogonal matrix F and positive definite symmetric matrix G. Now define g and J by

$$g(X_i, X_j) = G_{ij}, \quad JX_i = F_i^j X_j.$$

g is independent of the choice of k-orthonormal basis. For if $\{Y_1, \dots, Y_{2n}\}$ is another k-orthonormal basis, there is an orthogonal matrix P such that

$$B_{ij} = \Omega(Y_i, Y_j) = \Omega(P^k{}_i X_k, P^l{}_j X_l) = P^k{}_i P^l{}_j A_{kl} = (PAP^{-1})_{ij}.$$

If $B = \Phi\Gamma$ is the polarization of B, $\Phi\Gamma = PFP^{-1}PGP^{-1}$ and so by the uniqueness of the polar decomposition $\Phi = PFP^{-1}$ and $\Gamma = PGP^{-1}$. Thus, in particular, we see that g and J are globally defined. Also since A is skew-symmetric, $F^2 = -I$ and F is skew-symmetric. To see this note that $A^T = GF^T = -FG$ and hence applying F on the right, $G = -FFF^TGF$. But F^TGF is positive definite symmetric and so the uniqueness of the decomposition gives $-F^2 = I$. Finally $F = -F^{-1} = -F^T$.   ■

In particular given a symplectic manifold (M^{2n}, Ω) we say a Riemannian metric g is an *associated metric* if there exists an almost complex structure J such that $g(X, JY) = \Omega(X, Y)$. We remark that if g is an associated metric and one uses it as the starting metric k in the above polarization process, the process yields the metric g back again.

Theorem 4.4 *Let* (M^{2n+1}, η) *be a contact manifold and* ξ *its characteristic vector field. Then there exists an almost contact metric structure such that* $g(X, \phi Y) = d\eta(X, Y)$.

Proof. This time the proof is a two step process. First let k' be any Riemannian metric and define a new metric k by

$$k(X, Y) = k'(-X + \eta(X)\xi, -Y + \eta(Y)\xi) + \eta(X)\eta(Y).$$

Then $k(X, \xi) = \eta(X)$. Now polarize $d\eta$ on the contact subbundle $\mathcal{D}$ as in the symplectic case. This gives a metric g' and almost complex structure ϕ' on $\mathcal{D}$ such that $g'(X, \phi'Y) = d\eta(X, Y)$. Extending g' to a metric g agreeing with k in the direction ξ and extending ϕ' to a field of endomorphisms ϕ by requiring $\phi\xi = 0$, we have an almost contact metric structure (ϕ, ξ, η, g) such that $g(X, \phi Y) = d\eta(X, Y)$.   ■

As in the symplectic case, given a contact manifold (M^{2n+1}, η) we say a Riemannian metric g is an *associated metric* if there exists an almost contact metric structure such that $g(X, \phi Y) = d\eta(X, Y)$. In this case we also speak of a *contact metric structure*; other authors often use the phrase *contact Riemannian structure*. Working strictly with structure tensors, one may avoid the polarization process, by defining an associated metric as follows: Given a contact manifold (M^{2n+1}, η) with characteristic vector field ξ, a Riemannian metric g is an associated metric if there exists a tensor field of type $(1,1)$, ϕ such that

$$\phi^2 = -I + \eta \otimes \xi, \quad \eta(X) = g(X, \xi), \quad d\eta(X, Y) = g(X, \phi Y).$$

Finally we caution that it is possible to have a contact manifold (M^{2n+1}, η) with characteristic vector field ξ and an almost contact metric structure (ϕ, ξ, η, g), same ξ and η, without $g(X, \phi Y) = d\eta(X, Y)$; see Example 4.5.3 for an example.

In the course of this book we will give many properties of associated metrics; we give one simple property here as it is discussed in Example 4.5.5 and used periodically.

Theorem 4.5 *On a contact metric manifold the integral curves of ξ are geodesics.*

Proof. For a contact metric structure we have

$$0 = (\pounds_\xi \eta)(X) = \xi g(X, \xi) - g(\nabla_\xi X - \nabla_X \xi, \xi) = g(X, \nabla_\xi \xi)$$

so the integral curves of ξ are geodesics.   ∎

In our discussion above in both the symplectic and contact cases we started with an arbitrary Riemannian metric and obtained an associated metric. Thus we are led to believe that there are many associated metrics to a given symplectic or contact form. Indeed this is the case and we now show that the set $\mathcal{A}$ of all associated metrics is infinite dimensional by exhibiting a path of metrics in $\mathcal{A}$ determined by a C^∞ function with compact support. Such paths of associated metrics will be useful to us in the study of critical points of curvature functionals on $\mathcal{A}$. We give the construction in the symplectic case and then remark on the similarity with the contact case.

Let f be a C^∞ function with compact support contained in a neighborhood $\mathcal{U}$ of (M^{2n}, Ω) and $\{X_1, \dots, X_n, X_{1*}, \dots, X_{n*}\}$ a local J-basis. Let g be an associated metric. Make no change in g outside $\mathcal{U}$ and on $\mathcal{U}$ change g only in the planes spanned by $\{X_1, X_{1*}\}$ by the matrix

$$\begin{pmatrix} 1 + tf + \frac{1}{2}t^2 f^2 & \frac{1}{2}t^2 f^2 \\ \frac{1}{2}t^2 f^2 & 1 - tf + \frac{1}{2}t^2 f^2 \end{pmatrix}.$$

This defines a path of metrics g_t and it is easy to check that each g_t is an associated metric for the symplectic form Ω. In the contact case simply begin with a ϕ-basis and make the same construction.

Also, as already remarked in Section 1.1, it is evident that in the symplectic case, $\mathcal{A}$ may be thought of as the set of all almost Kähler metrics that have Ω as their fundamental 2-form.

On the other hand it is possible for a Riemannian metric g to be an associated metric for more than one symplectic structure. For example on a *hyper-Kähler manifold* one has, by definition, three independent global Kähler structures (J_a, g), $a = 1, 2, 3$, satisfying $J_1 J_2 + J_2 J_1 = 0$, $J_3 = J_1 J_2$. Thus the three fundamental 2-forms give three symplectic structures with g an associated metric for each of them.

We close this section by showing that all associated metrics have the same volume element and give the proof only in the symplectic case.

Theorem 4.6 Let (M^{2n}, Ω) be a symplectic manifold (resp. (M^{2n+1}, η) a contact manifold) and g an associated metric. Then

$$dV = \frac{(-1)^n}{2^n n!} \Omega^n, \ (resp. \ dV = \frac{(-1)^n}{2^n n!} \eta \wedge (d\eta)^n)).$$

Proof. Let $X_1, \dots, X_n, X_{1^*}, \dots, X_{n^*}$ be a J-basis and $\theta^1, \dots, \theta^n, \theta^{1^*}, \dots, \theta^{n^*}$ the dual basis. Then with respect to this dual basis $dV = \theta^1 \wedge \theta^{1^*} \wedge \theta^2 \wedge \theta^{2^*} \wedge$

$$\cdots \wedge \theta^n \wedge \theta^{n^*} \text{ and } \Omega = \sum_{i=1}^{n} (\theta^{i^*} \wedge \theta^i - \theta^i \wedge \theta^{i^*}) = -2 \sum_{i=1}^{n} \theta^i \wedge \theta^{i^*}. \text{ Thus}$$

$$\Omega^n = (-2)^n ((\theta^1 \wedge \theta^{1^*}) + \cdots + (\theta^n \wedge \theta^{n^*}))^n$$

$$= (-2)^n n! (\theta^1 \wedge \theta^{1^*}) \wedge \cdots \wedge (\theta^n \wedge \theta^{n^*}) = (-2)^n n! dV.$$

∎

4.3   Polarization of metrics as a projection

In the previous section we created associated metrics from an arbitrary metric by polarization. In this section we give some further properties of the set $\mathcal{A}$ and discuss how $\mathcal{A}$ sits in the set $\mathcal{N}$ of all Riemannian metrics with the same volume element. Restricting ourselves to $\mathcal{N}$ we will interpret the polarization process of constructing associated metrics from a given metric as a projection from $\mathcal{N}$ onto $\mathcal{A}$.

On a compact manifold M the set $\mathcal{M}$ of all Riemannian metrics may be given a Riemannian metric (Ebin [1970]): The tangent space, $T_g\mathcal{M}$, of $\mathcal{M}$ at a metric g is the space of symmetric tensor fields of type (0,2). For symmetric tensor fields S and T of type (0,2) we define a Riemannian metric $(.,.)$ by

$$(S, T)_g = \int_M S_{ij} T_{kl} g^{ik} g^{jl} dV_g.$$

For the set $\mathcal{N}$ of metrics with the same volume element, the geodesics in $\mathcal{N}$ were found by Ebin [1970] and are curves of the form ge^{St} where S is symmetric with $\operatorname{tr} S = 0$. $g_t = ge^{St}$ is computed by

$$g_t(X, Y) = g(X, e^{St} Y)$$

where here e^{St} acts on Y as a tensor field of type (1,1). Again and throughout we shall often denote a symmetric tensor field of type (0,2) and the corresponding tensor field of type (1,1) by the same letter. In fact in most of our computations we begin with a local orthonormal basis for some $g \in \mathcal{A}$ and regard S simply as its matrix of components.

As an aside, we remark that one can think of metrics with the same total volume on an n-dimensional manifold as being at a fixed distance from the zero tensor. Consider the path tg, $t \in [0, 1]$ from the zero tensor to the metric g; then since g may be considered as the tangent to the path at each point, we have the following entertaining computation.

$$|g| \equiv \int_0^1 (g, g)_{tg}^{1/2} dt = \int_0^1 \left(\frac{n}{t^2} \int_M \sqrt{t^n \det g} \, dx^1 \cdots dx^n \right)^{1/2} dt = 4 \sqrt{\frac{\mathrm{vol}_g M}{n}}.$$

4.3.1 Some linear algebra

For the projection result below, Theorem 4.8, we will need the polarization of a particular path in $GL(2n, \mathbb{R})$. Let $\mathcal{J}$ denote the matrix $\begin{pmatrix} 0 & -I \\ I & 0 \end{pmatrix}$ and let S be any symmetric $2n \times 2n$ matrix. First diagonalize S, say $Q^{-1}SQ = \Lambda$, $Q \in O(2n)$, Λ diagonal, and set $P = Qe^{-\frac{1}{2}\Lambda t}$. Our problem will be to polarize $P^T \mathcal{J} P$, i.e., find $F(t) \in O(2n)$, $G(t) \in H(2n)$ such that $P^T \mathcal{J} P = F(t)G(t)$.

Lemma 4.4 *Any symmetric $2n \times 2n$ matrix S can be uniquely written as $B + D$, where B and D are symmetric and $\mathcal{J}B - B\mathcal{J} = 0$ and $\mathcal{J}D + D\mathcal{J} = 0$.*

Proof. Setting $B = \frac{1}{2}(-\mathcal{J}S\mathcal{J} + S)$ and $D = \frac{1}{2}(\mathcal{J}S\mathcal{J} + S)$ the decomposition is immediate. If now $B + D = B' + D'$, $B - B' = D' - D$ and hence $\mathcal{J}(B - B') = (B - B')\mathcal{J}$ gives $\mathcal{J}(D' - D) = (D' - D)\mathcal{J} = -\mathcal{J}(D' - D)$. Therefore $\mathcal{J}(D' - D) = 0$ and so, since $\mathcal{J}$ is non-singular, $D' = D$ and in turn $B' = B$. ∎

Of course B and D need not commute with each other. We will see below that if $[B, D] = 0$, then $F(t) = Q^{-1}\mathcal{J}Q$ and $G(t) = Q^{-1}e^{-Bt}Q$ and conversely if either $F(t)$ or $G(t)$ have this simple form, B and D commute.

As a matter of notation, using the analyticity of the polar decomposition, we write

$$F(t) = \sum_{k=0}^{\infty} F^{(k)}t^k, \quad G(t) = \sum_{k=0}^{\infty} G^{(k)}t^k.$$

At $t = 0$, $P^T \mathcal{J} P = Q^{-1}\mathcal{J}Q$ which we denote by M. Since $\mathcal{J} \in O(2n)$, $F(0) = M$ and $G(0) = I$. The main lemma of this section is the following; due to its complexity we refer to the author's paper [1983] for its proof.

Lemma 4.5

$$F^{(k)} = \frac{1}{2}\sum_{j=1}^{k-1}\left(\frac{(-1)^k}{k!2^k}\binom{k}{j}Q^{-1}\mathcal{J}[(B-D)^{k-j},(B+D)^j]Q\right.$$

$$\left. -F^{(k-j)}G^{(j)} - MG^{(j)}F^{(k-j)}M + MF^{(k-j)}F^{(j)}\right),$$

$$G^{(k)} = MF^{(k)} + M\sum_{j=1}^{k-1}F^{(k-j)}G^{(j)}$$

$$+\frac{(-1)^k}{k!2^k}Q^{-1}\left(\sum_{j=0}^{k}\binom{k}{j}(B-D)^{k-j}(B+D)^j\right)Q.$$

From this lemma we see easily that

$$F^{(1)} = 0, \quad F^{(2)} = \frac{1}{4}Q^{-1}\mathcal{J}[B,D]Q.$$

Continuing we can find $F^{(k)}$ and $G^{(k)}$ as far as desired; in particular

$$G^{(1)} = -Q^{-1}BQ, \quad G^{(2)} = \frac{1}{2}Q^{-1}B^2Q,$$

$$G^{(3)} = Q^{-1}\left(-\frac{1}{6}B^3 - \frac{1}{24}(BD^2 - 2DBD + D^2B) - \frac{1}{8}(B^2D - 2BDB + DB^2)\right)Q,$$

$$G^{(4)} = Q^{-1}\left(\frac{1}{24}B^4 + \frac{1}{16}(B^3D - B^2DB - BDB^2 + DB^3)\right.$$

$$+\frac{1}{96}(2B^2D^2 - 7BDBD + 3DB^2D + 7BD^2B - 7DBDB + 2D^2B^2))Q.$$

Corollary 4.1 $[B,D] = 0$ *implies* $F(t) = Q^{-1}\mathcal{J}Q$ *and* $G(t) = Q^{-1}e^{-Bt}Q$. *Conversely either of these implies the commutativity.*

Proof. $[B,D] = 0$ implies that the bracket in $F^{(k)}$ vanishes and hence by induction $F^{(k)} = 0$ for $k > 0$. Again if B and D commute

$$\sum_{j=0}^{k}\binom{k}{j}(B-D)^{k-j}(B+D)^j = 2^kB^k$$

and hence $G(t) = Q^{-1}e^{-Bt}Q$.

Clearly $F(t) = Q^{-1}\mathcal{J}Q$ gives $F^{(2)} = \frac{1}{4}Q^{-1}\mathcal{J}[B,D]Q = 0$ and hence $[B,D] = 0$. Finally if $G(t) = Q^{-1}e^{-Bt}Q$,

$$QG^{(3)}Q^{-1} + \frac{1}{6}B^3 = 0, \quad QG^{(4)}Q^{-1} - \frac{1}{24}B^4 = 0.$$

Thus if we multiply the rest of the expression for $G^{(3)}$ on the left by $48B$ and separately on the right by $48B$ and add these two to 96 times the rest of $G^{(4)}$ we have

$$-3BDBD + 3DB^2D + 3BD^2B - 3DBDB = 0.$$

Thus $[B, D]^2 = 0$, but $[B, D]$ is skew-symmetric and hence $[B, D] = 0$. ∎

Lemma 4.6 $\mathcal{J}e^S = e^{-S}\mathcal{J}$ if and only if $S\mathcal{J} + \mathcal{J}S = 0$.

Proof. The sufficiency is clear, so we prove only the necessity. Suppose $SX = \lambda X$, then $\mathcal{J}e^S X = e^\lambda \mathcal{J}X = e^{-S}\mathcal{J}X$. Also let $\{e_i\}$ be an orthonormal eigenvector basis of S with $Se_i = \lambda_i e_i$. If now $\mathcal{J}X = Y^i e_i$,

$$e^{-S}\mathcal{J}X = e^\lambda \mathcal{J}X = e^\lambda Y^i e_i$$

and

$$e^{-S}\mathcal{J}X = e^{-S}Y^i e_i = \sum_i Y^i e^{-\lambda_i} e_i.$$

Thus for each i, either $\lambda = -\lambda_i$ or $Y^i = 0$ and hence

$$S\mathcal{J}X = \sum_i Y^i \lambda_i e_i = -\lambda Y^i e_i = -\lambda \mathcal{J}X,$$

but $\mathcal{J}SX = \lambda \mathcal{J}X$ giving $(S\mathcal{J} + \mathcal{J}S)X = 0$ for any eigenvector X and hence $S\mathcal{J} + \mathcal{J}S = 0$. ∎

4.3.2   Results on the set $\mathcal{A}$

First we will give a remark about general curves in $\mathcal{A}$. If $g_t = g + D^{(1)}t + D^{(2)}t^2 + \cdots, g \in \mathcal{A}$ is a path of metrics, we can easily obtain a sequence of necessary conditions for g_t to lie in $\mathcal{A}$. We adopt the convention that D will signify a tensor field of type (0,2) or type (1,1) related by the metric, it being clear from context which is meant. We give the details in the symplectic case; the reader can easily read ϕ instead of J for the contact case and follow the computation, showing also that $D^{(k)}\xi = 0$ by a small amount of further computation.

$$g(X, JY) = \Omega(X, Y) = g_t(X, J_tY) = g(X, J_tY) + \sum_{k=1}^{\infty} D^{(k)}(X, J_tY)t^k$$

from which

$$J = J_t + D^{(1)}J_t t + D^{(2)}J_t t^2 + \cdots.$$

Applying J_t on the right and J on the left we have

$$J_t = J(I + D^{(1)}t + D^{(2)}t^2 + \cdots).$$

Squaring this and comparing coefficients gives

$$JD^{(k)} + D^{(k)}J = -\sum_{j=1}^{k-1} D^{(j)}JD^{(k-j)}.$$

Using this last result repeatedly we see that

$$JD^{(k)} + D^{(k)}J = J(\text{polynomial in the } D^{(j)} \text{'s}, j < k).$$

In particular we have

$$JD^{(1)} + D^{(1)}J = 0, \quad JD^{(2)} + D^{(2)}J = JD^{(1)2}.$$

These equations yield immediately the following corollary.

Corollary 4.2 $\mathcal{A}$ *contains no line of metrics.*

Recall that in the contact case the construction of associated metrics was a two-step process, the first being the creation of a metric k such that $k(X, \xi) = \eta(X)$ from an arbitrary metric k' and the second the polarization of $d\eta$ restricted to the contact subbundle. The first step leads to a linear subspace $\mathcal{L}$ of $\mathcal{M}$. If k_0 and k_1 are two Riemannian metrics for which $k_i(X, \xi) = \eta(X)$, $i = 0, 1$, then $k = (1 - t)k_0 + tk_1$ for all t for which k is positive definite is also a Riemannian metric with this property. Moreover if k'_0 and k'_1 yield k_0 and k_1 respectively under the first step of the process, $(1 - t)k'_0 + tk'_1$ yields $k = (1 - t)k_0 + tk_1$.

We have also seen that all associated metrics have the same volume element. On a symplectic manifold of dimension 2, $\mathcal{A} = \mathcal{N}$ and on a contact manifold of dimension 3, $\mathcal{A} = \mathcal{N} \cap \mathcal{L}$. In higher dimensions $\mathcal{A}$ is a proper subset of $\mathcal{N}$ ($\mathcal{N} \cap \mathcal{L}$). We shall now show that $\mathcal{A}$ is totally geodesic in $\mathcal{N}$ and then study how polarization of a path $k_t \in \mathcal{N}$ acts as a projection onto $g_t \in \mathcal{A}$.

Theorem 4.7 $\mathcal{A}$ *is totally geodesic in* $\mathcal{N}$ ($\mathcal{N} \cap \mathcal{L}$) *in the sense that if* $g \in \mathcal{A}$ *and* D *is a symmetric tensor field satisfying* $DJ + JD = 0$ ($D\phi + \phi D = 0$ *and* $D\xi = 0$), *then* $g_t = ge^{Dt}$ *lies in* $\mathcal{A}$.

Proof. Let $J_t = Je^{Dt}$ ($\phi_t = \phi e^{Dt}$) and note that $Je^{Dt} = e^{-Dt}J$. Then

$$g_t(X, J_t Y) = g(X, e^{Dt}Je^{Dt}Y) = g(X, JY) = \Omega(X, Y) \; (= d\eta(X, Y))$$

and

$$J_t^2 = Je^{Dt}Je^{Dt} = J^2 = -I \; (\phi_t^2 = -I + \eta \otimes \xi).$$

∎

In particular the tangent space, $T_g\mathcal{A}$, of $\mathcal{A}$ at g is the set of all symmetric tensor fields that anti-commute with J (ϕ and annihilate ξ).

We now turn to the problem of understanding how the polarization of a geodesic in $\mathcal{N}$ gives rise to path $g_t \in \mathcal{A}$ and hence of viewing polarization as a projection of $\mathcal{N}$ onto $\mathcal{A}$ (the author [1983]). We suppress mentioning the contact case and for that case all tensors T should be understood as also satisfying $T\xi = 0$.

Let us start with a geodesic $k_t = ge^{St}$ in $\mathcal{N}$ where $g \in \mathcal{A}$ and S any symmetric tensor field of vanishing trace. As we have seen if S and J anti-commute, k_t is already the path g_t in $\mathcal{A}$. We shall see as a corollary that if S and J commute, k_t collapses to the metric g. (One can think of a matrix B which commutes with $\mathcal{J} = \begin{pmatrix} 0 & -I \\ I & 0 \end{pmatrix}$ as being orthogonal to a matrix D which anti-commutes with $\mathcal{J}$ by showing $\operatorname{tr} BD = 0$.) Writing S as $B + D$ we will see that the construction process takes k_t to ge^{Dt} if and only if B and D commute, but that g_t and ge^{Dt} agree through second order in general.

Let $\{X_i\}$ be a local g-orthonormal basis with respect to which J is given by the matrix $\begin{pmatrix} 0 & -I \\ I & 0 \end{pmatrix}$. The first problem is to construct a k_t-orthonormal basis. As remarked above we also denote by S the matrix of S with respect to the initial basis $\{X_i\}$. If $SX = \lambda X$, $e^{St}X = e^{\lambda t}X$ and hence the eigenvalues of e^{St} are analytic in t. Let Q be an orthogonal matrix diagonalizing S as in Subsection 4.3.1, i.e., $Q^{-1}SQ = \Lambda$, and set $\Delta = e^{-\frac{1}{2}\Lambda t}$ and $P = Q\Delta$ so that $P^T e^{St} P = I$. Let $X_i(t) = P_{ki}(t)X_k$, then

$$k_t(X_i(t), X_j(t)) = g(P_{ki}X_k, e^{St}P_{lj}X_l) = P_{ki}P_{lj}(e^{St})^m{}_l g_{km} = P^T e^{St} P = I.$$

Thus $\{X_i(t)\}$ is a k_t-orthonormal basis, but note that $X_i(0) = Q_{ki}X_k$. Now our job is to polarize $A(t) = \Omega(X_i(t), X_j(t)) = P^T JP$, giving $F(t)$ and $G(t)$ as in Lemma 4.5.

If we are to have an expression for g_t that we can compare more easily for each t, we should express $g(t)$ with respect to the original basis $\{X_i\}$. Now $G(t) = g_t(X_i(t), X_j(t)) = P_{ki}P_{lj}g_t(X_k, X_l) = P^T(g_t(X_k, X_l))P$, but as $P = Q\Delta$ we have

$$g_t(X_k, X_l) = Q\Delta^{-1}G(t)\Delta^{-1}Q^{-1}.$$

Theorem 4.8 *If $k_t = ge^{St}$ is a geodesic in $\mathcal{N}$ through $g \in \mathcal{A}$, then the path $g_t = g + D^{(1)}t + D^{(2)}t^2 + \cdots$ in $\mathcal{A}$ obtained by polarization is given with respect to the basis $\{X_i\}$ by*

$$D^{(l)} = \sum_{j=0}^{l} \frac{1}{j!2^j} \sum_{k=0}^{j} \binom{j}{k}(B+D)^{j-k}QG^{(l-j)}Q^{-1}(B+D)^k,$$

$G^{(l-j)}$ being given by Lemma 4.5.

The proof is by expansion of $g_t(X_k, X_l) = Q\Delta^{-1}G(t)\Delta^{-1}Q^{-1}$ using the series expansions of $\Delta^{-1} = e^{\frac{1}{2}\Lambda t}$ and $G(t)$ and noting that $\Lambda = Q^{-1}SQ$ (again see the author [1983]).

We remark that Q need not be unique in the above argument (e.g., if S does not have distinct eigenvalues), but each $G^{(k)}$ is of the form

$$Q^{-1} (\text{polynomial in } B \text{ and } D) \, Q$$

and therefore $D^{(k)}$ is independent of the orthogonal matrix Q diagonalizing S. Again we list the first few terms:

$$D^{(1)} = D, \quad D^{(2)} = \frac{D^2}{2}, \quad D^{(3)} = \frac{D^3}{6} - \frac{1}{12}(B^2 D - 2BDB + DB^2),$$

$$D^{(4)} = \frac{D^4}{24} + \frac{1}{96}(-4B^2D^2 + 5BDBD + 3BD^2B - 5DB^2D + 5DBDB - 4D^2B^2).$$

Corollary 4.3 $[B, D] = 0$ *if and only if* $g_t = ge^{Dt}$.

Proof. If $[B, D] = 0$, then as in Corollary 4.1

$$QG^{(k-j)}Q^{-1} = \frac{(-1)^{k-j}}{(k-j)!}B^{k-j}.$$

Thus

$$D^{(k)} = \sum_{j=0}^{k} \frac{1}{j!}(B+D)^j \frac{(-1)^{k-j}}{(k-j)!}B^{k-j}$$

$$= \frac{1}{k!}\sum_{j=0}^{k} \binom{k}{j}(-1)^{k-j}(B+D)^j B^{k-j} = \frac{1}{k!}D^k.$$

Conversely if $D^{(k)} = \frac{1}{k!}D^k$, multiply the rest of the expression for $D^{(3)}$ on both the left and right by $-48D$ and add these two to 96 times the rest of $D^{(4)}$ to yield $-3BDBD + 3DB^2D + 3BD^2B - 3DBDB = 0$ and in turn the conclusion as in Corollary 4.1. ∎

Corollary 4.4 *If* $k_t = ge^{St}$ *and* S *commutes with* J, $g_t = g$.

Proof. In this case $D = 0$ and the result follows from the previous corollary. ∎

Theorem 4.9 *Two metrics in* $\mathcal{A}$ *may be joined by a unique geodesic.*

Proof. Let g_0 and g_1 be two metrics in $\mathcal{A}$. From what has been said so far, the problem is to find S such that $g_1 = g_0e^S$ and $SJ_0 + J_0S = 0$. As before let

$\{X_i\}$ be a local g_0-orthonormal basis with respect to which J_0 is given by the matrix $\mathcal{J} = \begin{pmatrix} 0 & -I \\ I & 0 \end{pmatrix}$. Then

$$\mathcal{J} = \Omega(X_i, X_j) = g_1(X_i, J_1 X_j)$$

from which $g_1 = -\mathcal{J} J_1$ where here J_1 and g_1 are regarded as the matrices of J_1 and g_1 with respect to the basis $\{X_i\}$. In particular $-\mathcal{J} J_1$ is positive definite symmetric and hence there exists a unique real symmetric matrix S satisfying $e^S = -\mathcal{J} J_1$. Then $g_1 = g_0 e^S$ and, since $J_1^2 = -I$, $\mathcal{J} e^S \mathcal{J} e^S = -I$ from which $e^{-S} \mathcal{J} = \mathcal{J} e^S$. Lemma 4.6 then gives $S\mathcal{J} + \mathcal{J} S = 0$. ∎

4.4   Action of symplectic and contact transformations

We begin by showing that if f is a symplectomorphism or strict contact transformation and g an associated metric, then f^*g is also an associated metric.

Theorem 4.10 *Let (M, Ω) be a symplectic manifold, or respectively (M, η) a contact manifold and f a diffeomorphism satisfying $f^*\Omega = \Omega$, respectively $f^*\eta = \eta$. Then for any associated metric g, f^*g is also an associated metric.*

Proof. In the symplectic case define J^* by $(f^*g)(X, J^*Y) = \Omega(X, Y)$. Then

$$g(f_*X, f_*J^*Y) = \Omega(X, Y) = \Omega(f_*X, f_*Y) = g(f_*X, Jf_*Y)$$

and therefore $f_*J^* = Jf_*$. Now $f_*J^{*2}X = J^2f_*X = -f_*X$, so that J^* is an almost complex structure satisfying $(f^*g)(X, J^*Y) = \Omega(X, Y)$ and hence f^*g is an associated metric.

In the case of a strict contact transformation, first note that

$$\eta(f_*\xi) = (f^*\eta)(\xi) = \eta(\xi) = 1, \quad d\eta(f_*\xi, f_*X) = (f^*d\eta)(\xi, X) = d\eta(\xi, X) = 0$$

and therefore $f_*\xi = \xi$. Now define ϕ^* by $(f^*g)(X, \phi^*Y) = d\eta(X, Y)$ and proceed as in the symplectic case to get $f_*\phi^* = \phi f_*$ and in turn $\phi^{*2} = -I + \eta \otimes \xi$. Also it is easy to check that $(f^*g)(\xi, X) = \eta(X)$. ∎

There is a partial converse for the case where M is a compact symplectic manifold and the diffeomorphism belongs to the connected component of the identity of the diffeomorphism group; in general the converse is not true and we will also give a couple of counterexamples (cf. Apostolov and Draghici [1999]). The diffeomorphism group of a compact manifold will be denoted by $Diff$ and the connected component of the identity by $Diff_0$. The group of symplectomorphisms will be denoted by $\mathcal{S}$.

Theorem 4.11 *Let (M, Ω) be a compact symplectic manifold and g an associated metric. If for a diffeomorphism $f \in Diff_0$, f^*g is also an associated metric, then $f \in \mathcal{S}$.*

Proof. For the associated metrics g and f^*g let J and $\tilde{J}$ be the corresponding almost complex structures. Thus (J, g, Ω) and $(\tilde{J}, f^*g, \Omega)$ are almost Kähler structures and consider the action of f^{-1} on the second structure, i.e., setting $\tilde{J}^* = f_* \tilde{J} f^{-1}_*$, $(\tilde{J}^*, g, f^{-1*}\Omega)$ is again an almost Kähler structure with the same metric g. Since the fundamental 2-form of a compact almost Kähler manifold is harmonic, both Ω and $f^{-1*}\Omega$ are harmonic with respect to the metric g. Now $f \in Diff_0$ and so Ω and $f^{-1*}\Omega$ represent the same cohomology class in $H^2(M, \mathbb{R})$. Therefore $\Omega - f^{-1*}\Omega$ is exact, but since it is also harmonic, it must be zero by the Hodge decomposition. Thus $f^{-1*}\Omega = \Omega$ giving $f \in \mathcal{S}$. ■

In general the above theorem is not true and we give two counterexamples. First consider the diffeomorphism f of $\mathbb{R}^4$ given by $\bar{x}^1 = \frac{1}{2}(x^1 + x^2 + x^3 + x^4)$, $\bar{x}^2 = \frac{1}{2}(-x^1 + x^2 - x^3 + x^4)$, $\bar{x}^3 = \frac{1}{2}(x^1 + x^2 - x^3 - x^4)$, $\bar{x}^4 = \frac{1}{2}(-x^1 + x^2 + x^3 - x^4)$ and the symplectic form $\Omega = 2(d\bar{x}^1 \wedge d\bar{x}^3 + d\bar{x}^2 \wedge d\bar{x}^4)$. Let g be the standard Euclidean metric on $\mathbb{R}^4$ which is clearly an associated metric. Now it is easy to see that $\left(\frac{\partial \bar{x}^j}{\partial x^i}\right) \in SO(4)$ but not in $Sp(4, \mathbb{R})$. Thus $f^*g = g$ but $f^*\Omega \neq \Omega$; in fact $f^*\Omega = -\Omega$.

For a compact counterexample consider almost Kähler structures of the form $(M = M_1 \times M_2, g = g_1 + g_2, \Omega = \Omega_1 + \Omega_2)$ where (M_1, g_1, Ω_1) is any almost Kähler manifold and $(M_2 = S^1 \times S^1, g_2, \Omega_2)$ is the standard product Kähler structure on $S^1 \times S^1$. Let f be the diffeomorphism $id_{M_1} \times \psi$, where ψ is the map on $S^1 \times S^1$ that interchanges the two factors. Clearly, f is an isometry of g, but it is not a $\pm$-symplectomorphism, as $f^*\Omega = \Omega_1 - \Omega_2$.

In [1970] Ebin proved a "slice theorem" for the set of Riemannian metrics $\mathcal{M}$, i.e., given a metric $g \in \mathcal{M}$ there is a neighborhood of g in $\mathcal{M}$ that is the product of a neighborhood of g in the orbit of g under the action of the diffeomorphism group and a submanifold orthogonal to the orbit with respect to the inner product on $\mathcal{M}$. The tangent space to this submanifold or 'slice' at g is the kernel of the codifferential δ of g acting on second order symmetric tensor fields (Berger–Ebin [1969]). In Theorem 4.10 we saw that if $f \in \mathcal{S}$ and $g \in \mathcal{A}$, then $f^*g \in \mathcal{A}$ and we may consider the quotient space $\mathcal{A}/\mathcal{S}$. Considering the group of isometries that are also symplectic transformations, Smolentsev [1995] shows that the slice theorem of Ebin can be restricted to give a slice theorem for $\mathcal{A}$.

This is a good point to give a technical lemma for use in Chapter 10, to show the Berger–Ebin result that a symmetric tensor field D orthogonal to an orbit of $Diff$ is in the kernel of the codifferential and to show at least that a symmetric tensor field D is orthogonal to an orbit of $\mathcal{S}$ if and only if there exists a 2-form Ψ such that $(\delta D) \circ J = \delta \Psi$.

Lemma 4.7 *Let (M, g) be a compact orientable Riemannian manifold and for a vector field V, let $v(X) = g(V, X)$. Then $\int_M V^i \theta_i dV_g = 0$ for every closed 1-form θ if and only if $v = \delta \Psi$ for some 2-form Ψ. $\int_M v_i X^i dV_g = 0$ for every vector field X if and only if $v = 0$.*

Proof. Taking θ exact, say df, we have $(v, df) = 0$ and hence $(\delta v, f) = 0$ for every smooth function f where $(.,.)$ denotes the global inner product of differential forms. Therefore $\delta v = 0$ so that $v = \omega + \delta \Psi$ for some 2-form Ψ and harmonic 1-form ω. Now taking $\theta = \omega$, $(v, \theta) = 0$ gives $(\omega, \omega) = 0$. Thus $\omega = 0$ and $v = \delta \Psi$. Conversely for $v = \delta \Psi$ and θ closed, $(v, \theta) = (\delta \Psi, \theta) = (\Psi, d\theta) = 0$. For the second statement we already have $v = \delta \Psi$ but now let X be the contravariant form of $\delta \Psi$, then $0 = \int_M v_i X^i dV_g = (\delta \Psi, \delta \Psi)$ giving $v = 0$. ∎

We now look again at the diffeomorphism group, $Diff$, of M. For a vector field X on M, let f_t be its 1-parameter subgroup in $Diff$. Then $f_t(m)$ is the integral curve of X starting at $m \in M$, so in particular

$$\frac{d}{dt} f_t(m)\big|_{t=0} = X(m).$$

Conversely given a path f_t in $Diff$ with $f_0 = id$, we have for every $m \in M$, $\frac{d}{dt} f_t(m)|_{t=0} \in T_m M$. Thus the tangent space to $Diff$ at the identity may be viewed as the Lie algebra of vectors fields, $\mathfrak{X}$, on M.

Now consider the orbit of $\mathcal{O}_g$ of $g \in \mathcal{M}$ under $Diff$. We have remarked (Section 4.3) that the tangent space $T_g \mathcal{M}$ is the set of symmetric tensor fields of type $(0, 2)$ on M and ask what are the symmetric tensor fields that are tangent to the orbit. Let $\psi_g : Diff \longrightarrow \mathcal{M}$ be defined by $\psi_g(f) = f^* g$. Then $\psi_{g*} : \mathfrak{X} \longrightarrow T_g \mathcal{O}_g$, so given $X \in \mathfrak{X}$, let f_t be its 1-parameter subgroup. Then

$$\psi_{g*}(X) = \frac{d}{dt} f_t^* g = \pounds_X g.$$

Suppose now that a symmetric tensor field $D \in T_g \mathcal{M}$ is orthogonal to $\mathcal{O}_g$ at g. Then

$$0 = (\pounds_X g, D) = \int_M (\nabla_i X_j + \nabla_j X_i) D^{ij} dV_g$$

$$= 2 \int_M (\nabla_i X_j) D^{ij} dV_g = -2 \int_M (\nabla_i D^{ij}) X_j dV_g$$

for every vector field X and hence by Lemma 4.7, $\delta D = 0$.

Now let M be a compact symplectic manifold and g an associated metric with corresponding almost complex structure J. A symmetric tensor field D is orthogonal to the orbit of g under action of $\mathcal{S}$ if and only if there exists a 2-form Ψ such that $(\delta D) \circ J = \delta \Psi$. To see this first note that the tangent space to the orbit of g under the action of $\mathcal{S}$ at g is the set of tensor fields of the form $\mathcal{L}_X g$ where X is a symplectic vector field. As we have seen (Theorem 1.5) $X^i = J^{ik} \theta_k$ for some closed 1-form θ. Then as in the argument above

$$(\mathcal{L}_X g, D) = 2 \int_M (\delta D)_i J^{ik} \theta_k dV_g$$

and the result follows from Lemma 4.7.

4.5 Examples of almost contact metric manifolds

4.5.1 $\mathbb{R}^{2n+1}$

In Example 3.2.1 we considered $\mathbb{R}^{2n+1}$ with its usual contact structure $dz - \sum_{i=1}^{n} y^i dx^i$ and saw that the contact subbundle $\mathcal{D}$ is spanned by $\frac{\partial}{\partial x^i} + y^i \frac{\partial}{\partial z}, \frac{\partial}{\partial y^i}$, $i = 1, \ldots, n$. For normalization convenience, we take as the standard contact structure on $\mathbb{R}^{2n+1}$ the 1-form $\eta = \frac{1}{2}(dz - \sum_{i=1}^{n} y^i dx^i)$. The characteristic vector field is then $\xi = 2\frac{\partial}{\partial z}$ and the Riemannian metric

$$g = \eta \otimes \eta + \frac{1}{4} \sum_{i=1}^{n} ((dx^i)^2 + (dy^i)^2)$$

gives a contact metric structure on $\mathbb{R}^{2n+1}$. For reference purposes, we give the matrix of components of g, namely

$$\frac{1}{4} \begin{pmatrix} \delta_{ij} + y^i y^j & 0 & -y^i \\ 0 & \delta_{ij} & 0 \\ -y^j & 0 & 1 \end{pmatrix}.$$

The tensor field ϕ is given by the matrix

$$\begin{pmatrix} 0 & \delta_{ij} & 0 \\ -\delta_{ij} & 0 & 0 \\ 0 & y^j & 0 \end{pmatrix}$$

and the vector fields $X_i = 2\frac{\partial}{\partial y^i}$, $X_{n+i} = 2(\frac{\partial}{\partial x^i} + y^i \frac{\partial}{\partial z})$, $i = 1, \ldots, n$ and ξ form a ϕ-basis for the contact metric structure.

The Riemannian metric given here has the following properties. The vector field ξ is a Killing vector field, i.e., it generates a 1-parameter group of isometries. The sectional curvature of any plane section containing ξ is equal to 1.

The sectional curvature of a plane section spanned by a vector X orthogonal to ξ and ϕX is equal to -3; for this reason this example is often denoted $\mathbb{R}^{2n+1}(-3)$.

In dimension 3 this example is often identified with the Heisenberg group

$$H_{\mathbb{R}} = \left\{ \begin{pmatrix} 1 & y & z \\ 0 & 1 & x \\ 0 & 0 & 1 \end{pmatrix} \;\middle|\; x, y, z \in \mathbb{R} \right\};$$

left translation preserves η and g is a left invariant metric on $H_{\mathbb{R}}$.

We have already seen that associated metrics are not unique and in Section 7.2 we shall give another associated metric of the contact form η on $\mathbb{R}^{2n+1}$ which is less standard but which has some interesting and basic properties.

4.5.2 $M^{2n+1} \subset \tilde{M}^{2n+2}$ almost complex

We begin with a result of Tashiro [1963] that every C^∞ orientable hypersurface of an almost complex manifold has an almost contact structure. Let $(\tilde{M}^{2n+2}, J)$ be an almost complex manifold and $\iota : M^{2n+1} \longrightarrow \tilde{M}^{2n+2}$ a C^∞ orientable hypersurface. There exists a transverse vector field ν along M^{2n+1} such that $J\nu$ is tangent. For if $J\iota_*X$ is tangent for every tangent vector X, $J\iota_*X = \iota_* fX$ defines a $(1,1)$-tensor field f on M^{2n+1}. Applying J we have $f^2 = -I$ on M^{2n+1} making M^{2n+1} an almost complex manifold, a contradiction. Thus there exists a vector field ξ on M^{2n+1} such that $\nu = J\iota_*\xi$ is transverse.

Define a tensor field ϕ of type $(1,1)$ and a 1-form η on M^{2n+1} by

$$J\iota_*X = \iota_*\phi X + \eta(X)\nu; \qquad (*)$$

then applying J we have

$$-\iota_*X = \iota_*\phi^2 X + \eta(\phi X)\nu - \eta(X)\iota_*\xi$$

and hence $\phi^2 = -I + \eta \otimes \xi$ and $\eta \circ \phi = 0$. Taking $X = \xi$ in equation $(*)$ gives $\nu = \iota_*\phi\xi + \eta(\xi)\nu$ and hence $\phi\xi = 0$ and $\eta(\xi) = 1$. Therefore (ϕ, ξ, η) is an almost contact structure on M^{2n+1}.

If $\tilde{M}^{2n+2}$ is almost Hermitian with Hermitian metric $\tilde{g}$, set $g = \iota^*\tilde{g}$ and take ν to be a unit normal. Then $J\nu$ is tangent and defines ξ by $J\nu = -\xi$. Then again using equation $(*)$

$$g(X, Y) = \tilde{g}(J\iota_*X, J\iota_*Y) = g(\phi X, \phi Y) + \eta(X)\eta(Y)$$

and we see that (ϕ, ξ, η, g) is an almost contact metric structure.

We can construct the usual contact structure on an odd-dimensional sphere in this way. Let S_r^{2n+1} be a sphere of radius r in $\mathbb{C}^{2n+2}$ with its usual Kähler structure denoted as on $\tilde{M}^{2n+2}$ as above with $\tilde{\nabla}$ denoting the connection on $\mathbb{C}^{2n+2}$. With ν as the unit outer normal, η is the standard contact form (cf. Example 3.2.3). Since S_r^{2n+1} is an umbilical hypersurface, its second fundamental form is $\sigma(X,Y) = -\frac{1}{r}g(X,Y)\nu$. Thus, using the fact that J is parallel and the Gauss–Weingarten equations, we have

$$0 = (\tilde{\nabla}_X J)\xi = \tilde{\nabla}_X \nu - J(\nabla_X \xi + \sigma(X,\xi)) = \frac{1}{r}X - \phi\nabla_X\xi - \frac{1}{r}\eta(X)\xi$$

where ∇ is the Levi-Cività connection of g. Applying ϕ we have $\nabla_X \xi = -\frac{1}{r}\phi X$. This in turn yields

$$d\eta(X,Y) = \frac{1}{2}(g(\nabla_X\xi, Y) - g(\nabla_Y\xi, X)) = \frac{1}{r}g(X, \phi Y).$$

Thus for $r \neq 1$ g is not an associated metric, but this situation is easily rectified. The structure $\bar{\eta} = \frac{1}{r}\eta$, $\bar{\xi} = r\xi$, $\bar{\phi} = \phi$ and $\bar{g} = \frac{1}{r^2}g$ is a contact metric structure. Alternatively the metric $g' = \frac{1}{r}g + (1 - \frac{1}{r})\eta \otimes \eta$ is an associated metric for the induced contact form η on S_r^{2n+1}.

Of course in general one cannot expect the induced almost contact metric structure to be a contact metric structure. The condition for this when the ambient space is Kähler was obtained by Okumura [1966] and we have the following theorem.

Theorem 4.12 *Let M^{2n+1} be a hypersurface of a Kähler manifold $\tilde{M}^{2n+2}$, (ϕ, ξ, η, g) its induced almost contact metric structure and A its Weingarten map. Then (ϕ, ξ, η, g) is a contact metric structure if and only if $A\phi + \phi A = -2\phi$.*

Proof. From the Gauss–Weingarten equations we have on the one hand $\tilde{\nabla}_X\xi = \nabla_X\xi + \sigma(X,\xi)$ and on the other

$$\tilde{\nabla}_X\xi = -\tilde{\nabla}_X J\nu = JAX = \phi AX + \eta(AX)\nu.$$

Comparing gives $\nabla_X\xi = \phi AX$. Therefore

$$2d\eta(X,Y) = g(\nabla_X\xi, Y) - g(\nabla_Y\xi, X) = g((\phi A + A\phi)X, Y)$$

from which the result follows.   ∎

4.5.3 $S^5 \subset S^6$

First consider $\mathbb{R}^7$ as the imaginary part of the Cayley numbers $\mathbb{O}$ and define a vector product $\mathbf{u} \times \mathbf{v}$ by the imaginary part of $\mathbf{uv}$. Then

$$(\mathbf{u} \times \mathbf{v}) \times \mathbf{w} - (\mathbf{u} \cdot \mathbf{w})\mathbf{v} + (\mathbf{v} \cdot \mathbf{w})\mathbf{u} = -\mathbf{u} \times (\mathbf{v} \times \mathbf{w}) + (\mathbf{u} \cdot \mathbf{w})\mathbf{v} - (\mathbf{u} \cdot \mathbf{v})\mathbf{w}$$

both sides not being $\equiv 0$ as in dimension 3. Also

$$(\mathbf{u} \times \mathbf{v}) \cdot (\mathbf{u} \times \mathbf{w}) = (\mathbf{u} \cdot \mathbf{u})(\mathbf{v} \cdot \mathbf{w}) - (\mathbf{u} \cdot \mathbf{w})(\mathbf{v} \cdot \mathbf{u})$$

though again in dimension 7, this would not hold if the second $\mathbf{u}$ were replaced by a fourth vector.

The unit sphere $(S^6(1), \tilde{g})$ in $\mathbb{R}^7$ with outer unit normal N inherits an almost complex structure J defined by $JX = N \times X$. From the above vector identities

$$J^2 = N \times (N \times X) = -X,$$

$$\tilde{g}(JX, JY) = (N \times X) \cdot (n \times Y) = X \cdot Y = \tilde{g}(X, Y)$$

giving S^6 an almost Hermitian structure $(J, \tilde{g})$. One can also show that

$$(\tilde{\nabla}_X J)Y + (\tilde{\nabla}_Y J)X = 0,$$

and hence that the almost Hermitian structure is *nearly Kähler* (see also example 6.7.3 below).

Now consider the totally geodesic 5-sphere in $S^6(1) \subset \mathbb{R}^7$ defined by $x^7 = 0$ with $\nu = -\frac{\partial}{\partial x^7}$. Let (ϕ, ξ, η, g) be the induced almost contact metric structure; in particular

$$\xi = -J\nu = N \times \frac{\partial}{\partial x^7} = \sum_{i=1}^{6} x^i \frac{\partial}{\partial x^i} \times \frac{\partial}{\partial x^7}$$

$$= x^1 \frac{\partial}{\partial x^6} - x^2 \frac{\partial}{\partial x^5} - x^3 \frac{\partial}{\partial x^4} + x^4 \frac{\partial}{\partial x^3} + x^5 \frac{\partial}{\partial x^2} - x^6 \frac{\partial}{\partial x^1}.$$

η is the restriction of $x^1 dx^6 - x^6 dx^1 + x^5 dx^2 - x^2 dx^5 + x^4 dx^3 - x^3 dx^4$ to S^5 and hence is the usual contact form.

Compare this with the construction of (ϕ', ξ, η, g) on

$$S^5 \subset \mathbb{R}^6(x^7 = 0) \simeq \mathbb{C}^3 = \{x^2 + ix^5, x^3 + ix^4, x^6 + ix^1\};$$

$J' \frac{\partial}{\partial x^2} = \frac{\partial}{\partial x^5}$, etc., so viewing $\mathbb{C}^3 \subset \mathbb{C}^4 \simeq \mathbb{O}$, J' is just left multiplication by $\frac{\partial}{\partial x^7}$ considered as an imaginary unit in $\mathbb{O}$. Then for $X \perp \xi$,

$$\phi'X = J'X = \frac{\partial}{\partial x^7} X = \frac{\partial}{\partial x^7} \times X$$

since $\frac{\partial}{\partial x^7} \perp X$. Then $g(\phi X, \phi'X) = (N \times X) \cdot (\frac{\partial}{\partial x^7} \times X) = 0$. Therefore (ϕ, ξ, η, g) is an almost contact metric structure with η contact and ξ its characteristic vector field but is not a contact metric structure, $d\eta(X, Y) = g(X, \phi'Y) \neq g(X, \phi Y)$. The difference between ϕ and ϕ' will be seen again in Example 6.7.3 by comparison of their covariant derivatives.

4.5.4   The Boothby–Wang fibration

Let M^{2n+1} be a compact regular contact manifold and $\pi : M^{2n+1} \longrightarrow M^{2n}$ the Boothby–Wang fibration of M^{2n+1} over a symplectic manifold M^{2n} of integral class with symplectic form Ω. Let G be an associated metric to Ω and J the corresponding almost complex structure; in particular (M^{2n}, J, G) is almost Kählerian. As we have seen the contact form η can be viewed as a connection form on the principal circle bundle M^{2n+1}. Thus denoting the horizontal lift by $\tilde{\pi}$, we define a tensor field ϕ on M^{2n+1} by $\phi X = \tilde{\pi} J \pi_* X$. Then, since the characteristic vector field ξ is vertical, $\phi^2 = -I + \eta \otimes \xi$ and (ϕ, ξ, η) is an almost contact structure. Now define a Riemannian metric on M^{2n+1} by $g = \pi^* G + \eta \otimes \eta$. Since $d\eta = \pi^* \Omega$ we have

$$g(X, \phi Y) = G(\pi_* X, J\pi_* Y) \circ \pi = \Omega(\pi_* X, \pi_* Y) \circ \pi = \pi^* \Omega(X, Y) = d\eta(X, Y).$$

Clearly $\eta(X) = g(X, \xi)$ and hence (ϕ, ξ, η, g) is a contact metric structure on M^{2n+1}. It is also clear that ξ is a Killing vector field, i.e., ξ generates a 1-parameter group of isometries. A contact metric structure for which the characteristic vector field is a Killing vector field, is called a *K-contact structure*, a notion that we will discuss further in later chapters.

We can at this point give a topological result on compact regular contact manifolds. Since the characteristic class of the principal circle bundle $\pi : M^{2n+1} \longrightarrow M^{2n}$ is $[\Omega] \in H^2(M^{2n}, \mathbb{Z})$, the bundle is non-trivial and hence the Gysin sequence becomes

$$0 \to H^1(M^{2n}, \mathbb{R}) \xrightarrow{\pi^*} H^1(M^{2n+1}, \mathbb{R}) \to H^0(M^{2n}, \mathbb{R}) \xrightarrow{L} H^1(M^{2n}, \mathbb{R}) \to \cdots$$

where L is left exterior multiplication by Ω. Thus L is an isomorphism of $H^0(M^{2n}, \mathbb{R})$ into $H^2(M^{2n}, \mathbb{R})$ and therefore the map π^* is an onto isomorphism giving the following theorem of Tanno [1967a].

Theorem 4.13 *Let* $\pi : M^{2n+1} \longrightarrow M^{2n}$ *be the Boothby–Wang fibration of a compact regular contact manifold* M^{2n+1}. *Then the first Betti numbers of* M^{2n+1} *and* M^{2n} *are equal.*

In Example 3.2.7 we saw that the 5-dimensional torus carries a contact structure; we note however that it is not regular.

Theorem 4.14 *No torus* T^{2n+1} *can carry a regular contact structure.*

Proof. If T^{2n+1} admitted a regular contact structure, it would be a principal circle bundle over a symplectic manifold M^{2n} by the Boothby–Wang fibration. We have just seen that the first Betti number of the base $b_1(M^{2n})$ is equal to $b_1(T^{2n+1}) = 2n + 1$. On the other hand we have the homotopy sequence of the bundle

$$0 \to \pi_2(M^{2n}) \to \pi_1(S^1) \to \pi_1(T^{2n+1}) \to \pi_1(M^{2n}) \to 0$$

since $\pi_2(T^{2n+1}) = 0$. Now consider the universal covering space $\mathbb{R}^{2n+1}$ of T^{2n+1} and the lift of the fibration; each circle lifts to a line and hence the fibration of T^{2n+1} by circles has no null-homotopic fibres. Thus the map from $\pi_1(S^1)$ into $\pi_1(T^{2n+1})$ is non-trivial and hence $\pi_2(M^{2n}) = 0$. Then, by the exactness of the sequence, $\pi_1(M^{2n}) = \frac{\mathbb{Z}\oplus\cdots\oplus\mathbb{Z}}{\mathbb{Z}}$ and hence $b_1(M^{2n}) = 2n$, a contradiction. ∎

A proof that no torus can carry a K-contact structure can be found in Itoh [1997].

4.5.5 $M^{2n} \times \mathbb{R}$

Let (M^{2n}, J, G) be an almost Hermitian manifolds with local coordinates $x^1,\ldots,x^{2n}$ and let t be the coordinate on $\mathbb{R}$. Then on $M^{2n} \times \mathbb{R}$ set $\eta = f dt$, $\xi = \frac{1}{f}\frac{\partial}{\partial t}$ for some non-vanishing function f. Note that $d\eta = df \wedge dt$ and so $\eta \wedge d\eta \equiv 0$. Without stressing notation for simplicity, set $g = G + \eta \otimes \eta$ and define ϕ by $\phi\xi = 0$ and $\phi X = JX$ for X orthogonal to ξ. Then (ϕ,ξ,η,g) is an almost contact metric structure which is certainly not a contact metric structure. Generally in this example f is taken to be identically 1. However since for a contact metric structure the integral curves of ξ are geodesics, as we have seen, the question of whether for an almost contact metric structure the integral curves of ξ must be geodesics sometimes arises. That the integral curves of ξ need not be geodesics can be seen in this example by choosing f so that it is not independent of x^i. For then, using the standard formula for the Levi-Cività connection, $2g(\nabla_X Y, Z) = Xg(Y,Z) + Yg(X,Z) - Zg(X,Y) + g([X,Y],Z) + g([Z,X],Y) - g([Y,Z],X)$, we have

$$2g(\nabla_\xi\xi, \frac{\partial}{\partial x^i}) = g([\frac{\partial}{\partial x^i},\xi],\xi) - g([\xi,\frac{\partial}{\partial x^i}],\xi) = 2g((\frac{\partial}{\partial x^i}\frac{1}{f})\frac{\partial}{\partial t},\xi) = -\frac{2}{f}\frac{\partial f}{\partial x^i}.$$

4.5.6 Parallelizable manifolds

Let M^{2n+1} be a parallelizable manifold with $\{X_1,\ldots,X_{2n+1}\}$ a set of parallelizing vector fields. Define a Riemannian metric by $g(X_A,X_B) = \delta_{AB}$. Let $\xi = X_{2n+1}$ and η its covariant form with respect to g. Similarly let w^i be the covariant form of X_i, $i = 1,\ldots,n$ and w^{i^*} that of $X_{i^*} = X_{n+i}$. Then define ϕ by

$$\phi = \sum_{i=1}^{n}(w^i \otimes X_{i^*} - w^{i^*} \otimes X_i)$$

and it is easy to check that (ϕ,ξ,η,g) is an almost contact metric structure.

In particular any odd-dimensional Lie group carries an almost contact structure.

5

Integral Submanifolds and Contact Transformations

5.1 Integral submanifolds

Let M^{2n+1} be a contact manifold with contact form η. We have seen that $\eta = 0$ defines a $2n$-dimensional subbundle $\mathcal{D}$ called the *contact distribution* or *subbundle* and that since $\eta \wedge (d\eta)^n \neq 0$, $\mathcal{D}$ is non-integrable. This non-integrability was easily visualized, for example, in Example 3.2.6.

A submanifold M^r of M^{2n+1} is called an *integral submanifold* if $\eta(X) = 0$ for every tangent vector X. It is clear that for any pair of tangent vector fields we have

$$d\eta(X,Y) = \frac{1}{2}(X\eta(Y) - Y\eta(X) - \eta([X,Y])) = 0.$$

Then in terms of associated metrics, $g(X, \phi Y) = 0$ and for this reason integral submanifolds are often called *C-totally real submanifolds* . In particular ϕ maps tangent vectors to normal vectors; also since ξ is a normal vector, the dimension r can be at most n. On the other hand, by the Darboux theorem we have local coordinates $(x^1, \ldots, x^n, y^1, \ldots, y^n, z)$ with respect to which $\eta = dz - \sum_{i=1}^n y^i dx^i$. So $x^i = const., z = const.$ define an n-dimensional integral submanifold and we have the following theorem.

Theorem 5.1 *Let M^{2n+1} be a contact manifold with contact form η. Then there exist integral submanifolds of the contact subbundle $\mathcal{D}$ of dimension n but of no higher dimension.*

Continuing this theme we have the following result of Sasaki [1964].

Theorem 5.2 *Let $(x^i, y^i, z), i = 1, \ldots, n$ be local coordinates about a point $m = (x_0^i, i, y_0^i, z_0)$ such that $\eta = dz - \sum_{i=1}^n y^i dx^i$ on the coordinate neighbor-*

hood. In order that r linearly independent vectors $X_\lambda, \lambda = 1, \dots, r \le n$ at m with components $(a_\lambda^i, b_\lambda^i, c_\lambda)$ be tangent to an r-dimensional integral submanifold, it is necessary and sufficient that $\eta(X_\lambda) = 0$ and $d\eta(X_\lambda, Y_\mu) = 0$, that is $c_\lambda = \sum_i y_0^i a_\lambda^i$, and $\sum_i a_\lambda^i b_\mu^i = \sum_i a_\mu^i b_\lambda^i$.

Proof. Again the necessity is clear. To prove the sufficiency, set $c_{\lambda\mu} = \sum_i a_\lambda^i b_\mu^i$ and choose a sufficiently small neighborhood $\mathcal{U}$ of the origin of $\mathbb{R}^r$ with coordinates $(u^1, \dots, u^r)$ such that

$$x^i = x_0^i + \sum_\lambda a_\lambda^i u^\lambda, \quad y^i = y_0^i + \sum_\lambda b_\lambda^i u^\lambda,$$

$$z = z_0 + \sum_\lambda c_\lambda u^\lambda + \frac{1}{2} \sum_{\lambda,\mu} c_{\lambda\mu} u^\lambda u^\mu$$

defines a mapping ι of $\mathcal{U}$ into M^{2n+1}. Then $\dfrac{\partial x^i}{\partial u^\lambda} = a_\lambda^i$, $\dfrac{\partial y^i}{\partial u^\lambda} = b_\lambda^i$ and

$$\frac{\partial z}{\partial u^\lambda} = c_\lambda + \sum_\mu c_{\lambda\mu} u^\mu = \sum_i y_0^i \frac{\partial x^i}{\partial u^\lambda} + \sum_{i,\mu} \frac{\partial x^i}{\partial u^\lambda} \frac{\partial y^i}{\partial u^\mu} u^\mu = \sum_i y^i \frac{\partial x^i}{\partial u^\lambda}$$

and hence the mapping ι defines an integral submanifold of $\mathcal{D}$ tangent to $X_1, \dots, X_r$ and m. ∎

Finally as in the symplectic case we note the abundance of integral submanifolds of $\mathcal{D}$; more precisely we have the following result (Sasaki [1964]).

Theorem 5.3 *Given a vector $X \in \mathcal{D}$ at $m \in M^{2n+1}$ and any r, $1 \le r \le n$, there exists an r-dimensional integral submanifold M^r of $\mathcal{D}$ through m with X tangent to M^r.*

The proof is again immediate from the Darboux theorem, choosing the Darboux coordinates $(x^i, y^i, z), i = 1, \dots, n$ such that $X = \dfrac{\partial}{\partial y^1}(m)$.

In Chapter 1 we discussed a theorem of Weinstein (Theorem 1.4) that locally a symplectic manifold is the cotangent bundle of a Lagrangian submanifold. We now state an analogous theorem due to Lychagin [1977] (see also Kriegl–Michor [1997, p. 468]). Recall the Liouville form β on a cotangent bundle (Section 1.1, Examples 3.2.4, 3.2.5).

Theorem 5.4 *If L is an n-dimensional integral submanifold of a contact manifold (M^{2n+1}, η), then there exists an open neighborhood $\mathcal{U}$ of L in M^{2n+1}, an open neighborhood $\mathcal{V}$ of the zero section in $T^*L \times \mathbb{R}$ and a diffeomorphism $f : \mathcal{U} \longrightarrow \mathcal{V}$ such that $f|_L$ is the identity on L and $f^*(\beta - dt) = \eta$.*

5.2   Contact transformations

Recall that a diffeomorphism f of M^{2n+1} is a *contact transformation* if $f^*\eta = \tau\eta$ for some non-vanishing function τ and that f is a *strict contact transformation* if $\tau \equiv 1$. Clearly $f^*d\eta = d\tau \wedge \eta + \tau d\eta$ so if $d\eta$ is invariant, $(\tau - 1)d\eta = -d\tau \wedge \eta$ and hence $(\tau - 1)\eta \wedge d\eta = 0$ giving $\tau \equiv 1$. Thus f is strict if and only if $d\eta$ is invariant.

Theorem 5.5 f *is a contact transformation if and only if* $X \in \mathcal{D}$ *implies* $f_*X \in \mathcal{D}$.

Proof. $\eta(f_*X) = (f^*\eta)(X) = \tau\eta(X) = 0$ and conversely $0 = \eta(f_*X) = (f^*\eta)(X)$ implies that $f^*\eta$ is proportional to η. ∎

Theorem 5.6 *A diffeomorphism* f *is a contact transformation if and only if* f *maps* r*-dimensional integral submanifolds to* r*-dimensional integral submanifolds.*

Proof. If f is a contact transformation and M^r an integral submanifold, then for X tangent to M^r, $f_*X \in \mathcal{D}$ by the previous result and therefore $f(M^r)$ is an integral submanifold. Conversely given $X \in \mathcal{D}_m$, we have seen there exists an integral submanifold M^r through m with X tangent to M^r. Now since $f(M^r)$ is an integral submanifold, $f_*X \in \mathcal{D}$ and hence f is a contact transformation. ∎

If $\pounds_X\eta = \sigma\eta$ for some function σ, X is called an *infinitesimal contact transformation*. If $\sigma \equiv 0$ we say that X is a *strict infinitesimal contact transformation*.

Theorem 5.7 *A vector field* X *is an infinitesimal contact transformation if and only if there exists a function* f *on* M^{2n+1} *such that* $X = -\frac{1}{2}\phi\nabla f + f\xi$.

Proof. For the sufficiency we compute as follows.

$$(\pounds_X\eta)(Y) = (X \lrcorner \, d\eta)(Y) + df(Y) = 2d\eta(-\frac{1}{2}\phi\nabla f + f\xi, Y) + Yf$$

$$= -d\eta(\phi\nabla f, Y) + Yf = -g(\nabla f, Y) + \eta(\nabla f)\eta(Y) + Yf = (\xi f)\eta(Y).$$

Conversely $\pounds_X\eta = \sigma\eta$ implies $X\eta(Y) - \eta([X, Y]) = \sigma\eta(Y)$. Then setting $f = \eta(X)$ we have

$$2d\eta(X, Y) + Yf = \sigma\eta(Y)$$

or

$$-2g(\phi X, Y) + g(\nabla f, Y) = \sigma g(\xi, Y)$$

from which $-2\phi X + \nabla f = \sigma\xi$ and applying ϕ we have

$$X = -\frac{1}{2}\phi\nabla f + f\xi$$

as desired; note also that $\sigma = \xi f$ and hence we have the following corollary. ∎

Corollary 5.1 X *is strict if and only if* $\xi f = 0$.

As in the symplectic case (Theorem 1.6) we have the following theorem of Hatakeyama [1966].

Theorem 5.8 *Let* M^{2n+1} *be a compact contact manifold. Then for any two points* P, Q, *there exists a contact transformation mapping* P *to* Q. *If* M *is regular, there exists a strict contact transformation mapping* P *to* Q.

Proof. We first prove the result for a Darboux neighborhood $\mathcal{U}$ about $P = (0,0,0)$. Suppose $Q = (A^i, B^i, C)$ in this coordinate system and define a function f on $\mathcal{U}$ by $f = \sum(B^i x^i - A^i y^i) + C - \frac{1}{2}\sum A^i B^i$. Let X be the infinitesimal contact transformation generated by f (strictly speaking X is determined by $f \in C^\infty(M)$ such that on $\mathcal{U}$, f is as given and f vanishes outside some larger neighborhood). Writing $X = X^i \frac{\partial}{\partial x^i} + X^{i*} \frac{\partial}{\partial y^i} + X^0 \frac{\partial}{\partial z}$, we have since $\xi f = 0$ on $\mathcal{U}$, $d\eta(X, \frac{\partial}{\partial x^i}) = -\frac{1}{2}X^{i*}$. Now by Theorem 5.6

$$d\eta\left(X, \frac{\partial}{\partial x^i}\right) = d\eta\left(-\frac{1}{2}\phi\nabla f + f\xi, \frac{\partial}{\partial x^i}\right) = -\frac{1}{2}g\left(\phi\nabla f, \phi\frac{\partial}{\partial x^i}\right)$$

$$= -\frac{1}{2}\left(g\left(\nabla f, \frac{\partial}{\partial x^i}\right) - \eta(\nabla f)\eta\left(\frac{\partial}{\partial x^i}\right)\right) = -\frac{1}{2}\frac{\partial f}{\partial x^i} + \frac{1}{2}(\xi f)\eta\left(\frac{\partial}{\partial x^i}\right) = -\frac{1}{2}B^i$$

and hence $X^{i*} = B^i$. Similarly $d\eta(X, \frac{\partial}{\partial y^i}) = \frac{1}{2}A^i$ giving $X^i = A^i$. Now $f = \eta(X) = X^0 - y^i X^i = X^0 - y^i A^i$ and so $X^0 - y^i A^i = B^i x^i - y^i A^i + C - \frac{1}{2}\sum A^i B^i$. Therefore

$$X = A^i \frac{\partial}{\partial x^i} + B^i \frac{\partial}{\partial y^i} + \left(B^i x^i + C - \frac{1}{2}\sum A^i B^i\right)\frac{\partial}{\partial z}.$$

Thus the integral curves of X in $\mathcal{U}$ are given by $x^i = A^i t$, $y^i = B^i t$, $z = (C - \frac{1}{2}\sum A^i B^i)t + (\sum A^i B^i)\frac{t^2}{2}$. When $t = 1$ this curve is at Q. Thus the corresponding 1-parameter group f_t of X gives a diffeomorphism f_1 mapping P to Q. Now for P and Q in M^{2n+1}, the usual continuation argument gives the result.

Now for M regular, the result is a consequence of the following lemma.

Lemma 5.1 *If* M^{2n+1} *is a compact regular contact manifold and* f *a* C^∞ *function on a neighborhood* $\mathcal{U}$ *of* P *such that* $\xi f = 0$, *then there exists* $\tilde{f} \in C^\infty(M)$ *such that* $\xi \tilde{f} = 0$ *and on some neighborhood* $\mathcal{V}$ *of* P, $\tilde{f} \equiv f$.

Proof. Let $\pi : M^{2n+1} \longrightarrow M^{2n}$ be the principal circle bundle structure of M^{2n+1}. Since $\xi f = 0$, f is constant on fibres and therefore there exists a neighborhood $\mathcal{V}'$ about $\pi(P)$ and a function f' on $\mathcal{V}'$ such that $f = f' \circ \pi$ on $\pi^{-1}(\mathcal{V}')$. Now extend f' to a C^∞ function $\tilde{f}'$ on M^{2n} agreeing with f' on $\mathcal{V}'$. Then setting $\tilde{f} = \tilde{f}' \circ \pi$ we have the desired function. ∎

As in the symplectic case, Boothby [1969] mapped k points to k points; see also Kriegl–Michor [1997, p. 472].

5.3 Examples of integral submanifolds

One can readily cite some simple examples of integral submanifolds, e.g., an n-dimensional integral submanifold in $\mathbb{R}^{2n+1}$ given by $y^i = 0$, as already noted, and the fibres of $T_1^* M$ and $T_1 M$ with the contact structures given in Example 3.2.4. We give a few more examples here and we will give further discussion of integral submanifolds and additional examples from time to time.

5.3.1 $S^n \subset S^{2n+1}$

Consider the space $\mathbb{C}^{n+1}$ of $n + 1$ complex variables and let J be its usual almost complex structure. Let $S^{2n+1} = \{z \in \mathbb{C}^{n+1} | |z| = 1\}$. Then as we have seen we can give S^{2n+1} its usual contact structure as follows. For every $z \in S^{2n+1}$ and $X \in T_z S^{2n+1}$, $\xi = -Jz$ and ϕX is the tangential part of JX. Let g be the standard metric on S^{2n+1} and η the dual 1-form of ξ. Then (ϕ, ξ, η, g) is a contact metric structure on S^{2n+1}. Now let L be an $(n + 1)$-dimensional linear subspace of $\mathbb{C}^{n+1}$ passing through the origin and such that JL is orthogonal to L. Then since ξ is simply the action of J on the position vector, $S^n = S^{2n+1} \cap L$ is orthogonal to ξ and is therefore an n-dimensional integral submanifold of the contact structure on S^{2n+1}. Clearly S^n is a totally geodesic integral submanifold.

5.3.2 $T^2 \subset S^5$

The following imbedding of a 3-torus into the unit 5-sphere as a flat minimal submanifold is well known. Given in terms of its position vector $\mathbf{x} : T^3 \longrightarrow S^5 \subset E^6 \approx \mathbb{C}^3$, it is

$$\mathbf{x} = \frac{1}{3}(\cos u, \sin u, \cos v, \sin v, \cos w, \sin w).$$

Now imbedding T^2 in T^3 diagonally by $u + v + w = 0$ we have a flat minimal surface in S^5 given by

$$\mathbf{x} = \frac{1}{3}(\cos u, \sin u, \cos v, \sin v, \cos(u + v), -\sin(u + v)).$$

As we have seen the characteristic vector field on S^5 is given by

$$\xi = -J\mathbf{x} = \frac{1}{3}(-\sin u, \cos u, -\sin v, \cos v, \sin(u+v), \cos(u+v)).$$

Computing the tangent vectors $\mathbf{x}_u = \frac{\partial \mathbf{x}}{\partial u}$ and $\mathbf{x}_v = \frac{\partial \mathbf{x}}{\partial v}$ directly, it follows easily that $<\xi, \mathbf{x}_u> = <\xi, \mathbf{x}_v> = 0$ and hence that this torus in an integral surface of the contact structure on S^5.

Examples 5.3.1 and 5.3.2 show that S^5 contain both S^2 and T^2 as integral submanifolds. It is known for topological reasons (see e.g., Steenrod [1951, p. 144]) that S^5 does not admit a continuous field of 2-planes. Thus S^5 cannot be foliated by integral surfaces of its contact structure.

5.3.3 Legendre curves and Whitney spheres

Recall that a 1-dimensional integral submanifold of a contact manifold is called a *Legendre curve* and we begin with an elementary property of Legendre curves in the contact manifold $(\mathbb{R}^3, \eta = dz - ydx)$. The projection γ^* of a closed Legendre curve γ in $\mathbb{R}^3$ to the xy-plane must have self-intersections; moreover the algebraic (signed) area enclosed by γ^* is zero. Since $dz - ydx = 0$ along γ, this follows from the elementary formula for the area enclosed by a curve given by Green's theorem,

$$0 = -\int_\gamma dz = \int_{\gamma^*} -ydx = area,$$

the area being $+$ for γ^* traversed counterclockwise and $-$ for clockwise. Legendre curves and their projections are discussed further in Section 8.3. Here we note that one can think of the pair of γ and its projection γ^* in the following terms. Suppose that γ itself does not have self-intersections and regard γ^* as a Lagrangian submanifold in $\mathbb{R}^2 \cong \mathbb{C}$ with self-intersections; then think of going from γ^* to γ as a way of removing the singularity but preserving the "Lagrangian–Legendre" property.

For example the map of the circle $u^2 + v^2 = 1$ into $\mathbb{R}^2$ given by

$$(u, v) \longrightarrow (v, 2uv)$$

has a double point, viz., $(\pm 1, 0) \to (0, 0)$. On the other hand the map of the circle $u^2 + v^2 = 1$ into $(\mathbb{R}^3, \eta = dz - ydx)$ given by

$$(u, v) \longrightarrow (2uv, v, 2u - \frac{4}{3}u^3)$$

is an imbedding and is a Legendre curve.

A generalization of this example and an important Lagrangian submanifold of $\mathbb{R}^{2n} \cong \mathbb{C}^n$ is the Whitney sphere. We give two descriptions of the Whitney sphere. Let $\Omega = \sum_{i=1}^{n} dx^i \wedge dy^i$ be the standard symplectic form on $\mathbb{R}^{2n}$ and consider the sphere S^n in $\mathbb{R}^{n+1}$ given by $\sum_{i=0}^{n}(u^i)^2 = 1$ immersed in $\mathbb{R}^{2n}$ by

$$(u^0, \ldots, u^n) \longrightarrow (u^1, \ldots, u^n, 2u^0u^1, \ldots, 2u^0u^n).$$

Again notice the double point $(\pm1, 0, \ldots, 0)$ and it is easy to check that this immersed sphere is a Lagrangian submanifold of $\mathbb{R}^{2n}$ (cf. Weinstein [1977, p. 26], Morvan [1983]). In Section 1.2 we remarked that the sphere S^n can not be imbedded in $\mathbb{C}^n$ as a Lagrangian submanifold. In a related vein there are no umbilical, non-totally-geodesic, Lagrangian submanifolds isometrically immersed in any complex space-form (Chen–Ogiue [1974b]).

Now imbed $\sum_{i=0}^{n}(u^i)^2 = 1$ in the contact manifold $\mathbb{R}^{2n+1}$ by

$$(u^0, \ldots, u^n) \longrightarrow (2u^0u^1, \ldots, 2u^0u^n, u^1, \ldots, u^n, 2u^0 - \frac{4}{3}(u^0)^3)$$

giving an imbedded sphere as an integral submanifold of the standard contact structure. We refer to this sphere as a *contact Whitney sphere* .

The Whitney sphere is often presented in another form, which, though slightly more complicated, lends itself to a natural geometric characterization. For the Whitney sphere M^n as a Lagrangian submanifold of $\mathbb{R}^{2n} \cong \mathbb{C}^n$, the immersion is

$$(u^0, \ldots, u^n) \longrightarrow \frac{1}{1 + (u^0)^2}(u^1, \ldots, u^n, u^0u^1, \ldots, u^0u^n).$$

This submanifold satisfies the relation

$$|\mathbf{H}|^2 = \frac{n+2}{n^2(n-1)}\tau$$

where $\mathbf{H}$ is the mean curvature vector and τ the scalar curvature of M^n. This equality characterizes the Whitney sphere as a Lagrangian submanifold of $\mathbb{C}^n$. More precisely it was proven by Borrelli, Chen and Morvan [1995] and independently by Ros and Urbano [1998] that if M^n is a Lagrangian submanifold of $\mathbb{C}^n$, then $|\mathbf{H}|^2 \geq \frac{n+2}{n^2(n-1)}\tau$ with equality if and only if M^n is either totally geodesic or a (piece of a) Whitney sphere. Borrelli, Chen and Morvan [1995] and Ros and Urbano [1998] also gave the characterization that the second fundamental form σ of a Lagrangian submanifold in $\mathbb{C}^n$ is given by

$$\sigma(X, Y) = \frac{n}{n+2}(\tilde{g}(X, Y)\mathbf{H} + \tilde{g}(JX, \mathbf{H})JY + \tilde{g}(JY, \mathbf{H})JX)$$

if and only if the submanifold is either totally geodesic or a (piece of a) Whitney sphere.

In the contact manifold $\mathbb{R}^{2n+1}$ with its standard contact metric structure (Example 4.5.1) we also have a second presentation of the contact Whitney sphere as an imbedded sphere and an integral submanifold of the contact structure, namely

$$(u^0, \dots, u^n) \longrightarrow \frac{1}{1+(u^0)^2}\left(u^0 u^1, \dots, u^0 u^n, u^1, \dots, u^n, \frac{u^0}{1+(u^0)^2}\right).$$

For this contact Whitney sphere the analogues of the above results of Borrelli, Chen and Morvan, and Ros and Urbano were given by A. Carriazo and the author [toap].

5.3.4 Lift of a Lagrangian submanifold

Let $\pi : M^{2n+1} \longrightarrow M^{2n}$ be the Boothby–Wang fibration of a compact regular contact manifold M^{2n+1} over a symplectic manifold M^{2n} of integral class with symplectic form Ω and recall the details of Example 4.5.4. Let L be a Lagrangian submanifold of M^{2n} and consider the set of fibres over L. Then since $\phi X = \tilde{\pi} J \pi_* X$, this is a submanifold N^{n+1} of M^{2n+1} with the property that ϕ maps the tangent space into the normal space; such a submanifold is sometimes called an *anti-invariant submanifold*. If X and Y are horizontal tangent vector fields to N^{n+1}, then

$$0 = 2g(X, \phi Y) = 2d\eta(X, Y) = X\eta(Y) - Y\eta(X) - \eta([X, Y]) = -\eta([X, Y]).$$

Thus the horizontal distribution in N^{n+1} is integrable giving n-dimensional integral submanifolds of M^{2n+1}.

6

Sasakian and Cosymplectic Manifolds

6.1 Normal almost contact structures

Recall that almost contact manifolds were defined as manifolds with structural group $U(n) \times 1$ and hence can be thought of as odd-dimensional analogues of almost complex manifolds. We now consider almost contact manifolds which are, in the sense to be defined, analogous to complex manifolds.

As is well known an almost complex structure need not come from a complex structure. The celebrated theorem of Newlander and Nirenberg [1957] states that an almost complex structure J of class $C^{2n+\alpha}$ with vanishing Nijenhuis torsion is integrable, i.e., is the corresponding almost complex structure of a complex structure. The Nijenhuis torsion $[T, T]$ of a tensor field T of type (1,1) is the tensor field of type (1,2) given by

$$[T, T](X, Y) = T^2[X, Y] + [TX, TY] - T[TX, Y] - T[X, TY].$$

All manifolds under consideration are of class C^∞, so the theorem of Newlander and Nirenberg applies. For detailed studies of complex manifolds see for example Goldberg [1962], Kobayashi and Nomizu [1963–69, Chapter IX], Kobayashi and Wu [1983], Morrow and Kodaira [1971], Yano [1965].

Let M^{2n+1} be an almost contact manifold with structure tensors (ϕ, ξ, η) and consider the manifold $M^{2n+1} \times \mathbb{R}$. We denote a vector field on $M^{2n+1} \times \mathbb{R}$ by $(X, f\frac{d}{dt})$ where X is tangent to M^{2n+1}, t the coordinate on $\mathbb{R}$ and f a C^∞ function on $M^{2n+1} \times \mathbb{R}$. Define an almost complex structure J on $M^{2n+1} \times \mathbb{R}$ by

$$J(X, f\frac{d}{dt}) = (\phi X - f\xi, \eta(X)\frac{d}{dt});$$

that $J^2 = -I$ is easy to check. If now J is integrable, we say that the almost contact structure (ϕ, ξ, η) is *normal* (Sasaki and Hatakeyama [1961]).

As the vanishing of the Nijenhuis torsion of J is a necessary and sufficient condition for integrability, we seek to express the condition of normality in terms of the Nijenhuis torsion of ϕ. Since $[J, J]$ is a tensor field of type (1,2), it suffices to compute $[J, J]((X, 0), (Y, 0))$ and $[J, J]((X, 0), (0, \frac{d}{dt}))$ for vector fields X and Y on M^{2n+1}.

$$[J, J]((X, 0), (Y, 0)) = -([X, Y], 0) + \left[\left(\phi X, \eta(X)\frac{d}{dt}\right), \left(\phi Y, \eta(Y)\frac{d}{dt}\right)\right]$$

$$-J\left[\left(\phi X, \eta(X)\frac{d}{dt}\right), (Y, 0)\right] - J\left[(X, 0), \left(\phi Y, \eta(Y)\frac{d}{dt}\right)\right]$$

$$= (\phi^2[X, Y] - \eta([X, Y])\xi, 0) + \left([\phi X, \phi Y], (\phi X \eta(Y) - \phi Y \eta(X))\frac{d}{dt}\right)$$

$$- \left(\phi[\phi X, Y] + (Y \eta(X))\xi, \eta([\phi X, Y])\frac{d}{dt}\right)$$

$$- \left(\phi[X, \phi Y] - (X \eta(Y))\xi, \eta([X, \phi Y])\frac{d}{dt}\right)$$

$$= \left([\phi, \phi](X, Y) + 2d\eta(X, Y)\xi, ((\pounds_{\phi X}\eta)(Y) - (\pounds_{\phi Y}\eta)(X))\frac{d}{dt}\right).$$

$$[J, J]((X, 0), (0, \frac{d}{dt})) = \left[(\phi X, \eta(X)\frac{d}{dt}), (-\xi, 0)\right]$$

$$-J\left[(\phi X, \eta(X)\frac{d}{dt}), (0, \frac{d}{dt})\right] - J[(X, 0), (-\xi, 0)]$$

$$= (-[\phi X, \xi], (\xi \eta(X))\frac{d}{dt}) + (\phi[X, \xi], \eta([X, \xi])\frac{d}{dt})$$

$$= ((\pounds_\xi \phi)X, (\pounds_\xi \eta)(X)).$$

We are thus lead to define four tensors $N^{(1)}$, $N^{(2)}$, $N^{(3)}$, $N^{(4)}$ by

$$N^{(1)}(X, Y) = [\phi, \phi](X, Y) + 2d\eta(X, Y)\xi,$$

$$N^{(2)}(X, Y) = (\pounds_{\phi X}\eta)(Y) - (\pounds_{\phi Y}\eta)(X),$$

$$N^{(3)} = (\pounds_\xi \phi)X,$$

$$N^{(4)} = (\pounds_\xi \eta)(X).$$

Clearly the almost contact structure (ϕ, ξ, η) is normal if and only if these four tensors vanish. However the vanishing of $N^{(1)}$ implies the vanishing of $N^{(2)}$, $N^{(3)}$ and $N^{(4)}$, so that the normality condition is simply

$$[\phi, \phi](X, Y) + 2d\eta(X, Y)\xi = 0.$$

We now prove this and other properties of these tensors (cf. Sasaki and Hatakeyama [1961], [1962]).

Theorem 6.1 *For an almost contact structure (ϕ, ξ, η) the vanishing of $N^{(1)}$ implies the vanishing of $N^{(2)}$, $N^{(3)}$ and $N^{(4)}$.*

Proof. Setting $Y = \xi$ and recalling that $d\eta(X, \xi) = 0$, we have

$$0 = [\phi, \phi](X, \xi) = \phi^2[X, \xi] - \phi[\phi X, \xi] = \phi((\pounds_\xi \phi)X).$$

Applying ϕ and noting that $d\eta(\xi, \phi X) = 0$ implies $\eta([\xi, \phi X]) = 0$, we have $N^{(3)} = 0$. Moreover $(\pounds_\xi \eta)(\phi X) = 0$, but $(\pounds_\xi \eta)(\xi) = 0$ is immediate and hence $N^{(4)} = 0$. Finally applying η to

$$0 = [\phi, \phi](\phi X, Y) + 2d\eta(\phi X, Y)\xi$$

we have $\eta([\phi^2 X, \phi Y]) + \phi X \eta(Y) - \eta([\phi X, Y]) = 0$ which simplifies to $N^{(2)} = 0$. ∎

Theorem 6.2 *For a contact metric structure (ϕ, ξ, η, g), $N^{(2)}$ and $N^{(4)}$ vanish. Moreover $N^{(3)}$ vanishes if and only if ξ is a Killing vector field.*

Proof. We have already seen that $N^{(4)} = 0$. Now $N^{(2)}$ can be written

$$N^{(2)}(X, Y) = 2d\eta(\phi X, Y) - 2d\eta(\phi Y, X) = 2g(\phi X, \phi Y) - 2g(\phi Y, \phi X) = 0.$$

Turning to $N^{(3)}$, we note that since $d\eta$ is invariant under the action of ξ,

$$0 = (\pounds_\xi d\eta)(X, Y) = \xi g(X, \phi Y) - g([\xi, X], \phi Y) - g(X, \phi[\xi, Y])$$

$$= (\pounds_\xi g)(X, \phi Y) + g(X, (\pounds_\xi \phi)Y)$$

from which we see that $N^{(3)} = 0$ if and only if ξ is Killing. ∎

Next we establish a formula for the covariant derivative of ϕ for a general almost contact metric structure (ϕ, ξ, η, g).

Lemma 6.1 *For an almost contact metric structure (ϕ, ξ, η, g), the covariant derivative of ϕ is given by*

$$2g((\nabla_X \phi)Y, Z) = 3d\Phi(X, \phi Y, \phi Z) - 3d\Phi(X, Y, Z) + g(N^{(1)}(Y, Z), \phi X)$$

$$+ N^{(2)}(Y, Z)\eta(X) + 2d\eta(\phi Y, X)\eta(Z) - 2d\eta(\phi Z, X)\eta(Y).$$

Proof. Recall that the Levi-Cività connection ∇ of g is given by

$$2g(\nabla_X Y, Z) = Xg(Y, Z) + Yg(X, Z) - Zg(X, Y)$$

$$+g([X, Y], Z) + g([Z, X], Y) - g([Y, Z], X)$$

and that the coboundary formula for d on a 2-form Φ is

$$d\Phi(X, Y, Z) = \frac{1}{3}\{X\Phi(Y, Z) + Y\Phi(Z, X) + Z\Phi(X, Y)$$

$$-\Phi([X, Y], Z) - \Phi([Z, X], Y) - \Phi([Y, Z], X)\}.$$

Therefore

$$2g((\nabla_X\phi)Y, Z) = 2g(\nabla_X\phi Y, Z) + 2g(\nabla_X Y, \phi Z)$$

$$= Xg(\phi Y, Z) + \phi Yg(X, Z) - Zg(X, \phi Y)$$

$$+g([X, \phi Y], Z) + g([Z, X], \phi Y) - g([\phi Y, Z], X)$$

$$+Xg(Y, \phi Z) + Yg(X, \phi Z) - \phi Zg(X, Y)$$

$$+g([X, Y], \phi Z) + g([\phi Z, X], Y) - g([Y, \phi Z], X)$$

$$= X\Phi(Y, Z) + \phi Y(\Phi(\phi Z, X) + \eta(Z)\eta(X)) - Z\Phi(X, Y)$$

$$-\Phi([X, \phi Y], \phi Z) + \eta([X, \phi Y])\eta(Z)$$

$$+\Phi([Z, X], Y) - g(\phi[\phi Y, Z], \phi X) + \eta(X)\eta([Z, \phi Y])$$

$$+X\Phi(\phi Y, \phi Z) - Y\Phi(Z, X) - \phi Z(\Phi(\phi Y, X) + \eta(Y)\eta(X))$$

$$+\Phi([X, Y], Z) - \Phi([\phi Z, X], \phi Y) + \eta([\phi Z, X])\eta(Y)$$

$$-g(\phi[Y, \phi Z], \phi X) + \eta(X)\eta([\phi Z, Y])$$

$$+ \{\Phi([Y, Z], X) - g([Y, Z], \phi X)\}$$

$$- \{\Phi([\phi Y, \phi Z], X) - g([\phi Y, \phi Z], \phi X)\}$$

$$+ \{g(2d\eta(Y, Z)\xi, \phi X)\}$$

$$= 3d\Phi(X, \phi Y, \phi Z) - 3d\Phi(X, Y, Z) + g(N^{(1)}(Y, Z), \phi X)$$

$$+N^{(2)}(Y, Z)\eta(X) + 2d\eta(\phi Y, X)\eta(Z) - 2d\eta(\phi Z, X)\eta(Y).$$

∎

Corollary 6.1 *For a contact metric structure the formula of Lemma 6.1 becomes*

$$2g((\nabla_X\phi)Y, Z) = g(N^{(1)}(Y, Z), \phi X) + 2d\eta(\phi Y, X)\eta(Z) - 2d\eta(\phi Z, X)\eta(Y).$$

Taking $X = \xi$ in Corollary 6.1 we see that $\nabla_\xi \phi = 0$ for any contact metric structure. By choosing a ϕ-basis, Corollary 6.1 also yields

$$\nabla_i \phi^i{}_j = -2n\eta_j$$

on a contact metric manifold.

While our main interest is in contact manifolds, we mention, in regard to the normality of almost contact structures, papers of Sato [1977] and Geiges [1997b]. Sato proved that if a compact 3-dimensional normal almost contact manifold M is not homotopic to $S^1 \times S^2$, then $\pi_2(M) = 0$ and Geiges gives a complete classification.

6.2   The tensor field h

We have seen that on a contact metric manifold, $N^{(3)}$ vanishes if and only if ξ is Killing (Theorem 6.2) and a contact metric structure for which ξ is Killing is called a K-*contact structure* . For a general contact metric structure the tensor field $N^{(3)}$ enjoys many important properties and for simplicity we define a tensor field h on a contact metric manifold by

$$h = \frac{1}{2}\mathcal{L}_\xi \phi = \frac{1}{2}N^{(3)}.$$

The first property to note is immediate, namely $h\xi = 0$. We now give a number of important properties of h.

Lemma 6.2 *On a contact metric manifold, h is a symmetric operator,*

$$\nabla_X \xi = -\phi X - \phi h X,$$

h anti-commutes with ϕ and tr$h = 0$.

Proof. We have already seen that on a contact metric manifold, $\nabla_\xi \phi = 0$ and $\nabla_\xi \xi = 0$. Thus

$$g((\mathcal{L}_\xi \phi)X, Y) = g(\nabla_\xi \phi X - \nabla_{\phi X}\xi - \phi\nabla_\xi X + \phi\nabla_X \xi, Y)$$

$$= g(-\nabla_{\phi X}\xi + \phi\nabla_X \xi, Y)$$

which vanishes if either X or Y is ξ. For X and Y orthogonal to ξ, $N^{(2)} = 0$ becomes $\eta([\phi X, Y]) + \eta([X, \phi Y]) = 0$; continuing the computation we then have

$$g((\mathcal{L}_\xi \phi)X, Y) = \eta(\nabla_{\phi X}Y) + \eta(\nabla_X \phi Y)$$

$$= \eta(\nabla_Y \phi X) + \eta(\nabla_{\phi Y}X)$$

$$= g((\pounds_\xi \phi)Y, X).$$

For the second statement using Lemma 6.1 we have

$$2g((\nabla_X \phi)\xi, Z) = g(\phi^2[\xi, Z] - \phi[\xi, \phi Z], \phi X) - 2d\eta(\phi Z, X)$$

$$= -g(\phi(\pounds_\xi \phi)Z, \phi X) - 2g(\phi Z, \phi X)$$

$$= -g((\pounds_\xi \phi)Z, X) + \eta((\pounds_\xi \phi)Z)\eta(X) - 2g(Z, X) + 2\eta(Z)\eta(X)$$

$$= -g((\pounds_\xi \phi)X, Z) - 2g(X, Z) + 2g(\eta(X)\xi, Z)$$

and hence $-\phi \nabla_X \xi = -\frac{1}{2}(\pounds_\xi \phi)X - X + \eta(X)\xi$. Applying ϕ we then have

$$\nabla_X \xi = -\phi X - \phi h X.$$

To see the anti-commutativity note that

$$2g(X, \phi Y) = 2d\eta(X, Y) = g(\nabla_X \xi, Y) - g(\nabla_Y \xi, X)$$

$$= g(-\phi X - \phi h X, Y) - g(-\phi Y - \phi h Y, X).$$

Therefore $0 = g(-\phi h X, Y) - g(Y, h\phi X)$ giving $h\phi + \phi h = 0$.

An immediate consequence of this anti-commutativity is that if $hX = \lambda X$, then $h\phi X = -\lambda \phi X$, thus if λ is an eigenvalue of h so is $-\lambda$ and hence tr$h = 0$. ∎

As a result we get the following easy corollary.

Corollary 6.2 *On a contact metric manifold, $\delta \eta = 0$.*

Proof. Choosing an eigenvector basis $\{e_i\}$ of h we have

$$g(\nabla_{e_i}\xi, e_i) = g(-\phi e_i - \phi h e_i, e_i) = 0.$$

∎

Example 3.2.6 provides a nice illustration of the tensor field h. Let $\eta = \frac{1}{2}(\cos x^3 dx^1 + \sin x^3 dx^2)$ be the contact form on $\mathbb{R}^3$; then $\xi = 2(\cos x^3 \frac{\partial}{\partial x^1} + \sin x^3 \frac{\partial}{\partial x^2})$ is the characteristic vector field. The flat metric $g_{ij} = \frac{1}{4}\delta_{ij}$ is an associated metric and ϕ is given by

$$\phi = \begin{pmatrix} 0 & 0 & \sin x^3 \\ 0 & 0 & -\cos x^3 \\ -\sin x^3 & \cos x^3 & 0 \end{pmatrix}.$$

$\mathcal{D}$ is spanned by $X = -\sin x^3 \frac{\partial}{\partial x^1} + \cos x^3 \frac{\partial}{\partial x^2}$ and $\frac{\partial}{\partial x^3}$. Since the metric is Euclidean on $\mathbb{R}^3$, $\nabla_X \xi = 0$. Therefore by Lemma 6.2, $h\phi X = \phi X$, but $\phi X =$

$\frac{\partial}{\partial x^3}$ and so $h\frac{\partial}{\partial x^3} = \frac{\partial}{\partial x^3}$, i.e., $\frac{\partial}{\partial x^3}$ spans the $+1$ eigenspace of h and in turn X spans the -1 eigenspace.

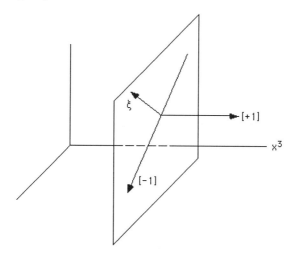

Conditions on $\nabla_\xi h$ arise frequently as we will see in later chapters, but we mention a couple at this point. Calvaruso and Perrone [2000] prove that a 3-dimensional contact metric manifold is locally homogeneous if and only if it is ball-homogeneous and $\nabla_\xi h = 2ah\phi$ where a is a constant. Moreover recalling the contact circles (Example 3.2.9) of Geiges and Gonzalo [1995], Calvaruso and Perrone prove that a compact orientable 3-manifold admits a taut contact circle if and only if it admits a locally homogeneous contact metric structure satisfying $\nabla_\xi h = 0$.

6.3 Definition of a Sasakian manifold

In this short section we give an important definition, namely that of a Sasakian manifold. A *Sasakian manifold* is a normal contact metric manifold. In some respects Sasakian manifolds may be viewed as odd-dimensional analogues of Kähler manifolds. This point of view is reflected in many of the examples and results on Sasakian manifolds that will be discussed. To begin with, the following theorem is analogous to an almost Hermitian manifold being Kähler if and only if the almost complex structure is parallel with respect to the Levi-Civ**ità** connection.

Theorem 6.3 *An almost contact metric structure* (ϕ, ξ, η, g) *is Sasakian if and only if*

$$(\nabla_X \phi)Y = g(X, Y)\xi - \eta(Y)X.$$

Proof. The necessity follows easily from Lemma 6.1, for if (ϕ, ξ, η, g) is a normal contact metric structure, $\Phi = d\eta$, $N^{(1)} = 0$ and $N^{(2)} = 0$, and hence

$$2g((\nabla_X \phi)Y, Z) = 2d\eta(\phi Y, X)\eta(Z) - 2d\eta(\phi Z, X)\eta(Y)$$

$$= 2(g(Y, X) - \eta(Y)\eta(X))\eta(Z) - 2g(Z, X) - \eta(Z)\eta(X))\eta(Y)$$

$$= 2g(g(X, Y)\xi - \eta(Y)X, Z).$$

Conversely, assuming $(\nabla_X \phi)Y = g(X, Y)\xi - \eta(Y)X$, setting $Y = \xi$ gives $-\phi\nabla_X\xi = \eta(X)\xi - X$ and hence $\nabla_X\xi = -\phi X$. Therefore

$$d\eta(X, Y) = \frac{1}{2}\big(g(\nabla_X\xi, Y) - g(\nabla_Y\xi, X)\big) = g(X, \phi Y)$$

showing that (ϕ, ξ, η, g) is a contact metric structure. Now

$$[\phi, \phi](X, Y) = (\phi\nabla_Y\phi - \nabla_{\phi Y}\phi)X - (\phi\nabla_X\phi - \nabla_{\phi X}\phi)Y$$

and straightforward substitution of the hypothesis simplifies this to

$$[\phi, \phi](X, Y) = -2d\eta(X, Y)\xi. \qquad \blacksquare$$

Recall (Example 4.5.4) that a contact metric structure (ϕ, ξ, η, g) is said to be *K-contact* if ξ is a Killing vector field. Since $\nabla_X\xi = -\phi X$ in the above proof and ϕ is skew-symmetric, we have the following corollary.

Corollary 6.3 *A Sasakian manifold is K-contact.*

In dimension 3 the converse is true, Corollary 6.5. For K-contact structures that are not Sasakian, see Example 6.7.2.

While we will see other examples of Sasakian manifolds in due course, let us show here that the standard contact metric structure on an odd-dimensional unit sphere is Sasakian. Recall that the standard contact metric on an odd-dimensional sphere was exhibited in Example 4.5.2 and in particular, if the radius is 1, the constant curvature metric is an associated metric. Using the notation there and the Kähler property of $\mathbb{C}^{2n+2}$, we have

$$0 = (\tilde{\nabla}_X J)Y = \tilde{\nabla}_X(\phi Y + \eta(Y)\nu) - J(\nabla_X Y - g(X, Y)\nu)$$

$$= \nabla_X \phi Y - g(X, \phi Y)\nu + (X\eta(Y))\nu + \eta(Y)X$$

$$-\phi\nabla_X Y - \eta(\nabla_X Y)\nu - g(X, Y)\xi$$

$$= (\nabla_X\phi)Y - g(X, Y)\xi + \eta(Y)X + ((\nabla_X\eta)(Y) - g(X, \phi Y))\nu.$$

Taking the tangential part we see that $(\nabla_X\phi)Y = g(X, Y)\xi - \eta(Y)X$ and hence that the structure is Sasakian.

We close this section with a couple of diagrams indicating some analogies between almost Hermitian manifolds and almost contact metric manifolds. Let $(M^{2n} J, g)$ be an almost Hermitian manifold and let Ω denote the fundamental 2-form. Then we have the following schematic array of structures.

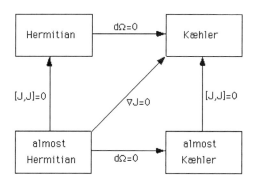

Recall that S^6 carries an almost Hermitian structure that is neither Hermitian nor almost Kähler (cf. Example 4.5.3 and Example 6.7.3 below). The well known Calabi–Eckmann manifolds $S^{2p+1} \times S^{2q+1}, p, q \geq 1$ are Hermitian manifolds (see also Section 6.6 below) which are not Kähler for the topological reason that the second Betti number of a compact Kähler manifold is non-zero. As noted in Section 1.1 there are many compact symplectic (and hence almost Kähler) manifolds with no Kähler structure. Also the tangent bundle of a non-flat Riemannian manifold carries an almost Kähler structure which is not Kählerian (Dombrowski [1962],Tachibana and Okumura [1962], see also Section 9.1 below). Finally there are many well known Kähler manifolds.

The corresponding diagram for almost contact metric manifolds is the following, the notion of a K-contact manifold being intermediate between a contact metric manifold and a Sasakian manifold.

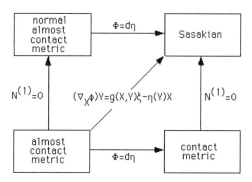

We have already seen in Example 4.5.3 that S^5 carries an almost contact metric structure which is not a contact metric structure and we will see in Example 6.7.3 that the structure is not normal. Cosymplectic manifolds as discussed in Section 6.5 are examples of normal almost contact metric manifolds which are not Sasakian. Since the first Betti number of a compact Sasakian manifold is even (including zero), the 3-dimensional and 5-dimensional tori have no Sasakian structures though they have the contact structures presented in Examples 3.2.6 and 3.2.7. Also we will see in Section 9.2 that the tangent sphere bundles are not in general Sasakian. Finally we just noted that the odd-dimensional spheres are Sasakian and other examples are given in Section 6.7. Also in the Geiges [1997b] classification of compact 3-dimensional normal almost contact manifolds, he identifies those which are normal contact (Sasakian). A similar treatment can be found in Belgun [2000].

6.4   CR-manifolds

In this section we discuss some aspects of the relation of almost contact structures and contact metric structures to CR-structures. Let N be an n-dimensional C^∞ manifold and $T^{\mathbb C}N$ its complexified tangent bundle, i.e., $T_p^{\mathbb C}N = T_pN \otimes_{\mathbb R} \mathbb C \simeq T_pN \oplus iT_pN$. Let $\mathcal H$ be a C^∞ complex subbundle of complex dimension l. A CR-*manifold*, as introduced by Greenfield [1968], of real dimension n and CR-*dimension l* is a pair $(N, \mathcal H)$ such that $\mathcal H_p \cap \bar{\mathcal H}_p = 0$ and $\mathcal H$ is involutive, i.e., for vector fields $X, Y \in \mathcal H$, $[X, Y] \in \mathcal H$. Then there exists a unique subbundle $\mathcal D$ of TN such that $\mathcal D^{\mathbb C} = \mathcal H \oplus \bar{\mathcal H}$, and a unique bundle map $\mathcal J : \mathcal D \longrightarrow \mathcal D$ such that $\mathcal J^2 = -I$ and $\mathcal H = \{X - i\mathcal J X | X \in \mathcal D\}$.

Now let $(M, \mathcal H)$ be a CR-manifold with M of real dimension $2n + 1$ and $\mathcal H$ of complex dimension n. Consider F_x the space of all covectors $f \in T_x^*M$ such that $\mathcal D \subseteq \ker f$. This defines a real line bundle $F \subset T^*M$. If M is orientable, then $F \longrightarrow M$ admits a global nowhere vanishing section η which is called a *pseudo-Hermitian structure* and $(M, \mathcal H, \eta)$ is called a *pseudo-Hermitian manifold*. The *Levi form* of $(M, \eta, \mathcal H)$ is defined by

$$L_\eta(X, Y) = -d\eta(X, \mathcal J Y), \quad X, Y \in \mathcal D.$$

$(M, \eta, \mathcal H)$ is *non-degenerate* if L_η is non-degenerate. In this case M has a natural volume form $\eta \wedge (d\eta)^n$; thus η is a contact form and its characteristic vector field ξ is transversal to $\mathcal D$. $(M, \mathcal H, \eta)$ is said to be *strongly pseudo-convex* if L_η is positive definite. Using the direct sum decomposition $TM = \mathcal D \oplus \{\xi\}$ we may extend L_η to a Riemannian metric g_η on M, called the *Webster metric*, (see e.g., S. Dragomir [1995]) by $g_\eta(\xi, \xi) = 1$, $g_\eta(\xi, X) = 0$ for $X \in \mathcal D$ and $g_\eta(X, Y) = L_\eta(X, Y)$ for $X, Y \in \mathcal D$. Moreover we may extend $\mathcal J$ to a tensor field ϕ on M by $\phi\xi = 0$ and $\phi X = JX$ for $X \in \mathcal{D}$. Therefore a strongly pseudo convex CR manifold $(M, \mathcal{H}, \eta)$, carries a contact metric structure $(\phi, \xi, \eta, g_\eta)$.

In [1978] Bejancu defined the notion of a CR-submanifold of an almost Hermitian manifold. A submanifold N of an almost Hermitian manifold (M, J, g) is a CR-*submanifold* if there exists a holomorphic (J-invariant) subbundle $\mathcal{D}$ with $\mathcal{D}_p \neq \{0\}, T_pN$ such that $\mathcal{D}^\perp$ is totally real, i.e., $J\mathcal{D}_p^\perp \subset T_p^\perp N$. Clearly, real hypersurfaces are CR-submanifolds.

Let P denote the projection from TN to $\mathcal{D}$ and Q the projection from TN to $\mathcal{D}^\perp$. Now since $\mathcal{D}$ is J-invariant, set $\mathcal{J} = JP$ and define a complex subbundle $\mathcal{H}$ of $T^{\mathbb{C}}N$ by $\mathcal{H} = \{X - i\mathcal{J}X | X \in \mathcal{D}\}$. The following lemma is immediate.

Lemma 6.3 $\mathcal{J}(X - i\mathcal{J}X) = JPX - i\mathcal{J}JPX.$

Now suppose that the ambient space (M, J, g) is Hermitian, i.e., the almost complex structure J is integrable. We then have the following lemma and the theorem of Blair and Chen [1979].

Lemma 6.4 *Let N be a CR-submanifold of a Hermitian manifold (M, J, g). Then for $X, Y \in \mathcal{D}$, $Q([JX, Y] + [X, JY]) = 0$.*

Proof. Since M is Hermitian, $[J, J] = 0$ and hence

$$0 = [J, J](JX, Y) = -[JX, Y] - [X, JY] + J([X, Y] - [JX, JY]),$$

but $[X, Y]$ and $[JX, JY]$ are tangent to N, so $J([X, Y] - [JX, JY])$ has no $\mathcal{D}^\perp$-component. Therefore $[JX, Y] + [X, JY]$ has no $\mathcal{D}^\perp$-component. ∎

Theorem 6.4 *Let M be a Hermitian manifold and N a CR-submanifold, then N is a CR-manifold.*

Proof. Let $X, Y \in \mathcal{D}$. Then

$$[X - i\mathcal{J}X, Y - i\mathcal{J}Y] = [X, Y] - [JX, JY] - i[JX, Y] - i[X, JY]$$

$$= -J[JX, Y] - J[X, JY] - iP[JX, Y] - iP[X, JY]$$

by virtue of $[J, J] = 0$ and Lemma 6.4. Continuing the computation using Lemma 6.4 again and then Lemma 6.3,

$$[X - i\mathcal{J}X, Y - i\mathcal{J}Y] = -\mathcal{J}[JX, Y] - \mathcal{J}[X, JY] + i\mathcal{J}^2[JX, Y] + i\mathcal{J}^2[X, JY]$$

$$= -\mathcal{J}([JX, Y] - i\mathcal{J}[JX, Y]) - \mathcal{J}([X, JY] - i\mathcal{J}[X, JY]) \in \mathcal{H}.$$

∎

Turning to the case of almost contact structures, consider an almost contact manifold M^{2n+1} with structure tensors (ϕ, ξ, η). Since $\phi^2 = -I + \eta \otimes \xi$ and $\phi\xi = 0$, the eigenvalues of ϕ are 0 and $\pm i$ each with multiplicity n; in particular

ϕ is an almost complex structure on the subbundle $\mathcal{D}$ defined by $\eta = 0$. Thus the complexification of $\mathcal{D}_p$ is decomposable as $\mathcal{D}'_p \oplus \mathcal{D}''_p$ where $\mathcal{D}'_p = \{X - i\phi X | X \in \mathcal{D}_p\}$ and $\mathcal{D}''_p = \{X + i\phi X | X \in \mathcal{D}_p\}$, the eigenspaces of $\pm i$ respectively.

Lemma 6.5 $\phi(X - i\phi X) \in \mathcal{D}'_p$.

Proof. $\phi(\phi(X - i\phi X)) = \phi(i(X - i\phi X)) = i\phi(X - i\phi X)$.   ∎

We now prove a Theorem of Ianus [1972] that a normal almost contact manifold is a CR-manifold. The converse is not true and in Theorem 6.6 we will obtain a necessary and sufficient condition for a contact metric manifold to be a CR-manifold.

Theorem 6.5 If $(M^{2n+1}, \phi, \xi, \eta, g)$ is a normal almost contact manifold, then $(M^{2n+1}, \mathcal{D}')$ is a CR-manifold.

Proof. First of all $\bar{\mathcal{D}}'_p = \mathcal{D}''_p$ and $\mathcal{D}'_p \cap \mathcal{D}''_p = 0$; thus it remains to show that $[X - i\phi X, Y - i\phi Y] \in \mathcal{D}'$ for $X, Y \in \mathcal{D}$. By the normality, $[\phi, \phi] + 2d\eta \otimes \xi = 0$, so that

$$0 = -[X, Y] + [\phi X, \phi Y] - \phi[\phi X, Y] - \phi[X, \phi Y].$$

Also since $N^{(2)} = 0$, $(\mathcal{L}_{\phi X}\eta)(Y) - (\mathcal{L}_{\phi Y}\eta)(X) = 0$ from which

$$\eta([\phi X, Y] + [X, \phi Y]) = 0.$$

Now

$$[X - i\phi X, Y - i\phi Y] = [X, Y] - [\phi X, \phi Y] - i[\phi X, Y] - i[X, \phi Y]$$

$$= -\phi[\phi X, Y] - \phi[X, \phi Y] + i\phi^2[\phi X, Y] - i\eta([\phi X, Y])\xi$$

$$+ i\phi^2[X, \phi Y] - i\eta([X, \phi Y])\xi$$

$$= -\phi([\phi X, Y] - i\phi[\phi X, Y]) - \phi([X, \phi Y] - i\phi[X, \phi Y]) \in \mathcal{D}'.$$

∎

On a contact metric manifold M^{2n+1}, $(M^{2n+1}, \mathcal{D}')$ might be CR without the structure being normal. We present an important result of Tanno [1989] giving a necessary and sufficient condition for a contact metric manifold to be a CR-manifold.

Theorem 6.6 Let (M^{2n+1}, η, g) be a contact metric manifold. Then the pair $(M^{2n+1}, \mathcal{D}')$ is a (strongly pseudo-convex) CR-manifold if and only if

$$(\nabla_X \phi)Y = g(X + hX, Y)\xi - \eta(Y)(X + hX). \qquad (*)$$

Proof. The strong pseudo-convexity refers to the positive definiteness of the Levi form

$$L(X,Y) = -d\eta(X, \phi|_{\mathcal{D}}Y), \quad X, Y \in \mathcal{D}.$$

We must show the equivalence of $(*)$ and

$$[X - i\phi X, Y - i\phi Y] \in \mathcal{D}' \text{ for } X, Y \in \mathcal{D}. \tag{†}$$

Since $N^{(2)} = 0$, $\eta([X,Y] - [\phi X, \phi Y]) = 0$ for $X, Y \in \mathcal{D}$ from which we can see that (†) is equivalent to

$$\phi[X,Y] - \phi[\phi X, \phi Y] - [\phi X, Y] - [X, \phi Y] = 0 \quad X, Y \in \mathcal{D}. \tag{‡}$$

Now replacing X, Y by $\phi X, \phi Y$ we have an expression for our condition in terms of general vector fields X, Y, viz.,

$$\phi[\phi X, \phi Y] - \phi[-X + \eta(X)\xi, -Y + \eta(Y)\xi]$$

$$-[-X + \eta(X)\xi, \phi Y] - [\phi X, -Y + \eta(Y)\xi] = 0.$$

Applying ϕ and changing the sign we get

$$[\phi, \phi](X,Y) - \eta([\phi X, \phi Y])\xi - \phi^2[\eta(X)\xi, Y] - \phi^2[X, \eta(Y)\xi]$$

$$+\phi[\eta(X)\xi, \phi Y] + \phi[\phi X, \eta(Y)\xi] = 0.$$

Expressing the Nijenhuis torsion and Lie brackets in terms of covariant derivatives and noting that $-\eta([\phi X, \phi Y]) = 2d\eta(\phi X, \phi Y) = 2d\eta(X, Y)$, we have

$$(\phi\nabla_Y\phi - \nabla_{\phi Y}\phi)X - (\phi\nabla_X\phi - \nabla_{\phi X}\phi)Y + 2d\eta(X,Y)\xi$$

$$+\eta(X)(\phi^2\nabla_Y\xi - \phi\nabla_{\phi Y}\xi) - \eta(Y)(\phi^2\nabla_X\xi - \phi\nabla_{\phi X}\xi) = 0.$$

Now recall that since $g(X, \phi Y) = d\eta(X, Y)$, the cyclic sum on $\{X, Y, Z\}$ in $g((\nabla_X\phi)Y, Z)$ must vanish. Thus taking the inner product of our condition with a vector field Z and using this cyclic sum property twice we obtain

$$g(\phi(\nabla_Y\phi)X, Z) - g(\phi(\nabla_X\phi)Y, Z) + g((\nabla_X\phi)Z, \phi Y) + g((\nabla_Z\phi)\phi Y, X)$$

$$-g((\nabla_Y\phi)Z, \phi X) - g((\nabla_Z\phi)\phi X, Y) + 2d\eta(X,Y)\eta(Z)$$

$$+\eta(X)\big(-g((\nabla_Y\xi, Z) + g((\nabla_{\phi Y}\xi, \phi Z)\big)$$

$$-\eta(Y)\big(-g((\nabla_X\xi, Z) + g((\nabla_{\phi X}\xi, \phi Z)\big) = 0.$$

Also since $\phi^2 = -I + \eta \otimes \xi$, we have $(\nabla_X\phi)\phi Y + \phi(\nabla_X\phi)Y = g(\nabla_X\xi, Y)\xi + \eta(Y)\nabla_X\xi$. Using this we obtain

$$2g((\nabla_Z\phi)\phi Y, X) - g(\nabla_Z\xi, Y)\eta(X) - \eta(Y)g(\nabla_Z\xi, X)$$
$$+\eta(X)g(\nabla_{\phi Y}\xi, \phi Z) - \eta(Y)g(\nabla_{\phi X}\xi, \phi Z) = 0.$$

Lemma 6.5 then yields $-(\nabla_Z\phi)\phi Y = \big(g(\phi Z, Y) + g(\phi hZ, Y)\big)\xi$. Finally replacing Y by ϕY, $(\nabla_Z\phi)Y = g(Z + hZ, Y)\xi - \eta(Y)(Z + hZ)$ as desired. ∎

In contrast to the fact that not every 3-dimensional contact metric manifold is Sasakian, every 3-dimensional contact metric manifold is a strongly pseudo-convex CR-manifold as we see in the following corollary.

Corollary 6.4 *A 3-dimensional contact metric manifold is a strongly pseudo-convex CR-manifold; in particular on a 3-dimensional contact metric manifold*

$$(\nabla_X\phi)Y = g(X + hX, Y)\xi - \eta(Y)(X + hX).$$

Proof. The left-hand side of (‡) when restricted to $\mathcal{D}$ is just $-\phi[\phi, \phi]$ and hence it is enough to verify (‡) on the basis $\{X, Y = \phi X\}$ of $\mathcal{D}$ which is straightforward. ∎

Corollary 6.5 *A 3-dimensional K-contact manifold is Sasakian.*

Proof. Since being K-contact is equivalent to a contact metric manifold satisfying $h = 0$, the result follows from Corollary 6.4 and Theorem 6.3. ∎

Remark. In the literature there is also the following definition of a CR-structure which does not include the integrability. A CR-*structure* on a manifold M is a contact form η together with a complex structure J on the contact subbundle $\mathcal{D}$ (see e.g., Chern–Hamilton [1985]). Let (M, η, J) be a CR-structure in this sense and set

$$\mathcal{H} = \{X - iJX | X \in \mathcal{D}\} = \{Z \in \mathcal{D}^{\mathbb{C}} | JZ = iZ\}.$$

If the Levi form L_η is Hermitian, then (M, η, J) is called a *non-degenerate pseudo-Hermitian manifold* (the condition is equivalent to the partial integrability condition: $[X, Y] \in \mathcal{D}^{\mathbb{C}}$ for $X, Y \in \mathcal{H}$). (M, η, J) is said to be *integrable* if $\mathcal{H}$ is involutive, i.e., $[X, Y] \in \mathcal{D}$ for $X, Y \in \mathcal{D}$.

If L_η is positive definite, (M, η, J) is called a *strongly pseudo-convex CR-manifold*. So the notion, in this sense, of a strongly pseudo-convex CR-structure on M is equivalent to our notion of a contact metric structure (ϕ, ξ, η, g) by the relations $g = L_\eta + \eta \otimes \eta$, $J = \phi|_{\mathcal{D}}$ where L_η also denotes its natural extension to a (0,2) tensor field on M. Using this definition of CR-structure, Theorem 6.6 can be given in the following form: Let (M, η, g) be a contact metric manifold and (η, J) the corresponding strongly pseudo-convex CR-structure; then (η, J) is integrable if and only if the condition $(*)$ of Theorem 6.6 holds. In particular (M, η, g) is Sasakian if and only if $h = 0$ and (η, J) is integrable.

6.5 Cosymplectic manifolds and remarks on the Sasakian definition

There are two notions of cosymplectic structure in the literature (in fact counting "Hermitian cosymplectic" $(\delta\Omega = 0)$ there are at least three). P. Libermann [1959] defines a cosymplectic manifold to be a $(2n + 1)$-dimensional manifold admitting a closed 1-form η and a closed 2-form Φ such that $\eta \wedge \Phi^n$ is a volume element. As before on the subbundle defined by $\eta = 0$, Φ may be polarized to give an almost contact metric structure (ϕ, ξ, η, g) for which both η and the fundamental 2-form $\Phi(X, Y) = g(X, \phi Y)$ are closed. In [1967] the author defined a *cosymplectic structure* to be a normal almost contact metric structure (ϕ, ξ, η, g) with both η and Φ closed.

Corresponding to Theorem 6.3 for Sasakian structures we have the following result.

Theorem 6.7 *An almost contact metric structure (ϕ, ξ, η, g) is cosymplectic if and only if ϕ is parallel.*

Proof. Since $N^{(1)} = 0$ implies $N^{(2)} = 0$, that a cosymplectic manifold satisfies $\nabla_X \phi = 0$ follows immediately from Lemma 6.1. Conversely $\nabla_X \phi = 0$ implies that $d\Phi = 0$ and $N^{(1)} = 2d\eta \otimes \xi$. Now

$$N^{(2)}(Y, \xi) = (\mathcal{L}_{\phi Y}\eta)(\xi) = -\eta([\phi Y, \xi])$$

$$= -g(\xi, \nabla_{\phi Y}\xi - \nabla_\xi \phi Y) = g(\xi, \phi \nabla_\xi Y) = 0;$$

so setting $Z = \xi$ in Lemma 6.1, $d\eta(\phi Y, X) = 0$ for all X. Therefore $d\eta = 0$ and in turn $N^{(1)} = 0$. ∎

Since $d\eta = 0$ on a cosymplectic manifold, it is clear that the subbundle defined by $\eta = 0$ is integrable; moreover one can show that η is parallel. Also the projection map to the tangent spaces of the integral submanifolds, $-\phi^2 = I - \eta \otimes \xi$, is parallel. Thus from Theorem 6.7 we see that locally a cosymplectic manifold is the product of a Kähler manifold and an interval, the complex structure being the restriction of ϕ to integral submanifolds. There are however cosymplectic manifolds which are not globally the product of a Kähler manifold and a 1-dimensional manifold; a 3-dimensional example was given by Chinea, de Leon and Marrero [1993] and higher dimensional examples were given by Marrero and Padron [1998].

We have remarked that a Sasakian manifold is sometimes viewed as an odd-dimensional analogue of a Kähler manifold. In view of the present theorem the same can be said of cosymplectic manifolds. Moreover recalling the almost complex structure $J(X, f\frac{d}{dt}) = (\phi X - f\xi, \eta(X)\frac{d}{dt})$ on $M^{2n+1} \times \mathbb{R}$ defined at the beginning of this chapter, consider the product metric $G = g + dt^2$. An easy calculation shows that

$$G\left(J\left(X, f_1\frac{d}{dt}\right), J\left(Y, f_2\frac{d}{dt}\right)\right) = G\left(\left(X, f_1\frac{d}{dt}\right), \left(Y, f_2\frac{d}{dt}\right)\right)$$

so that $(M^{2n+1} \times \mathbb{R}, J, G)$ is an almost Hermitian manifold. Denote by $\bar{\nabla}$ the Levi-Cività connection of G. If now one assumes that this structure is Kähler, then another straightforward computation gives

$$0 = (\bar{\nabla}_{(X,0)}J)(Y, 0) = \left((\nabla_X\phi)Y, (\nabla_X\eta)(Y)\frac{d}{dt}\right)$$

from which we see that M^{2n+1} is cosymplectic.

Thus we have the question of the relation between the metric structure on $M^{2n+1} \times \mathbb{R}$ and the Sasakian condition. This was discussed by Tashiro [1963] and Oubina [1985]. Let $G = e^{2\rho}G'$ be a conformal change of a Kähler metric, then, as is well known, the respective connections are related by

$$D_X Y = D'_X Y + (X\rho)Y + (Y\rho)X - G'(X,Y)P$$

where $P = \mathbf{grad}'\rho$ and we have

$$(D_X J)Y = (JY\rho)X - G'(X, JY)P - (Y\rho)JX + G'(X,Y)JP.$$

Now suppose that the product metric G on $M^{2n+1} \times \mathbb{R}$ is conformally equivalent to a Kähler metric G' with $\rho = -t$; then $\mathbf{grad}'\rho = -e^{-2t}\frac{d}{dt}$. Thus for X, Y vector fields on M^{2n+1}

$$(D_{(X,0)}J)(Y, 0) = -\eta(Y)(X, 0) + g(X, \phi Y)\left(0, \frac{d}{dt}\right) + g(X,Y)(\xi, 0)$$

on the one hand and on the other

$$(D_{(X,0)}J)(Y, 0) = D_{(X,0)}\left(\phi Y, \eta(Y)\frac{d}{dt}\right) - J(\nabla_X Y, 0)$$

$$= \left(\nabla_X\phi Y, (X\eta(Y))\frac{d}{dt}\right) - \left(\phi\nabla_X Y, \eta(\nabla_X Y)\frac{d}{dt}\right).$$

Comparing the components, we see that $(\nabla_X\phi)Y = g(X,Y)\xi - \eta(Y)X$ and hence that M^{2n+1} is Sasakian. Also $(\nabla_X\eta)(Y) = g(X, \phi Y)$, since ξ is now Killing.

Conversely it is clear that if M^{2n+1} is Sasakian, then $(D_{(X,0)}J)(Y, 0)$ is given as above. Therefore making the inverse conformal change

$$(D'_{(X,0)}J)(Y, 0) = 0.$$

Moreover

$$(D_{(X,0)}J)\left(0,\frac{d}{dt}\right) = D_{(X,0)}(-\xi,0) = (-\nabla_X\xi,0) = (\phi X,0)$$

and this is equal to

$$(D'_{(X,0)}J)\left(0,\frac{d}{dt}\right) - g(X,\xi)\frac{d}{dt} + \left(\phi X, \eta(X)\frac{d}{dt}\right).$$

Therefore $(D'_{(X,0)}J)(0,\frac{d}{dt}) = 0$ and similarly

$$(D'_{(0,\frac{d}{dt})}J)(Y,0) = 0, \quad (D'_{(0,\frac{d}{dt})}J)\left(0,\frac{d}{dt}\right) = 0.$$

The idea of using a deformed metric on $M \times \mathbb{R}$ to characterize Sasakian manifolds can be done in another way which is taken as the definition of a Sasakian manifold in Boyer and Galicki [1999]. Let (M^m, g) be a Riemannian manifold and $\mathbb{R}_+$ the positive reals. Then (M^m, g) is Sasakian if and only if the holonomy group of $(\mathbb{R}_+ \times M^m, dr^2 + r^2 g)$ reduces to a subgroup of $U(\frac{m+1}{2})$. Thus $(\mathbb{R}_+ \times M^m, dr^2 + r^2 g)$ is Kähler and $m = 2n+1$, $n \geq 1$. Boyer and Galicki are particularly interested in defining 3-Sasakian manifolds in an analogous way, see Chapter 13 and Boyer, Galicki and Mann [1994].

6.6   Products of almost contact manifolds

We continue our discussion in the last section with a look at products of almost contact manifolds. However we stress only the statement of results rather than the details of their proofs; the interested reader may consult the references. Let M_1 and M_2 be almost contact manifolds with almost contact structures $(\phi_i, \xi_i, \eta_i), i = 1, 2$. A. Morimoto [1963] defined an almost complex structure J on the product $M_1 \times M_2$ by

$$J(X_1, X_2) = (\phi_1 X_1 - \eta_2(X_2)\xi_1, \phi_2 X_2 + \eta_1(X_1)\xi_2);$$

that $J^2(X_1, X_2) = -(X_1, X_2)$ is an easy calculation. Morimoto then proved the following theorem.

Theorem 6.8 *J is integrable if and only if both (ϕ_1, ξ_1, η_1) and (ϕ_2, ξ_2, η_2) are normal.*

An interesting corollary of this is a result of Calabi and Eckmann [1953] that the product of two odd-dimensional spheres is a complex manifold.

Corollary 6.6 *$S^{2p+1} \times S^{2q+1}$ is a complex manifold.*

Turning again to metric considerations, let M_1 and M_2 be almost contact metric manifolds with almost contact metric structures $(\phi_i, \xi_i, \eta_i, g_i)$, $i = 1, 2$. Capursi [1984] studied the product metric $G = g_1 + g_2$ and it is again an easy calculation that

$$G(J(X_1, X_2), J(Y_1, Y_2)) = G((X_1, X_2), (Y_1, Y_2)).$$

The result of Capursi is the following.

Theorem 6.9 $(M_1 \times M_2, J, G)$ *is Kähler if and only if both* $(M_1, \phi_1, \xi_1, \eta_1, g_1)$ *and* $(M_2, \phi_2, \xi_2, \eta_2, g_2)$ *are cosymplectic.*

In view of this and the observation of Tashiro and Oubina in the last section on how to obtain the Sasakian condition from the structure on $M \times \mathbb{R}$, one might ask how to get both M_1 and M_2 Sasakian out of the almost Hermitian structure (J, G) on $M_1 \times M_2$. To the author's knowledge this is an open question. On the other hand a special almost contact metric structure introduced by Kenmotsu [1972] seems to play a role here.

An almost contact metric manifold (M, ϕ, ξ, η, g) is called a *Kenmotsu manifold* if it satisfies

$$(\nabla_X \phi)Y = g(\phi X, Y)\xi - \eta(Y)\phi X.$$

Kenmotsu [1972] gave a local characterization of this structure.

Theorem 6.10 *Every point of a Kenmotsu manifold has a neighborhood which is a warped product* $(-\epsilon, \epsilon) \times_f V$ *where* $f(t) = ce^t$ *and* V *is Kähler.*

Returning to the product space $M_1 \times M_2$, define an almost complex structure by

$$J(X_1, X_2) = (\phi_1 X_1 - e^{-2\mu}\eta_2(X_2)\xi_1, \phi_2 X_2 + e^{-2\mu}\eta_1(X_1)\xi_2).$$

Again $J^2(X_1, X_2) = -(X_1, X_2)$ is an easy calculation. Let $G = e^{2\rho}g_1 + e^{2\tau}g_2$, then G is Hermitian if and only if $\mu = \frac{1}{2}(\rho - \tau)$. Blair and Oubina [1990] then noted the following.

Theorem 6.11 *Let M_1 and M_2 be almost contact metric manifolds and U a coordinate neighborhood on M_2 such that $\xi_2 = \frac{\partial}{\partial t}$. Consider the change of metric $G = e^{2\rho}g_1 + e^{2\tau}g_2$ on $M_1 \times U$ where $\rho = \log(k - e^{-t})$ and $\tau = -t$ for some constant k. Then $(M_1 \times U, J, G)$ is Kähler if and only if the structure on M_1 is Sasakian and the structure on U is Kenmotsu.*

A variation of the above result is that if $(M_1 \times U, J, G)$ is Kähler for the conformal change $\rho = \tau = -t$, then M_1 is Sasakian and the structure $(\phi_2, -\xi_2, -\eta_2, g_2)$ is Kenmotsu.

The fact that this theorem is local in regard to the second manifold M_2 is not unnatural. Even for $M_1 \times \mathbb{R}$, the 1-dimensional case for M_2, note that the Hopf manifold $S^{2n+1} \times S^1$ is locally conformally Kähler but not globally conformally Kähler.

We close this section with a remark on a generalization of these structures. In the classification of Gray and Hervella [1980] of almost Hermitian manifolds there appears a class, $\mathcal{W}_4$, of Hermitian manifolds which are closely related to locally conformally Kähler manifolds. Again consider $M_1 \times \mathbb{R}$ with the almost complex structure $J(X, f\frac{d}{dt}) = (\phi X - f\xi, \eta(X)\frac{d}{dt})$ and product metric G. Oubina [1985] introduced the notion of a *trans-Sasakian* structure as an almost contact metric structure (ϕ, ξ, η, g) for which the almost Hermitian manifold $(M_1 \times \mathbb{R}, J, G)$ belongs to the class $\mathcal{W}_4$. This may be expressed by the condition

$$(\nabla_X \phi)Y = \alpha(g(X,Y)\xi - \eta(Y)X) + \beta(g(\phi X, Y)\xi - \eta(Y)\phi X)$$

for functions α and β on M and the trans-Sasakian structure is said to be of type (α, β). If β but not α (respectively α but not β) vanishes, the structure is α-*Sasakian* (resp. β-*Kenmotsu*) (see also Janssens and Vanhecke [1981]). Marrero [1992] showed that a trans-Sasakian manifold of dimension ≥ 5 is either α-Sasakian, β-Kenmotsu or cosymplectic. He also showed that if M is a 3-dimensional Sasakian manifold with structure tensors (ϕ, ξ, η, g), f a positive non-constant function and $\bar{g} = fg + (1-f)\eta \otimes \eta$, then $(\phi, \xi, \eta, \bar{g})$ is trans-Sasakian of type $(\frac{1}{f}, \frac{1}{2}\xi(\log f))$.

6.7   Examples

6.7.1 $\mathbb{R}^{2n+1}$

In Example 4.5.1 we gave explicitly an associated almost contact metric structure (ϕ, ξ, η, g) to the Darboux contact form $\eta = \frac{1}{2}(dz - \sum_{i=1}^{n} y^i dx^i)$ on $\mathbb{R}^{2n+1}$. From the matrix expression for ϕ given in Example 4.5.1 it is easy to check that $[\phi, \phi] + 2d\eta \otimes \xi = 0$ and hence that this contact metric structure is Sasakian.

6.7.2   *Principal circle bundles*

In Example 4.5.4 we saw that a compact regular contact manifold M^{2n+1} carries a K-contact structure (ϕ, ξ, η, g), defined in terms of the almost Kähler structure (J, G) of the base manifold M^{2n}. Since $\mathcal{L}_\xi \phi = N^{(3)} = 0$,

$$[\phi, \phi](\xi, X) + 2d\eta(\xi, X)\xi = \phi^2[\xi, X] - \phi[\xi, \phi X] = 0.$$

Now $\phi X = \tilde{\pi} J \pi_* X$, so for projectable horizontal vector fields X and Y,

$$[\phi, \phi](X, Y) + 2d\eta(X, Y)\xi = \tilde{\pi} J^2 \pi_*[X, Y] + [\tilde{\pi} J \pi_* X, \tilde{\pi} J \pi_* Y]$$

$$-\tilde{\pi}J\pi_*[\tilde{\pi}J\pi_*X,Y] - \tilde{\pi}J\pi_*[X,\tilde{\pi}J\pi_*Y] + d\eta(X,Y)\xi$$

$$= \tilde{\pi}J^2[\pi_*X,\pi_*Y] + \tilde{\pi}[J\pi_*X,J\pi_*Y] + \eta([\tilde{\pi}J\pi_*X,\tilde{\pi}J\pi_*Y])\xi$$

$$-\tilde{\pi}J[J\pi_*X,\pi_*Y] - \tilde{\pi}J[\pi_*X,J\pi_*Y] + d\eta(X,Y)\xi$$

$$= \tilde{\pi}[J,J](\pi_*X,\pi_*Y) - 2(\Omega(J\pi_*X,J\pi_*Y)\circ\pi - \Omega(\pi_*X,\pi_*Y)\circ\pi)\xi$$

$$= \tilde{\pi}[J,J](\pi_*X,\pi_*Y).$$

Thus we see that the K-contact structure (ϕ,ξ,η,g) is Sasakian if and only if the base manifold $(M^{2n}J,G)$ is Kählerian (Hatakeyama [1963]). If $(M^{2n}J,G)$ is only almost Kähler, then (ϕ,ξ,η,g) is only K-contact.

Similar to the Boothby–Wang fibration of compact regular contact manifolds, A. Morimoto [1964] obtained a fibration of compact normal almost contact manifolds with ξ regular. First however, let $\pi : M^{2n+1} \longrightarrow M^{2n}$ be a principal circle bundle over a complex manifold M^{2n} and suppose there exists a connection form η such that $d\eta = \pi^*\Psi$ where Ψ is a form of bidegree $(1,1)$ on M^{2n}. Then we again define $\phi X = \tilde{\pi}J\pi_*X$ where J is the almost complex structure on M^{2n} and $\tilde{\pi}$ the horizontal lift with respect to η. Let ξ be a vertical vector field with $\eta(\xi) = 1$. Then (ϕ,ξ,η) is an almost contact structure. Noting that $\mathcal{L}_\xi\eta = 0$, since η is a connection form, and $\mathcal{L}_\xi\phi = 0$ by the definition of ϕ, a computation of $N^{(1)}$ similar to the one just given shows that M^{2n+1} is a normal almost contact manifold (Morimoto [1963]).

Conversely we give the following theorem of Morimoto [1964] and just sketch its proof since the major ideas have already been given.

Theorem 6.12 *Let M^{2n+1} be a compact normal almost contact manifold with structure tensors (ϕ,ξ,η) and suppose that ξ is a regular vector field. Then M^{2n+1} is the bundle space of a principal circle bundle $\pi : M^{2n+1} \longrightarrow M^{2n}$ over a complex manifold M^{2n}. Moreover η is a connection form and the 2-form Ψ on M^{2n} such that $d\eta = \pi^*\Psi$ is of bidegree $(1,1)$.*

Proof. In the proof of the Boothby–Wang Theorem (Section 3.3) we defined the period function λ of the vector field ξ and showed that λ was constant on M^{2n+1} which we then took to be 1. The argument (cf. Tanno [1965]) required only that $\eta(\xi) = 1$ and $\mathcal{L}_\xi\eta = N^{(4)} = 0$. Thus we again have a circle bundle structure as in the Boothby–Wang fibration with η a connection form. Now since $\mathcal{L}_\xi\phi = N^{(3)} = 0$, ϕ is projectable and we can define an almost complex structure J on M^{2n} by $JX = \pi_*\phi\tilde{\pi}X$ where $\tilde{\pi}$ denotes the horizontal lift with respect to η. That $J^2 = -I$ is immediate and

$$\tilde{\pi}[J,J](X,Y) = -[\tilde{\pi}X,\tilde{\pi}Y] + \eta([\tilde{\pi}X,\tilde{\pi}Y])\xi + [\phi\tilde{\pi}X,\phi\tilde{\pi}Y] - \eta([\phi\tilde{\pi}X,\phi\tilde{\pi}Y])\xi$$

$$-\phi[\phi\tilde{\pi}X,\tilde{\pi}Y] - \phi[\tilde{\pi}X,\phi\tilde{\pi}Y]$$

$$= [\phi,\phi](\tilde{\pi}X,\tilde{\pi}Y) + 2d\eta(\phi\tilde{\pi}X,\phi\tilde{\pi}Y)\xi$$

$$= [\phi,\phi](\tilde{\pi}X,\tilde{\pi}Y) + 2d\eta(\tilde{\pi}X,\tilde{\pi}Y)\xi = 0,$$

the last equality following from $N^{(1)} = 0$ and the next to last from $N^{(2)} = 0$. Finally $\Psi(JX,JY)\circ\pi = d\eta(\phi\tilde{\pi}X,\phi\tilde{\pi}Y) = d\eta(\tilde{\pi}X,\tilde{\pi}Y) = \Psi(X,Y)\circ\pi$ showing that Ψ is of bidegree $(1,1)$. ∎

6.7.3 A non-normal almost contact structure on S^5

In Example 4.5.3 we saw that S^5 inherits from the almost Hermitian structure on S^6, an almost contact metric structure different than the standard one. Recall that the almost Hermitian structure $(J,\tilde{g})$ on S^6 given in Example 4.5.3 is a nearly Kähler structure, i.e., $(\tilde{\nabla}_X J)X = 0$ for all vector fields X. The geometric meaning of this condition is that geodesics are holomorphically planar curves. A curve γ on an almost Hermitian manifold is *holomorphically planar* if the holomorphic section determined by its tangent field is parallel along the curve.

We will now show that the induced almost contact metric structure (ϕ,ξ,η,g) on S^5 satisfies a similar condition, namely $(\nabla_X\phi)X = 0$. This is an immediate consequence of the following theorem of the author [1971]; for notation see Example 4.5.2.

Theorem 6.13 *Let M^{2n+1} be a hypersurface of a nearly Kähler manifold $\tilde{M}^{2n+2}$. Then the induced almost contact metric structure (ϕ,ξ,η,g) satisfies $(\nabla_X\phi)X = 0$ if and only if the second fundamental form σ is proportional to $(\eta\otimes\eta)\nu$.*

Proof. From equation $(*)$ in Example 4.5.2 and $\tilde{\nabla}_X Y = \nabla_X Y + \sigma(X,Y)$ we have

$$(\nabla_X\Phi)(Y,Z) = (\tilde{\nabla}_X\Omega)(Y,Z) + \tilde{g}(\sigma(X,Y),\nu)\eta(Z) + \tilde{g}(\sigma(X,Z),\nu)\eta(Y)$$

where Ω is the fundamental 2-form of the nearly Kähler structure. Interchanging X and Z and adding we obtain

$$(\nabla_X\Phi)(Y,Z) + (\nabla_Z\Phi)(Y,X) = -2\tilde{g}(\sigma(X,Z),\nu)\eta(Y)$$

$$+\tilde{g}(\sigma(X,Y),\nu)\eta(Z) + +\tilde{g}(\sigma(Z,Y),\nu)\eta(X).$$

Now if σ is proportional to $(\eta\otimes\eta)\nu$, $(\nabla_X\phi)X = 0$. Conversely if $(\nabla_X\phi)X = 0$,

$$0 = -2\tilde{g}(\sigma(X,Z),\nu)\eta(Y) + \tilde{g}(\sigma(X,Y),\nu)\eta(Z) + \tilde{g}(\sigma(Z,Y),\nu)\eta(X).$$

Setting $Y = \xi$ gives $2\tilde{g}(\sigma(X, Z), \nu) = \tilde{g}(\sigma(X, \xi), \nu)\eta(Z) + \tilde{g}(\sigma(Z, \xi), \nu)\eta(X)$ but now setting $X = \xi$ we have $\tilde{g}(\sigma(\xi, Z), \nu) = \tilde{g}(\sigma(\xi, \xi), \nu)\eta(Z)$ and consequently $\tilde{g}(\sigma(X, Z), \nu) = \tilde{g}(\sigma(\xi, \xi), \nu)\eta(X)\eta(Z)$. ∎

A cosymplectic version of this theorem was given by Goldberg in [1968a] (see also Okumura [1965]). An almost contact metric structure (ϕ, ξ, η, g) satisfying $(\nabla_X \phi)X = 0$ is called a *nearly cosymplectic structure*. As a further justification of this name we remark without proof that a normal nearly cosymplectic manifold is cosymplectic (see the author's paper [1971] for details).

Proposition 6.1 *On a nearly cosymplectic manifold ξ is a Killing vector field.*

Proof. Clearly $(\nabla_\xi \phi)\xi = 0$ from which one easily obtains $\nabla_\xi \xi = 0$. Now differentiating the compatibility condition, $g(\phi X, \phi Y) = g(X, Y) - \eta(X)\eta(Y)$, with respect to ξ we find

$$g((\nabla_\xi \phi)X, \phi Y) + g(\phi X, (\nabla_\xi \phi)Y) = 0.$$

The nearly cosymplectic condition then gives

$$g((\nabla_X \phi)\xi, \phi Y) + g(\phi X, (\nabla_Y \phi)\xi) = 0$$

which easily simplifies to

$$g(\nabla_X \xi, Y) + g(\nabla_Y \xi, X) = 0. \qquad ∎$$

Proposition 6.2 *On a normal nearly cosymplectic manifold, $d\eta = 0$.*

Proof. Since the structure is normal, $N^{(1)}$ and $N^{(2)}$ vanish, thus setting $Y = X$ and $Z = \xi$ in Lemma 6.1 we have $d\eta(X, \phi X) = 0$ for all vector fields X. Linearizing this we obtain $d\eta(X, \phi Y) + d\eta(Y, \phi X) = 0$, but from the vanishing of $N^{(2)}$, $d\eta(X, \phi Y) = -d\eta(\phi X, Y)$ and hence $d\eta(X, \phi Y) = 0$. Also $d\eta(X, \xi) = \frac{1}{2}(g(\nabla_X \xi, \xi) - g(\nabla_\xi \xi, X)) = 0$ and so we obtain $d\eta = 0$. ∎

Turning now to S^5 as a totally geodesic hypersurface of S^6 (Example 4.5.3), its induced structure is nearly cosymplectic by Theorem 6.13. This structure is not normal, for if it were, then by our two propositions η is respectively co-closed and closed and hence harmonic, contradicting the vanishing of the first Betti number of S^5.

6.7.4 $M^{2n+1} \subset \tilde{M}^{2n+2}$

We have already seen that the odd-dimensional spheres are Sasakian manifolds and that this structure may be obtained both by viewing the sphere as a hypersurface in $\mathbb{C}^{n+1}$ or by considering the Hopf fibration $\pi : S^{2n+1} \longrightarrow \mathbb{C}P^n$ as a special case of the Boothby–Wang fibration. In Theorem 4.12 we saw the condition for the induced almost contact metric structure on a hypersurface of a Kähler manifold to be contact metric. Now similar to Theorem 6.13 we give the condition for the hypersurface to be Sasakian; this is a result of Tashiro [1963] and we will omit the proof as it is similar to the proof of Theorem 6.13.

Theorem 6.14 *Let M^{2n+1} be a hypersurface of a Kähler manifold $\tilde{M}^{2n+2}$. Then the induced almost contact metric structure (ϕ, ξ, η, g) is Sasakian if and only if the second fundamental form $\sigma = (-g + \beta(\eta \otimes \eta))\nu$, for some function β.*

A contact metric structure (ϕ, ξ, η, g) is said to be *nearly Sasakian* if $(\nabla_X \phi)Y + (\nabla_Y \phi)X = 2g(X,Y)\xi - \eta(X)Y - \eta(Y)X$. The above theorem also holds for this structure on a hypersurface of a nearly Kähler manifold (Blair, Showers, and Yano [1976]). Similar to the structure on S^5 obtained in Examples 4.5.3 and 6.7.3, consider S^5 as an umbilical hypersurface of the unit sphere S^6 at a "latitude" of 45° so that $\sigma(X,Y) = -g(X,Y)\nu$. Then the induced almost contact metric structure is nearly Sasakian but not Sasakian.

For other results on hypersurfaces of Kähler manifolds see Okumura [1964a, 1964b, 1966], Vernon [1987].

6.7.5 Brieskorn manifolds

In this section we will show that the Brieskorn manifolds admit Sasakian structures which are often non-regular.

Consider $\mathbb{C}^{n+1}$ with coordinates $z = (z_0, \dots, z_n)$ and let $(a_0, \dots, a_n)$ be an $(n+1)$-tuple of positive integers. A polynomial of the form $P(z) = z_0^{a_0} + \cdots + z_n^{a_n}$ is called a *Brieskorn polynomial* and let $V^{2n}(a_0, \dots, a_n)$, or just V^{2n} denote the zero set of P. Then $\Sigma^{2n-1}(a_0, \dots, a_n) = V^{2n} \cap S^{2n+1}(1)$ or just Σ^{2n-1} is called a *Brieskorn manifold*.

For $w \in \mathbb{C}$ define $f_w : \mathbb{C}^{n+1} \longrightarrow \mathbb{C}^{n+1}$ by

$$f_w(z) = \left(e^{\frac{mw}{a_0}} z_0, \dots, e^{\frac{mw}{a_n}} z_n\right)$$

where m is the least common multiple of $(a_0, \dots, a_n)$. $G_{\mathbb{C}} = \{f_w | w \in \mathbb{C}\}$ is an Abelian group of diffeomorphisms of $\mathbb{C}^{n+1}$. Let $G_{\mathbb{R}}$ be the 1-parameter subgroup given by $\{f_s | s \in \mathbb{R}\}$. The $\mathbb{R}$-action leaves V^{2n} invariant and differentiating with respect to s at $s = 0$, we see that the vector field which generates $G_{\mathbb{R}}$ is

$$\mathbf{a} = \left(\frac{m}{a_\alpha} z_\alpha\right).$$

Now let $G_S = \{f_{it}|t \in \mathbb{R}\}$; $f_{it} = f_{i(t+2\pi)}$ and is an S^1-action on $\mathbb{C}^{n+1}$ leaving Σ^{2n-1} invariant. The vector field which generates G_S is

$$\mathbf{b} = i\mathbf{a} = \left(\frac{m}{a_\alpha} iz_\alpha\right).$$

The almost complex structure J on $\mathbb{C}^{n+1}$ restricts to V^{2n} and $\xi = -\mathbf{b} = -J\mathbf{a}$ is a vector field tangent to Σ^{2n-1}. Moreover for an arbitrary vector field tangent to Σ^{2n-1}, decomposing JX as

$$JX = \phi X + \eta(X)\mathbf{a}$$

gives Σ^{2n-1}, an almost contact structure (ϕ, ξ, η) as in Example 4.5.2; see Sasaki and Takahashi [1976] and Sasaki [1985]. (In Sasaki [1985], the structure tensors have the opposite sign from our construction here.) Sasaki and Takahashi prove the following.

Theorem 6.15 *The almost contact structure (ϕ, ξ, η) on the Brieskorn manifold Σ^{2n-1} is normal.*

Recalling the theorem of Morimoto, Theorem 6.8, we have the following result of Brieskorn and Van de Ven [1968].

Corollary 6.7 *Let Σ^{2n-1} and Σ^{2m-1} be Brieskorn manifolds. Then $\Sigma^{2n-1} \times \mathbb{R}$ and $\Sigma^{2n-1} \times \Sigma^{2m-1}$ are complex manifolds.*

In regard to non-regularity, Sasaki and Takahashi prove the following.

Theorem 6.16 *The almost contact structure (ϕ, ξ, η) on the Brieskorn manifold $\Sigma^{2n-1}(a_0, \ldots, a_n)$, is non-regular if and only if there exist three positive integers a_λ, a_μ, a_ν among the positive integers $a_0, \ldots, a_n$ such that the least common multiple of a_λ, a_μ is not equal to the least common multiple of a_λ, a_μ, a_ν.*

Now let H be the inner product of $\mathbf{a}$ with the position vector $z \in \mathbb{C}^{n+1}$, i.e., denoting the inner product on $\mathbb{C}^{n+1}$ by $<,>$, $H = <\mathbf{a}, z> = m\sum_{\alpha=0}^{n} \frac{z_\alpha \bar{z}_\alpha}{a_\alpha}$. Taking the inner product of $JX = \phi X + \eta(X)\mathbf{a}$ with the position vector z we have $H\eta(X) = <JX, z> = -<X, Jz>$. Thus if $\iota : \Sigma^{2n-1} \longrightarrow \mathbb{C}^{n+1}$ denotes the imbedding and $\omega = \frac{i}{2}\sum_{\alpha=0}^{n}(\bar{z}_\alpha dz_\alpha - z_\alpha d\bar{z}_\alpha)$, $\eta = \frac{\iota^*\omega}{H}$. η and $\iota^*\omega$ are contact forms on Σ^{2n-1}. This was shown by Sasaki and Hsu [1976] and simpler proofs given by Abe [1977] and Vaisman [1978].

Turning to the question of an associated metric for η, set

$$g(X, Y) = \frac{1}{H}(<X, Y> - \eta(X)<Y, \xi> - \eta(Y)<X, \xi>$$

$$+\eta(X)\eta(Y) < \xi, \xi >) + \eta(X)\eta(Y).$$

Note that g is not the induced metric from the imbedding $\iota : \Sigma^{2n-1} \longrightarrow \mathbb{C}^{n+1}$. Sasaki [1985] showed that (ϕ, ξ, η, g) is a Sasakian structure on Σ^{2n-1}.

Thus we see that there are in fact many non-regular Sasakian manifolds. Much of the above goes over to more general spaces. For discussion of these generalizations, see the references mentioned as well as Abe [1976], Abe and Erbacher [1975], Lutz and Meckert [1976] and Sato [1977]. An earlier example of a non-regular Sasakian manifold was given by Tanno [1969].

6.8 Topology

In the 1960s a great deal of work was done on the topology of compact Sasakian and to a lesser extent cosymplectic manifolds. The idea was to see how much a compact Sasakian manifold must be like a sphere. In the case of a compact Kähler manifold, the even-dimensional Betti numbers are different from zero and the odd-dimensional Betti numbers are even, properties which are certainly enjoyed by complex projective space (see e.g., Goldberg [1962, Chapter V]). Furthermore the Betti numbers b_p of a compact Kähler manifold of positive constant holomorphic curvature are equal to 1 for p even and vanish for p odd (see e.g., Goldberg [1962, Chapter VI]). More recently there has been renewed interest in the cosymplectic case. For brevity we will just mention a few results and the interested reader can consult the references for details.

In [1965] Tachibana proved that the first Betti number of a compact Sasakian manifold M^{2n+1} is zero or even. This is proved by first showing that on a compact K-contact manifold, a harmonic 1-form ω is orthogonal to the contact form η. Then letting $\tilde{\omega} = \omega \circ \phi$ and computing the Laplacian of $\tilde{\omega}$ one obtains the harmonicity of $\tilde{\omega}$ as well. Thus the number of independent harmonic 1-forms is even. The computation uses the fact that on a Sasakian manifold the Ricci operator commutes with ϕ, a fact that we will prove in our subsequent discussions of curvature. More generally the p-th Betti number is even for p odd and $1 \leq p \leq n$ and by duality for p even and $n + 1 \leq p \leq 2n$ (Fujitani [1966], see also Blair and Goldberg [1967]).

Considerable attention has been given to the vanishing of the second Betti number under some curvature restrictions as well as being isometric to the unit sphere under stronger conditions. A compact Sasakian manifold of strictly positive curvature has vanishing second Betti number (Moskal [1966], Tachibana and Ogawa [1966], Tanno [1968], Goldberg [1967] in the regular case and Goldberg [1968b] in the regular case with non-negative curvature). A compact, simply-connected Sasakian–Einstein space of strictly positive curvature is isometric to the unit sphere (Moskal [1966]). Pinching theorems have been obtained by Tanno [1968] including an analogue of holomorphic pinching.

More recently Goldberg [1986] showed that a compact simply-connected regular Sasakian manifold M of strictly positive curvature is homeomorphic to a sphere. If, in addition, M has constant scalar curvature, then Goldberg had shown earlier [1967] that M is isometric to a sphere, but not necessarily with a constant curvature metric (cf. the metrics on the sphere in Example 7.4.1).

Allowing some negative curvature, Tanno [1968] showed that if M^{2n+1} is a compact K-contact manifold with sectional curvature greater than $\frac{-3}{2n-1}$, then $b_1 = 0$. Similarly if the Ricci tensor ρ is such that $\rho + 2g$ is positive definite, then $b_1 = 0$. By duality in dimension 3, one also has $b_2 = 0$.

In dimension 5, Perrone [1989] showed that if M^5 is a compact simply-connected regular Sasakian manifold with $b_2 = 0$ and with scalar curvature $\tau > -4$, then M^5 is homeomorphic to a sphere. If, in addition, M^5 has constant scalar curvature, M is isometric to a sphere (but not necessarily with a constant curvature metric).

A classical result on the topology of a compact Kähler manifold M^{2n} is the monotonicity of the Betti numbers, namely $b_p \leq b_{p+2}$, $p \leq n-1$; this is proved using the idea of effective harmonic forms (see Goldberg [1962, Chapter V]). We briefly describe the idea of effective forms in the almost contact context. Let $(M^{2n+1}, \phi, \xi, \eta, g)$ be a compact almost contact manifold and, as before, we denote the fundamental 2-form by Φ. Define two operators $\mathbf{L}$ and Λ acting on differential forms α by

$$\mathbf{L}\alpha = \alpha \wedge \Phi, \quad \Lambda\alpha = *\mathbf{L} * \alpha$$

where $*$ denotes the Hodge star isomorphism ($\mathbf{L}\alpha = \alpha \wedge \Omega$ in the symplectic case). A p-form α is said to be *effective* if $\Lambda\alpha = 0$. We now have the following Proposition (see Blair and Goldberg [1967], Chinea, de Leon and Marrero [1993]).

Proposition 6.3 *On an almost contact metric manifold of dimension $2n+1$, every p-form α with $p \leq n+1$ may be written uniquely as a sum*

$$\alpha = \sum_{k=0}^{r} \mathbf{L}^k \beta_{p-2k}$$

where the β_{p-2k}'s, $0 \leq k \leq r$, are effective forms of degree $p - 2k$ and $r = \left[\frac{p}{2}\right]$.

In the cosymplectic case we cite the recent work of Chinea, de Leon and Marrero [1993]. They prove that on a compact cosymplectic manifold M^{2n+1}, $b_0 \leq b_1 \leq \cdots \leq b_n = b_{n+1}$ and $b_{n+1} \geq b_{n+2} \geq \cdots \geq b_{2n+1}$. Moreover the differences $b_{2p+1} - b_{2p}$ with $0 \leq p \leq n$ are even and so in particular the first Betti number of M^{2n+1} is odd. They also prove a strong Lefschetz property for a compact cosymplectic manifold and show that such a manifold is formal (i.e., the homotopy type of the differential graded algebra of differential forms is the same as the homotopy type of the cohomology ring).

This topological work involves generalizing from Kähler geometry notions of the bidegree of differential forms, effective harmonic forms, etc. For studies of the *tridegree* of differential forms and applications, see Chinea, de Leon and Marrero [1997], Moskal [1977] and Fujitani [1966]. A p-form α is said to be *coeffective* if $\mathbf{L}\alpha = 0$. Coeffective cohomology was studied in the symplectic case by Bouche [1990] and in the almost cosymplectic case by Chinea, de Leon and Marrero [1995]. In both papers the relation between the coeffective and the de Rham cohomologies of the manifolds is discussed (for the almost contact case see also M. Fernández, R. Ibánez and M. de Leon [1997], [1998]).

7
Curvature of Contact Metric Manifolds

7.1 Basic curvature properties

In this chapter we discuss many aspects of the curvature of contact metric manifolds. We begin with some preliminaries concerning the tensor field h. Let M^{2n+1} be a contact metric manifold with structure tensors (ϕ, ξ, η, g) and $h = \frac{1}{2}\mathcal{L}_\xi\phi$ as before. Recall that in Lemma 6.2 we saw that $\nabla_X\xi = -\phi X - \phi hX$.

Proposition 7.1 *On a contact metric manifold M^{2n+1} we have the formulas*

$$(\nabla_\xi h)X = \phi X - h^2\phi X - \phi R_{X\xi}\xi,$$

$$\frac{1}{2}(R_{\xi X}\xi - \phi R_{\xi\phi X}\xi) = h^2 X + \phi^2 X.$$

Proof. We compute $R_{\xi X}\xi = \nabla_\xi\nabla_X\xi - \nabla_X\nabla_\xi\xi - \nabla_{[\xi,X]}\xi$ using $\nabla_X\xi = -\phi X - \phi hX$; thus

$$R_{\xi X}\xi = \nabla_\xi(-\phi X - \phi hX) + \phi[\xi, X] + \phi h[\xi, X].$$

Applying ϕ and recalling that $\nabla_\xi\phi = 0$ we have

$$\phi R_{\xi X}\xi = \nabla_\xi(X + hX) - \eta(\nabla_\xi(X + hX))\xi - [\xi, X] + \eta([\xi, X])\xi - h[\xi, X]$$

$$= (\nabla_\xi h)X + \nabla_X\xi + h\nabla_X\xi.$$

Using $\nabla_X\xi = -\phi X - \phi hX$ and $\phi h + h\phi = 0$ (Lemma 6.2) this becomes

$$\phi R_{\xi X}\xi = (\nabla_\xi h)X - \phi X + h^2\phi X$$

which is the first formula.

Now from the first formula we have

$$R_{\xi X}\xi = h^2 X + \phi^2 X - \phi(\nabla_\xi h)X$$

and

$$\phi R_{\xi \phi X}\xi = -h^2 X - \phi^2 X - \phi(\nabla_\xi h)X;$$

subtracting then yields the second formula.   ∎

Corollary 7.1 *On a contact metric manifold M^{2n+1} the Ricci curvature in the direction ξ is given by*

$$Ric(\xi) = 2n - \mathrm{tr}h^2.$$

Proof. Choosing X to be a unit vector orthogonal to ξ, the inner product of X with the second formula yields the following formula for sectional curvatures

$$K(\xi, X) + K(\xi, \phi X) = 2(1 - g(h^2 X, X)).$$

Therefore if $\{X_1, \dots, X_n, \phi X_1, \dots, \phi X_n, \xi\}$ is a ϕ-basis, then summing over $\{X_1, \dots, X_n\}$ yields the result.   ∎

Recall that a K-contact structure is a contact metric structure for which the vector field ξ is Killing and that this is the case if and only if the symmetric operator h vanishes. Thus from the above corollary we have the following immediate result (Blair [1977]).

Theorem 7.1 *A contact metric manifold M^{2n+1} is K-contact if and only if $Ric(\xi) = 2n$.*

With regard to sectional curvature we have an earlier result obtained by Hatakeyama, Ogawa and Tanno [1963].

Theorem 7.2 *A contact metric manifold is K-contact if and only if the sectional curvature of all plane sections containing ξ are equal to 1. Moreover, on a K-contact manifold,*

$$R_{X\xi}\xi = X - \eta(X)\xi.$$

Proof. In view of the above the sufficiency is clear. Conversely if the structure is K-contact, then since $\nabla_X \xi = -\phi X$, we have for X orthogonal to ξ,

$$R_{\xi X}\xi = \nabla_\xi(-\phi X) + \phi[\xi, X] = -\phi \nabla_X \xi = \phi^2 X = -X.$$

The second statement is easily obtained.   ∎

Furthermore for the Ricci operator Q acting on ξ we have the following.

Proposition 7.2 *On a K-contact metric manifold M^{2n+1} $Q\xi = 2n\xi$.*

Proof. Since ξ is Killing, it is affine and therefore

$$\nabla_X \nabla_Y \xi - \nabla_{\nabla_X Y} \xi = R_{X\xi} Y.$$

From this we have $(\nabla_X \phi)Y = R_{X\xi}Y$ but in Section 6.1 we saw that on a contact metric manifold $\nabla_i \phi^i_j = -2n\eta_j$. Thus letting $\{X_A\}$ be a local orthonormal basis of the contact subbundle,

$$Q\xi = \sum R_{\xi X_A} X_A = \sum (\nabla_{X_A} \phi) X_A = 2n\xi.$$

■

We noted in Theorem 7.2 that on a K-contact manifold $R_{X\xi}\xi = X - \eta(X)\xi$. On a Sasakian manifold we have the following stronger result.

Proposition 7.3 *On a Sasakian manifold*

$$R_{XY}\xi = \eta(Y)X - \eta(X)Y.$$

Proof.

$$\begin{aligned} R_{XY}\xi &= -\nabla_X \phi Y + \nabla_Y \phi X + \phi[X,Y] \\ &= -(\nabla_X \phi)Y + (\nabla_Y \phi)X \\ &= \eta(Y)X - \eta(X)Y. \end{aligned}$$

■

As converses of these results one often sees propositions of the following type (Hatakeyama, Ogawa and Tanno [1963]).

Proposition 7.4 *Let (M^{2n+1}, g) be a Riemannian manifold admitting a unit Killing field ξ such that $R_{X\xi}\xi = X$ for X orthogonal to ξ. Then M^{2n+1} is a K-contact manifold.*

Proof. Let $\eta(X) = g(X, \xi)$ and $\phi X = -\nabla_X \xi$. Since ξ is unit Killing $\nabla_\xi \xi = 0$ and

$$\nabla_X \nabla_Y \xi - \nabla_{\nabla_X Y} \xi = R_{X\xi} Y. \qquad (*)$$

Thus for X orthogonal to ξ

$$\phi^2 X = \nabla_{\nabla_X \xi} \xi = R_{\xi X}\xi = -X$$

and $\phi\xi = 0$. Therefore $\phi^2 = -I + \eta \otimes \xi$. Moreover

$$d\eta(X,Y) = \frac{1}{2}(g(\nabla_X \xi, Y) - g(\nabla_Y \xi, X)) = -g(\nabla_Y \xi, X) = g(X, \phi Y).$$

Therefore (ϕ, ξ, η, g) is a contact metric structure on M^{2n+1}. ∎

An interesting variation is a result of Rukimbira [1995b] that if a Riemannian manifold admits a unit Killing field ξ such that the sectional curvatures of plane sections containing ξ is positive, then it also admits a K-contact structure but with a possibly different metric.

Proposition 7.5 *If in Proposition 7.4* $R_{XY}\xi = g(\xi, Y)X - g(X, \xi)Y$, *then* M^{2n+1} *is Sasakian.*

Proof. From equation $(*)$ in the proof of Proposition 7.4, $(\nabla_X \phi)Y = R_{\xi X}Y$ and hence

$$g((\nabla_X \phi)Y, Z) = g(R_{\xi X}Y, Z) = g(R_{YZ}\xi, X) = g(\eta(Z)Y - \eta(Y)Z, X). \quad ∎$$

These results start with a Riemannian structure and construct the desired structure. However given a contact metric structure we have the following proposition.

Proposition 7.6 *A contact metric structure is Sasakian if and only if*

$$R_{XY}\xi = \eta(Y)X - \eta(X)Y.$$

Proof. For X orthogonal to ξ, $R_{\xi X}\xi = -X$ and so from the second equation of Proposition 1 we have

$$\frac{1}{2}(-X - \phi(-\phi X)) = h^2 X + \phi^2 X = h^2 X - X.$$

Therefore $h^2 = 0$, but h is a symmetric operator and so $h = 0$. Thus ξ is Killing and the result follows as above. ∎

Finally for future use we establish some additional lemmas on Sasakian manifolds. The reader will recognize these curvature properties as being analogous to well-known curvature properties of Kähler manifolds. Let M^{2n+1} be a Sasakian manifold with structure tensors (ϕ, ξ, η, g) and define a tensor field P of type $(0,4)$ by

$$P(X, Y, Z, W) = d\eta(X, Z)g(Y, W) - d\eta(X, W)g(Y, Z)$$

$$-d\eta(Y, Z)g(X, W) + d\eta(Y, W)g(X, Z).$$

Lemma 7.1 *On a Sasakian manifold we have*

a) $g(R_{XY}Z, \phi W) + g(R_{XY}\phi Z, W) = -P(X, Y, Z, W).$

For X, Y, Z, W orthogonal to ξ we have

$$\text{b)} \quad g(R_{\phi X \, \phi Y}\phi Z, \phi W) = g(R_{XY}Z, W)$$

and

$$\text{c)} \quad g(R_{X\phi X}Y, \phi Y) = g(R_{XY}X, Y) + g(R_{X\phi Y}X, \phi Y) - 2P(X, Y, X, \phi Y).$$

Proof. A direct computation or the Ricci identity shows that

$$(\nabla_X \nabla_Y \Phi - \nabla_Y \nabla_X \Phi - \nabla_{[X,Y]}\Phi)(Z, W) = -g(R_{XY}Z, \phi W) - g(R_{XY}\phi Z, W).$$

Computing the left-hand side using $(\nabla_X \phi)Y = g(X, Y)\xi - \eta(Y)X$ yields a). Using a) and the definition of P we obtain b). Finally applying the first Bianchi identity to $g(R_{X\phi X}Y, \phi Y)$ and using a) we obtain c). ∎

Lemma 7.2 *On a Sasakian manifold $Q\phi = \phi Q$.*

Proof. Choosing a ϕ-basis $\{X_i, X_{n+i} = \phi X_i, \xi\}$ we have for X and Y orthogonal to ξ,

$$g(Q\phi X, \phi Y) = \sum_{A=1}^{2n} g(R_{\phi X \, X_A}X_A, \phi Y) + g(R_{\phi X \, \xi}\xi, \phi Y)$$

$$= \sum_{A=1}^{2n} g(R_{\phi X \, \phi X_A}\phi X_A, \phi Y) + g(X, Y)$$

$$= g(QX, Y)$$

where we have used b) from the previous lemma. We already know that $Q\xi = 2n\xi$ and hence the Ricci operator Q commutes with ϕ on a Sasakian manifold. ∎

7.2 Curvature of contact metric manifolds

Before giving our main curvature results, we present some rather complicated lemmas from the paper [1979] of Olszak.

Lemma 7.3 *On a contact metric manifold*

$$(\nabla_X \phi)Y + (\nabla_{\phi X}\phi)\phi Y = 2g(X, Y)\xi - \eta(Y)(X + hX + \eta(X)\xi).$$

Proof. Using Corollary 6.1 or by direct differentiation of $\nabla_Y \xi = -\phi Y - \phi hY$ we obtain

$$(\nabla_X \Phi)(\phi Y, Z) - (\nabla_X \Phi)(Y, \phi Z) = -\eta(Y)g(X + hX, \phi Z) - \eta(Z)g(X + hX, \phi Y) \tag{$*$}$$

and replacing Z by ϕZ and using Corollary 6.1 again we obtain

$$(\nabla_X \Phi)(\phi Y, \phi Z) + (\nabla_X \Phi)(Y, Z) = \eta(Y)g(X + hX, Z) - \eta(Z)g(X + hX, Y). \tag{$**$}$$

Now since $d\Phi = 0$ we have

$$(\nabla_X \Phi)(Y, Z) + (\nabla_Y \Phi)(Z, X) + (\nabla_Z \Phi)(X, Y)$$

$$+ (\nabla_{\phi X} \Phi)(\phi Y, Z) + (\nabla_{\phi Y} \Phi)(Z, \phi X) + (\nabla_Z \Phi)(\phi X, \phi Y)$$

$$+ (\nabla_{\phi X} \Phi)(Y, \phi Z) + (\nabla_Y \Phi)(\phi Z, \phi X) + (\nabla_{\phi Z} \Phi)(\phi X, Y)$$

$$- (\nabla_X \Phi)(\phi Y, \phi Z) - (\nabla_{\phi Y} \Phi)(\Phi Z, X) - (\nabla_{\phi Z} \Phi)(X, \phi Y) = 0.$$

Now $(*)$ and $(**)$ give

$$(\nabla_{\phi X} \Phi)(Z, \phi Y) + (\nabla_X \Phi)(Z, Y)$$

$$= 2\eta(Z)g(X, Y) - \eta(Y)g(X + hX, Z) - \eta(X)\eta(Y)\eta(Z)$$

from which the result follows. ∎

Lemma 7.4 *The curvature tensor of a contact metric manifold satisfies*

$$g(R_\xi X Y, Z) = -(\nabla_X \Phi)(Y, Z) - g(X, (\nabla_Y \phi h)Z) + g(X, (\nabla_Z \phi h)Y),$$

$$g(R_\xi X Y, Z) - g(R_\xi X \phi Y, \phi Z) + g(R_\xi \phi X Y, \phi Z) + g(R_\xi \phi X \phi Y, Z)$$

$$= 2(\nabla_{hX} \Phi)(Y, Z) - 2\eta(Y)g(X + hX, Z) + 2\eta(Z)g(X + hX, Y).$$

Proof. Differentiating $\nabla_Z \xi = -\phi Z - \phi hZ$ we have

$$R_{YZ}\xi = -(\nabla_Y \phi)Z + (\nabla_Z \phi)Y - (\nabla_Y \phi h)Z + (\nabla_Z \phi h)Y$$

which, since $d\Phi = 0$, yields the first formula. Now set

$$A(X, Y, Z) = -(\nabla_X \Phi)(Y, Z) + (\nabla_X \Phi)(\phi Y, \phi Z)$$

$$- (\nabla_{\phi X} \Phi)(Y, \phi Z) + (\nabla_{\phi X} \Phi)(\phi Y, Z)$$

and

$$B(X, Y, Z) = -g(X, (\nabla_Y \phi h)Z) + g(X, (\nabla_{\phi Y} \phi h)\phi Z)$$

$$- g(\phi X, (\nabla_Y \phi h)\phi Z) - g(\phi X, (\nabla_{\phi Y} \phi h)Z).$$

Then by the first formula, the left side of the second formula is $A(X, Y, Z) + B(X, Y, Z) - B(X, Z, Y)$. Now by Lemma 7.3 and equation $(**)$ in its proof we have

$$A(X, Y, Z) = 2g(X, Y)\eta(Z) - 2g(X, Z)\eta(Y).$$

Also it is straightforward to show that $\eta((\nabla_{\phi Y} h)Z) = g(-Y + hY, hZ)$. Now rewrite B as

$$B(X, Y, Z) = -g(X, (\nabla_Y \phi)hZ) + g(X, h(\nabla_Y \phi)Z)$$

$$+g(X, h\phi(\nabla_{\phi Y}\phi)Z) + g(X, \phi(\nabla_{\phi Y}\phi)hZ) + \eta(X)\eta((\nabla_{\phi Y}h)Z).$$

Then using Lemma 7.3 again

$$B(X, Y, Z) = 2g(hX, (\nabla_Y \phi)Z) + 2\eta(Z)g(hX, Y) - 2\eta(X)g(hY, hZ).$$

Finally computing $A(X, Y, Z) + B(X, Y, Z) - B(X, Z, Y)$ and using $d\Phi = 0$ we have the result. ∎

Let ρ denote the Ricci tensor and τ the scalar curvature. In addition we define the *-*Ricci tensor* ρ^* and *-*scalar curvature* τ^* by contracting the curvature tensor by ϕ instead of the metric. Precisely

$$\rho_{ij}^* = R_{iklm}\phi^{kl}\phi_j{}^m, \quad \tau^* = \rho^*{}_i^i.$$

These notions have their origin in almost Hermitian geometry and we shall see their almost Hermitian analogues in Section 10.2. We now have an important proposition due to Olszak [1979].

Proposition 7.7 *On a contact metric manifold* M^{2n+1},

$$\tau^* - \tau + 4n^2 = \mathrm{tr}h^2 + \frac{1}{2}(|\nabla\phi|^2 - 4n) \geq 0$$

with equality if and only if M^{2n+1} *is Sasakian.*

Proof. We have seen that $\nabla_i \phi^i{}_j = -2n\eta_j$ and $\nabla_k \xi^j = -\phi^j{}_k - \phi^j{}_m h^m{}_k$. Therefore using $\phi^2 = -I + \eta \otimes \xi$ and basic properties of h (Section 6.4) we have

$$\phi^{kj}\nabla_k\nabla_i\phi^i{}_j = -4n^2. \qquad (*)$$

Differentiation of $\phi^2 = -I + \eta \otimes \xi$ yields $\phi^{kj}\nabla_t\phi_{kj} = 0$. Since $d\Phi = 0$, we have $\phi^{kj}\nabla_k\phi_{tj} = \phi^{kj}(-\nabla_t\phi_{jk} - \nabla_j\phi_{kt})$ from which we obtain $\phi^{kj}\nabla_k\phi^t{}_j = 0$. In turn $\phi^{kj}\nabla_t\nabla_k\phi^t{}_j = -(\nabla_t\phi^{kj})(\nabla_k\phi^t{}_j)$. Using $d\Phi = 0$ on the second factor on the right and simplifying we have

$$\phi^{kj}\nabla_t\nabla_k\phi^t{}_j = -\frac{1}{2}|\nabla\phi|^2. \qquad (**).$$

From $(*)$ and $(**)$

$$\phi^{kj}(\nabla_k\nabla_t\phi^t{}_j - \nabla_t\nabla_k\phi^t{}_j) = \frac{1}{2}|\nabla\phi|^2 - 4n^2.$$

Therefore by Corollary 7.1

$$\frac{1}{2}|\nabla\phi|^2 - 4n^2 = \phi^{kj}(R_{kta}{}^t\phi^a{}_j - R_{ktj}{}^a\phi^t{}_a) = (g^{ka} - \xi^k\xi^a)(-\rho_{ka}) + \tau^*$$

$$= \tau^* - \tau + (2n - \mathrm{tr}h^2)$$

which is the desired formula.

Now $(\nabla_i\phi_{jk} - g_{ik}\eta_j + g_{ij}\eta_k)(\nabla^i\phi^{jk} - g^{ik}\xi^j + g^{ij}\xi^k) \geq 0$ is equivalent to $|\nabla\phi|^2 - 4n \geq 0$ giving the inequality and by Theorem 6.3 equality holds if and only if the structure is Sasakian. ∎

We now prove the following important result of Olszak [1979].

Theorem 7.3 *If a contact metric manifold M^{2n+1} is of constant curvature c and dimension ≥ 5, then $c = 1$ and the structure is Sasakian.*

Proof. Recall from Proposition 7.1 that, $\frac{1}{2}(R_{\xi X}\xi - \phi R_{\xi \phi X}\xi) = h^2X + \phi^2X$; thus if $R_{XY}Z = c(g(Y,Z)X - g(X,Z)Y)$, then $\frac{c}{2}(\eta(X)\xi - X + \phi^2X) = h^2X + \phi^2X$. Therefore $h^2X + (c-1)\phi^2X$ and hence $\mathrm{tr}h^2 = 2n(1-c)$. Now from Lemma 7.4

$$c(g(X,Y)\eta(Z) - \eta(Y)g(X,Z)) - 0 - c\eta(Y)g(\phi X, \phi Z) + c\eta(Z)g(\phi X, \phi Y)$$

$$= 2(\nabla_{hX}\Phi)(Y,Z) - 2\eta(Y)g(X + hX, Z) + 2\eta(Z)g(X + hX, Y).$$

Therefore

$$(\nabla_{hX}\Phi)(Y,Z) = (1-c)(\eta(Y)g(X,Z) - \eta(Z)g(X,Y))$$

$$+\eta(Y)g(hX,Z) - \eta(Z)g(hX,Y).$$

Replacing X by hX, we have

$$-g((\nabla_{(c-1)(-X+\eta(X)\xi)}\phi)Y, Z) = (1-c)(\eta(Y)g(hX,Z) - \eta(Z)g(hX,Y))$$

$$+\eta(Y)g((c-1)(-X + \eta(X)\xi), Z) - \eta(Z)g((c-1)(-X + \eta(X)\xi), Y)$$

and hence

$$(\nabla_X\phi)Y = g(X + hX, Y)\xi - \eta(Y)(X + hX).$$

Using this to compute $|\nabla\phi|^2$ and using $\mathrm{tr}h^2 = 2n(1-c)$ we obtain

$$|\nabla\phi|^2 = 4n(2-c).$$

On the other hand $\tau = 2n(2n+1)c$ and $\tau^* = 2nc$ as is easily checked. Now from Olszak's formula in Proposition 7.7,

$$\tau - \tau^* - 4n^2 = -\mathrm{tr}h^2 - \frac{1}{2}|\nabla\phi|^2 + 2n \leq 0$$

with equality if and only if the structure is Sasakian. Computing both sides of this formula we find that $4n^2(c-1) = 4n(c-1)$; thus if $n > 1$, $c = 1$ and hence $\tau - \tau^* - 4n^2 = 0$ giving M^{2n+1} Sasakian. ∎

Earlier the present author [1976] showed that in dimension ≥ 5, there are no flat associated metrics. Thus while the 5-torus carries a contact structure (Example 3.2.7), the flat metric is not an associated metric. In dimension 3, the only constant curvature cases are constant curvature 0 and 1 as we will note below. The non-existence of flat associated metrics does raise the question as to whether there are contact metric manifolds of everywhere non-positive curvature, except for the flat 3-dimensional case. If the manifold is compact and we ask for strictly negative curvature, we can answer this question in the negative using the following deep result of A. Zeghib [1995] on geodesic plane fields. Recall that a plane field on a Riemannian manifold is said to be *geodesic* if any geodesic tangent to the plane field at some point is everywhere tangent to it.

Theorem 7.4 *A compact negatively curved Riemannian manifold has no C^1 geodesic plane field (of non-trivial dimension).*

Since for any contact metric structure the integral curves of ξ are geodesics (Theorem 4.5), ξ determines a geodesic line field to which we can apply the theorem of Zeghib as was pointed out by Rukimbira [1998]. Thus we have the following corollary.

Corollary 7.2 *On a compact contact manifold, there is no associated metric of strictly negative curvature.*

The author conjectures that this and a bit more is true locally, viz., that except for the flat 3-dimensional case, any contact metric manifold has some positive sectional curvature. The fact that hyperbolic space has many 1-dimensional totally geodesic foliations does not violate such a conjecture, since by the above theorem of Olszak the hyperbolic metric cannot be an associated metric of any contact structure.

Along the line of the influence of a contact structure on the curvature of its associated metrics we also make the following remark. In [1941] Myers proved that a complete Riemannian manifold for which $Ric \geq \delta > 0$ is compact. In [1981] Hasegawa and Seino generalized Myers' theorem for a K-contact manifold by proving that a complete K-contact manifold for which $Ric \geq$

$-\delta > -2$ is compact (they state their result in the Sasakian case, but their proof uses only the K-contact property). As we have seen in the K-contact case, all sectional curvatures of plane sections containing ξ are equal to 1 and hence there is a lot of positive curvature from the outset. In an attempt to weaken the K-contact requirement in this result, Blair and Sharma [1990a] considered a contact metric manifold M^{2n+1} for which ξ is an eigenvector field of the Ricci operator, equivalently $\text{div} h\phi$ is proportional to η. In this case if $Ric \geq -\delta > -2$ and the sectional curvatures of plane sections containing ξ are $\geq \epsilon > \delta' \geq 0$ where

$$\delta' = 2\sqrt{n(\delta - 2\sqrt{2\delta} + n + 2)} - (\delta - 2\sqrt{2\delta} + 1 + 2n),$$

then M^{2n+1} is compact.

The flat case was investigated further by Rukimbira [1998] who showed that a compact flat contact metric manifold is isometric to the quotient for a flat 3-torus by a finite cyclic group of isometries of order 1,2,3,4 or 6.

In Example 3.2.6 and Section 6.2 we saw explicitly a flat contact metric structure on $\mathbb{R}^3$ and in turn on the 3-dimensional torus T^3. As in Section 6.2 let $\eta = \frac{1}{2}(\cos x^3 dx^1 + \sin x^3 dx^2)$ be the contact form; $g_{ij} = \frac{1}{4}\delta_{ij}$ is an associated metric and the non-zero eigenvalues of h were ± 1. It is interesting to study this example in Darboux coordinates and to generalize it to higher dimensions. By the Darboux Theorem there exist coordinates (x, y, z) such that the contact form η is given by $\frac{1}{2}(dz - ydx)$. Consider the map $f : \mathbb{R}^3 \longrightarrow \mathbb{R}^3$ given by

$$x^1 = z\cos x - y\sin x, \quad x^2 = -z\sin x - y\cos x, \quad x^3 = -x.$$

Then $\frac{1}{2}(dz - ydx) = f^*\eta$ and $g_0 = f^*g$ is a flat associated metric for the Darboux form $\eta_0 = \frac{1}{2}(dz - ydx)$. The metric g_0 is given by

$$g_0 = \frac{1}{4}\begin{pmatrix} 1 + y^2 + z^2 & z & -y \\ z & 1 & 0 \\ -y & 0 & 1 \end{pmatrix}.$$

Now consider the Darboux form $\eta = \frac{1}{2}(dz - \sum y^i dx^i)$ on $\mathbb{R}^{2n+1}$.

$$g = \frac{1}{4}\begin{pmatrix} \delta_{ij} + y^i y^j + \delta_{ij} z^2 & \delta_{ij} z & -y^i \\ \delta_{ij} z & \delta_{ij} & 0 \\ -y^j & 0 & 1 \end{pmatrix}$$

is an associated metric quite different from the Sasakian metric given in Examples 4.5.1 and 6.7.1. Direct computation shows that $h\frac{\partial}{\partial y^i} = -\frac{\partial}{\partial y^i}$. Therefore -1 is an eigenvalue of h of multiplicity n and hence in turn $+1$ is also an eigenvalue of multiplicity n. This metric enjoys the following curvature properties:

$R_{\xi X}\xi = 0$ for every X and $R_{XY}\xi = 0$ for $X, Y \in \{+1\}$. However $R_{XY}\xi \neq 0$ in general, for example in dimension 5,

$$R_{\frac{\partial}{\partial x^1} \frac{\partial}{\partial x^2}}\xi = \frac{1}{2}(-y^2\frac{\partial}{\partial x^1} + y^1\frac{\partial}{\partial x^2} + y^2 z\frac{\partial}{\partial y^1} - y^1 z\frac{\partial}{\partial y^2}).$$

We now show that the condition $R_{XY}\xi = 0$ for all X, Y has a strong and interesting implication for a contact metric manifold, namely that it is locally the product of Euclidean space E^{n+1} and a sphere of constant curvature $+4$ (Blair [1977]).

Theorem 7.5 *A contact metric manifold M^{2n+1} satisfying $R_{XY}\xi = 0$ is locally isometric to $E^{n+1} \times S^n(4)$ for $n > 1$ and flat for $n = 1$.*

Proof. $R_{XY}\xi = 0$ implies, by Proposition 7.1, that $h^2 + \phi^2 = 0$ and hence that the non-zero eigenvalues of h are ± 1, each with multiplicity n. For $X, Y \in \{-1\}$,

$$0 = -\nabla_{[X,Y]}\xi = \phi[X, Y] + \phi h[X, Y]$$

and $\eta([X, Y]) = -2d\eta(X, Y) = -2g(X, \phi Y) = 0$. Thus $\{-1\}$ is integrable. Also $0 = -\nabla_{[X,\xi]}\xi$, so that $[X, \xi] \in \{-1\}$. Therefore $\{-1\} \oplus \{\xi\}$ is integrable. Choose coordinates $(u^0, \ldots, u^{2n})$ such that $\frac{\partial}{\partial u^0}, \ldots, \frac{\partial}{\partial u^n} \in \{-1\} \oplus \{\xi\}$. Let $X_i = \frac{\partial}{\partial u^{n+i}} + \sum_{j=0}^{n} f_i^j \frac{\partial}{\partial u^j}$ where the f_i^j's are functions chosen so that $X_i \in \{+1\}$. Now $[\frac{\partial}{\partial u^k}, X_i] \in \{-1\} \oplus \{\xi\}$, $k = 0, \ldots, n$ and hence

$$0 = \nabla_{[\frac{\partial}{\partial u^k}, X_i]}\xi = \nabla_{\frac{\partial}{\partial u^k}}\nabla_{X_i}\xi - \nabla_{X_i}\nabla_{\frac{\partial}{\partial u^k}}\xi = -2\nabla_{\frac{\partial}{\partial u^k}}\phi X_i$$

from which

$$\nabla_{\phi X_j}\phi X_i = 0.$$

We therefore see that the integral submanifolds of $\{-1\} \oplus \{\xi\}$ are totally geodesic and flat.

From the second formula of Lemma 7.4, we have for X, Y, Z orthogonal to ξ that $g((\nabla_X \phi)Y, Z) = 0$. Also for $X, Y \in \{+1\}$,

$$0 = R_{XY}\xi = -2\nabla_X\phi Y + 2\nabla_Y\phi X - \nabla_{[X,Y]}\xi$$

$$= -2(\nabla_X\phi)Y + 2(\nabla_Y\phi)X - 2\phi[X, Y] + \phi[X, Y] + \phi h[X, Y].$$

Taking the inner product with $Z \in \{+1\}$, gives $g(-\phi[X, Y] - h\phi[X, Y], Z) = 0$ and hence $g(\phi[X, Y], Z) = 0$. Thus $[X, Y]$ is orthogonal to $\{-1\}$. Also $\eta([X, Y]) = 0$ and therefore $\{+1\}$ is integrable.

Now take $X \in \{-1\}$ and $Y \in \{+1\}$; then

$$0 = R_{XY}\xi = -2\nabla_X\phi Y + \phi[X, Y] + \phi h[X, Y]$$

$$= -2(\nabla_X \phi)Y - \phi \nabla_X Y - \phi \nabla_Y X - h\phi \nabla_X Y + h\phi \nabla_Y X.$$

Taking the inner product with $Z \in \{-1\}$, we have $g(\phi \nabla_Y X, Z) = 0$ and hence that $\nabla_Y X$ is orthogonal to $\{+1\}$. Also $\nabla_Y \xi = -2\phi Y$ is orthogonal to $\{+1\}$. Therefore the integral submanifolds of $\{+1\}$ are totally geodesic. Moreover we are now at the point that M^{2n+1} has a local Riemannian product structure.

For $X \in \{+1\}$,

$$g((\nabla_X \phi)Y, \xi) = -g((\nabla_X \phi)\xi, Y) = g(\phi \nabla_X \xi, Y)$$

$$= g(\phi(-2\phi X), Y) = 2g(X, Y)$$

and hence by the second formula of Lemma 7.4, $(\nabla_X \phi)Y = 2g(X, Y)\xi$ for $X, Y \in \{+1\}$. Now for $X, Y, Z, W \in \{+1\}$,

$$g(\nabla_X \nabla_Y \phi Z, \phi W) - g(\nabla_X \nabla_Y Z, W)$$

$$= g(\nabla_X(2g(Y, Z)\xi + \phi \nabla_Y Z), \phi W) - g(\nabla_X \nabla_Y Z, W)$$

$$= -4g(Y, Z)g(X, W)$$

using the property we noted above that $g((\nabla_X \phi)Y, Z) = 0$ for X, Y, Z orthogonal to ξ. Using this property again to treat the bracket terms we have

$$g(R_{XY} \phi Z, \phi W) - g(R_{XY} Z, W) = -4(g(Y, Z)g(X, W) - g(X, Z)g(Y, W)),$$

but $g(R_{XY} \phi Z, \phi W) = 0$ by virtue of the Riemannian product structure. Therefore the integral submanifolds of $\{+1\}$ are locally isometric to $S^n(4)$.

In dimension 3, the integrability of $\{-1\}$ and $\{+1\}$ is immediate and the rest of the proof goes through giving that M^3 is flat. ∎

As a generalization of both $R_{XY}\xi = 0$ and the Sasakian case, $R_{XY}\xi = \eta(Y)X - \eta(X)Y$, consider

$$R_{XY}\xi = \kappa(\eta(Y)X - \eta(X)Y) + \mu(\eta(Y)hX - \eta(X)hY)$$

for constants κ and μ. We say this is a contact metric structure for which ξ belongs to the (κ, μ)-*nullity distribution* or we refer to this condition as the (κ, μ)-*nullity condition* or the contact metric manifold as a (κ, μ)-*manifold*. Despite the technical appearance of this condition there are good reasons for considering it; we first mention them here referring to Blair, Koufogiorgos, and Papantoniou [1995] for details and then give a classification theorem due to Boeckx [2000].

Theorem 7.6 *A contact metric manifold M for which ξ belongs to the (κ, μ)-nullity distribution is a strongly pseudo-convex CR-manifold.*

Theorem 7.7 *Let M^{2n+1} be a contact metric manifold for which ξ belongs to the (κ, μ)-nullity distribution. Then $\kappa \leq 1$. If $\kappa = 1$, the structure is Sasakian. If $\kappa < 1$, the (κ, μ)-nullity condition determines the curvature of M^{2n+1} completely.*

We remark that when $\kappa = 1$, the proof shows that $h = 0$ and hence μ is indeterminate in this case. When $\kappa < 1$, the non-zero eigenvalues of h are $\pm\sqrt{1-\kappa}$ each with multiplicity n. Let λ be the positive eigenvalue. Explicit formulas for the curvature tensor may be found in Blair, Koufogiorgos, and Papantoniou [1995]. We mention only that for a unit vector $X \in [\lambda]$, $\phi X \in [-\lambda]$ and we have

$$K(X, \xi) = \kappa + \lambda\mu, \quad K(\phi X, \xi) = \kappa - \lambda\mu, \quad K(X, \phi X) = -(\kappa + \mu).$$

Thus, turning to the sign of the curvature question, if a (κ, μ)-manifold were of negative curvature, then $\lambda > 1$ by Corollary 7.1 and $\kappa \pm \lambda\mu < 0$, giving $\lambda^2 - 1 > \lambda|\mu|$. $K(X, \phi X) = -(\kappa + \mu) < 0$ would then give $\lambda|\mu| < \lambda^2 - 1 < \mu$, a contradiction.

Theorem 7.8 *Let M be a 3-dimensional contact metric manifold for which ξ belongs to the (κ, μ)-nullity distribution. Then M is either Sasakian or locally isometric to one of the unimodular Lie groups $SU(2)$, $SL(2, \mathbb{R})$. $E(2)$, $E(1,1)$ with a left invariant metric.*

Theorem 7.9 *The standard contact metric structure on the tangent sphere bundle $T_1 M$ (see Section 9.2) satisfies the (κ, μ)-nullity condition if and only if the base manifold M is of constant curvature. In particular if M has constant curvature c, then $\kappa = c(2 - c)$ and $\mu = -2c$.*

Finally given a contact metric structure (ϕ, ξ, η, g), consider the deformed structure

$$\bar{\eta} = a\eta, \quad \bar{\xi} = \frac{1}{a}\xi, \quad \bar{\phi} = \phi, \quad \bar{g} = ag + a(a - 1)\eta \otimes \eta$$

where a is a positive constant. Such a deformation is called a *D-homothetic deformation*, since the metrics restricted to the contact subbundle $\mathcal{D}$ are homothetic. This deformation was introduced by Tanno in [1968] and has many applications. While such a change preserves the state of being contact metric, K-contact, Sasakian or strongly pseudo-convex CR, it destroys a condition like $R_{XY}\xi = 0$ or $R_{XY}\xi = \kappa(\eta(Y)X - \eta(X)Y)$. However the form of the (κ, μ)-nullity condition is preserved under a $\mathcal{D}$-homothetic deformation with

$$\bar{\kappa} = \frac{\kappa + a^2 - 1}{a^2}, \quad \bar{\mu} = \frac{\mu + 2a - 2}{a}.$$

Given a non-Sasakian (κ, μ)-manifold M, Boeckx [2000] introduced an invariant

$$I_M = \frac{1 - \frac{\mu}{2}}{\sqrt{1 - \kappa}}$$

and showed that for two non-Sasakian (κ, μ)-manifolds $(M_i, \phi_i, \xi_i, \eta_i, g_i)$, $i = 1, 2$, we have $I_{M_1} = I_{M_2}$ if and only if up to a $\mathcal{D}$-homothetic deformation, the two spaces are locally isometric as contact metric manifolds. Thus we know all non-Sasakian (κ, μ)-manifolds locally as soon as we have for every odd dimension $2n + 1$ and for every possible value of the invariant I, one (κ, μ)-manifold (M, ϕ, ξ, η, g) with $I_M = I$. From Theorem 7.9 we see that for the standard contact metric structure on the tangent sphere bundle of a manifold of constant curvature c, $I = \frac{1+c}{|1-c|}$. Therefore as c varies over the reals, I takes on every value > -1. Boeckx now gives an example for any odd dimension and value of $I \le -1$; his construction is as follows.

Let $\mathfrak{g}$ be a $(2n + 1)$-dimensional Lie algebra. Introduce a basis for $\mathfrak{g}$, $\{\xi, X_1, \ldots, X_n, Y_1, \ldots, Y_n\}$, and for real numbers α and β define the Lie bracket by

$$[\xi, X_1] = -\frac{\alpha\beta}{2}X_2 - \frac{\alpha^2}{2}Y_1, \quad [\xi, X_2] = \frac{\alpha\beta}{2}X_1 - \frac{\alpha^2}{2}Y_2, \quad [\xi, X_i] = -\frac{\alpha^2}{2}Y_i, \ i \ge 3,$$

$$[\xi, Y_1] = \frac{\beta^2}{2}X_1 - \frac{\alpha\beta}{2}Y_2, \quad [\xi, Y_2] = \frac{\beta^2}{2}X_2 + \frac{\alpha\beta}{2}Y_1, \quad [\xi, Y_i] = \frac{\beta^2}{2}X_i, \ i \ge 3,$$

$$[X_1, X_i] = \alpha X_i, \ i \ne 1, \quad [X_i, X_j] = 0, \ i, j \ne 1,$$

$$[Y_2, Y_i] = \beta Y_i, \ i \ne 2, \quad [Y_i, Y_j] = 0, \ i, j \ne 2,$$

$$[X_1, Y_1] = -\beta X_2 + 2\xi, \quad [X_1, Y_i] = 0, \ i \ge 2,$$

$$[X_2, Y_1] = \beta X_1 - \alpha Y_2, \quad [X_2, Y_2] = \alpha Y_1 + 2\xi, \quad [X_2, Y_i] = \beta X_i, \ i \ge 3,$$

$$[X_i, Y_1] = -\alpha Y_i, \ i \ge 3, \quad [X_i, Y_2] = 0, \ i \ge 3,$$

$$[X_i, Y_j] = \delta_{ij}(-\beta X_2 + \alpha Y_1 + 2\xi), \ i, j \ge 3.$$

The associated Lie group G is not unimodular for dim $\mathfrak{g} \ge 5$ and not both α and β equal to zero, since tr $\mathrm{ad}_{X_1} = (n-1)\alpha$ and tr $\mathrm{ad}_{Y_2} = (n-1)\beta$. Now define a metric on G by left translation of the basis $\{\xi, X_1, \ldots, X_n, Y_1, \ldots, Y_n\}$, taken as orthogonal at the identity. Then taking η as the metric dual of ξ and defining ϕ by $\phi\xi = 0$, $\phi X_i = Y_i$ and $\phi Y_i = -X_i$, we have a contact metric structure on G. Now for the present purpose suppose that $\beta^2 > \alpha^2$. G is a non-Sasakian (κ, μ)-manifold and

$$I_G = -\frac{\beta^2 + \alpha^2}{\beta^2 - \alpha^2} \le -1;$$

thus for appropriate choices of $\beta > \alpha \ge 0$, I_G attains any value ≤ -1.

Returning to Theorem 7.8, consider the Lie algebra

$$[\xi, X] = -\frac{\alpha^2}{2}Y, \quad [\xi, Y] = \frac{\beta^2}{2}X, \quad [X, Y] = 2\xi$$

which corresponds to a unimodular Lie group. Boeckx [2000] points out that for appropriate values of α and β we obtain left invariant contact metric structures with values of the invariant I_G as follows: $I_{SU(2)} > 1$, $I_{E(2)} = 1$, $-1 < I_{SL(2,\mathbb{R})} < 1$, $I_{E(1,1)} = -1$ and $I_{SL(2,\mathbb{R})} < -1$ (see also Blair, Koufogiorgos, and Papantoniou [1995]).

Finally in [2000] Koufogiorgos and Tsichlias considered the question of contact metric manifolds for which ξ belongs to the (κ, μ)-nullity distribution but κ and μ are functions rather than constants. They showed that in dimensions ≥ 5, κ and μ must be constant and in dimension 3 gave an example where κ and μ are not constants; this case is studied further in Koufogiorgos and Tsichlias [toap].

If on a (κ, μ)-manifold, $\mu = 0$, the contact metric manifold is said to be one for which ξ belongs to the κ-*nullity distribution*. In general the κ-*nullity distribution* of a Riemannian manifold (M, g) is the subbundle $N(k)$ defined by

$$N_p(\kappa) = \{Z \in T_pM | R_{XY}Z = \kappa\big((g(Y, Z)X - g(X, Z)Y\big) \,\forall\, X, Y \in T_pM\}.$$

In dimension 3, Blair, Koufogiorgos, and Sharma [1990] showed that ξ belonging to $N(\kappa)$ is equivalent to the Ricci operator Q commuting with ϕ and equivalent to the contact metric manifold being η-*Einstein*, i.e.,

$$Q = aI + b\eta \otimes \xi$$

for some functions a and b. In dimensions ≥ 5 it is known that for any η-Einstein K-contact manifold, a and b are constants. The main result of the author's papers with Koufogiorgos and Sharma [1990] and with H. Chen [1992] is that a 3-dimensional contact metric manifold for which ξ belongs to the κ-nullity distribution is either Sasakian, flat or locally isometric to a left-invariant metric on the Lie groups $SU(2)$ or $SL(2, \mathbb{R})$.

An interesting side question associated with this case is the dimension of the κ-nullity distribution itself. Clearly if a vector Z belongs to $N(\kappa)$, then the sectional curvatures of all plane sections containing Z are equal to κ. In particular on any Riemannian manifold, $N(\kappa)$ is non-trivial for at most one value of κ. It is known that $N(\kappa)$ is an integrable subbundle with totally geodesic leaves of constant curvature κ, see e.g., Tanno [1978a]. In unpublished work, Baikoussis, Koufogiorgos and the author showed that if the characteristic vector field ξ on a contact metric manifold M^{2n+1}, $n > 1$ belongs to $N(\kappa)$, then

($\kappa \leq 1$ and) if $\kappa < 1$ and $\kappa \neq 0$, $dimN(\kappa) = 1$. The corresponding result for $n = 1$ is an unpublished result of R. Sharma. If $\kappa = 0$, M^{2n+1} is locally $E^{n+1} \times S^n(4)$, as we have seen, and ξ is tangent to the Euclidean factor giving $dimN(0) = n + 1$. If $\kappa = 1$, the structure is Sasakian. This leaves the question of the dimension of $N(1)$ on a Sasakian manifold; F. Gouli–Andreou and the author conjecture that it must be either $2n + 1$ (and M^{2n+1} is of constant curvature), or 3 (and M^{2n+1} has a Sasakian 3-structure, see Chapter 13) or 1.

We just mentioned η-Einstein contact metric manifolds and a few remarks about this and the Einstein case are in order. First note that on a Sasakian–Einstein manifold $Q\xi = 2n\xi$ and $Q = \frac{\tau}{2n+1}I$ and hence the scalar curvature is $\tau = 2n(2n + 1) > 0$. An important recent result is the following Theorem of Boyer and Galicki [2001]. This remarkable result should be compared with the Goldberg conjecture that a compact almost Kähler–Einstein space is Kähler (see Section 10.2). The Goldberg conjecture is true for non-negative scalar curvature (Sekigawa [1987]). Since Sasakian–Einstein manifolds have positive scalar curvature, one might hope for a Goldberg conjecture type result in contact geometry and Boyer and Galicki achieved the following.

Theorem 7.10 *A compact K-contact Einstein manifold is Sasakian.*

There are many Sasakian–Einstein manifolds as shown by Boyer and Galicki [2000] including the well known Sasakian–Einstein structure on $S^3 \times S^2$ (see Tanno [1978b] and Section 9.2). This paper of Boyer and Galicki contains a number of nice results on Sasakian–Einstein manifolds.

It is also known that a contact metric Einstein manifold of dimension ≥ 5 with ξ belonging to the κ-nullity distribution is Sasakian (Tanno [1988]).

If M is a compact η-Einstein K-contact manifold with Ricci tensor, $\rho = ag + b\eta \otimes \eta$, and if $a \geq -2$, then M is Sasakian (S. Morimoto [1992], Boyer and Galicki [2001]).

Turning to other curvature conditions we first prove the following result of Tanno [1967b] on K-contact manifolds; the corresponding result in the Sasakian case was given by Okumura [1962a].

Theorem 7.11 *A locally symmetric K-contact manifold is of constant curvature $+1$ and Sasakian.*

Proof. Since the structure is K-contact, $R_{\xi X}\xi = -X + \eta(X)\xi$ (cf. Proposition 7.1 with $h = 0$). Differentiation yields

$$R_{-\phi Y\, X}\xi + R_{\xi X}(-\phi Y) = -g(X, \phi Y)\xi - \eta(X)\phi Y.$$

Replacing Y by ϕY we have

$$R_{YX}\xi - \eta(Y)R_{\xi X}\xi + R_{\xi X}Y - \eta(Y)R_{\xi X}\xi = g(X,Y)\xi - 2\eta(Y)\eta(X)\xi + \eta(X)Y$$

and simplifying

$$R_{YX}\xi + R_{\xi X}Y = g(X,Y)\xi - 2\eta(Y)X + \eta(X)Y.$$

Differentiating this gives

$$-R_{YX}\phi Z - R_{\phi Z X}Y = -g(X,Y)\phi Z - 2g(Z,\phi Y)X + g(Z,\phi X)Y.$$

Replacing Z by ϕZ and simplifying

$$R_{YX}Z + R_{ZX}Y = g(X,Y)Z - 2g(Z,Y)X + g(Z,X)Y.$$

Thus finally $2g(R_{YX}X,Y) = 2|X|^2|Y|^2 - 2g(X,Y)^2$ giving constant curvature $+1$ and that the structure is Sasakian (cf. Theorem 7.3).   ∎

In [1989] the author showed that the standard contact metric structure of the tangent sphere bundle (or unit tangent bundle but with a homothetic change of metric) $T_1 M$ of a Riemannian manifold M is locally symmetric if and only if either M is flat in which case $T_1 M$ is locally isometric to $E^{n+1} \times S^n(4)$ or M is 2-dimensional and of constant curvature $+1$ in which case $T_1 M$ is locally isometric to $T_1 S^2 \sim \mathbb{R}P^3 \sim SO(3)$. We will prove this result in Section 9.2. We remark that even though $T_1 S^3$ is topologically $S^3 \times S^2$, the product metric is not an associated metric to the natural contact structure (again see Chapter 9 or Tanno [1978b] for an Einstein associated metric).

The above results raise the question of the classification of all locally symmetric contact metric manifolds and one might conjecture that the only two possibilities are contact metric manifolds which are locally isometric to $S^{2n+1}(1)$ or $E^{n+1} \times S^n(4)$. In dimension 3, Blair and Sharma [1990b] showed that a locally symmetric contact metric manifold is of constant curvature 0 or 1. In dimension 5, A. M. Pastore [1998] showed that indeed local isometry with $S^5(1)$ or $E^3 \times S^2(4)$ are the only two possibilities. More generally she showed that if M^{2n+1} is a locally symmetric contact metric manifold, then M^{2n+1} admits a foliation whose leaves are totally geodesic and locally isometric to the Riemannian product $E^{m+1} \times S^m(4)$, where m is the multiplicity of the $+1$ eigenvalue of the operator h. In [1994] K. Bang showed that a locally symmetric contact metric manifold with $R_{X\xi}\xi = 0$ is locally isometric to $E^{n+1} \times S^n(4)$; this answered positively a question posed by D. Perrone [1992a, p. 148]. Also in dimension > 3 Sharma and Koufogiorgos [1991] showed that a locally symmetric contact metric manifold such that the sectional curvatures of plane sections containing ξ are equal to a constant c is either of constant curvature $+1$ and Sasakian or $c = 0$. In the same paper they showed that if ξ belongs to the κ-nullity distribution and the contact metric structure has parallel Ricci tensor, then the contact metric manifold is either Sasakian and Einstein or locally isometric to $E^{n+1} \times S^n(4)$.

Similarly one can ask the question: When is a contact metric structure conformally flat? In [1962a] Okumura showed that a conformally flat Sasakian manifold of dimension ≥ 5 is of constant curvature $+1$ and in [1963,1967b] Tanno extended this result to the K-contact case and for dimensions ≥ 3.

Theorem 7.12 *A conformally flat K-contact manifold is of constant curvature $+1$ and Sasakian.*

In a similar vein, Ghosh, Koufogiorgos and Sharma [2001] have shown that a conformally flat contact metric manifold of dimension ≥ 5 with a strongly pseudo-convex CR-structure is of constant curvature $+1$.

In dimension ≥ 5, as we have seen, a contact metric structure of constant curvature must be of constant curvature $+1$ and in dimension 3 a contact metric structure of constant curvature must be of constant curvature 0 or 1. For simplicity set $R_{X\xi}\xi = lX$, then l is a symmetric operator. K. Bang [1994] showed that in dimension ≥ 5 there are no conformally flat contact metric structures with $l = 0$, even though there are many contact metric manifolds satisfying $l = 0$, see Theorem 9.14. Bang's result was extended to dimension 3 and in fact generalized by Gouli-Andreou and Xenos [1999] who showed that in dimension 3 the only conformally flat contact metric structures satisfying $\nabla_\xi l = 0$ (equivalent to $\nabla_\xi h = 0$, Perrone [1992a]) are those of constant curvature 0 or 1. More recently Calvaruso, Perrone and Vanhecke [1999] showed that in dimension 3 the only conformally flat contact metric structures for which ξ is an eigenvector of the Ricci tensor are those of constant curvature 0 or 1. In the case of the standard contact metric structure on the tangent sphere bundle, the metric is conformally flat if and only if the base manifold is a surface of constant Gaussian curvature 0 or 1 (Theorem 9.7) as was shown by Blair and Koufogiorgos [1994]. In view of these strong curvature results one may ask if there are any conformally flat contact metric structures which are not of constant curvature. We devote the rest of this section to showing the local existence and giving some additional remarks.

We will work in $\mathbb{R}^3$, primarily with cylindrical coordinates (r, θ, z). Let $\eta = \frac{1}{2}(\alpha dr + \beta rd\theta + \gamma dz)$ be a contact form on $\mathbb{R}^3$. Then

$$d\eta = \frac{1}{2}\big((\beta + r\beta_r - \alpha_\theta)dr \wedge d\theta + (\gamma_r - \alpha_z)dr \wedge dz + (\gamma_\theta - r\beta_z)d\theta \wedge dz\big).$$

If g is a conformally flat metric, we may write it as $ds^2 = \frac{1}{4}e^{2\sigma}(dr^2 + r^2 d\theta^2 + dz^2)$. If g is also an associated metric, the characteristic vector field is given by

$$\xi = 2e^{-2\sigma}\left(\alpha\frac{\partial}{\partial r} + \beta\frac{1}{r}\frac{\partial}{\partial \theta} + \gamma\frac{\partial}{\partial z}\right)$$

and $\eta(\xi) = 1$ gives $e^{2\sigma} = \alpha^2 + \beta^2 + \gamma^2$.

Much of our analysis will actually be done with respect to the Euclidean metric on $\mathbb{R}^3$ and we denote the Euclidean length of a vector field $\mathbf{B}$ by $|\mathbf{B}|$. In particular if $\mathbf{B} = \alpha\frac{\partial}{\partial r} + \beta\frac{1}{r}\frac{\partial}{\partial\theta} + \gamma\frac{\partial}{\partial z}$, $|\mathbf{B}| = e^{2\sigma}$. Now $d\eta(\xi, X) = 0$ for all X gives

$$\frac{1}{r}\beta(\beta + r\beta_r - \alpha_\theta) + \gamma(\gamma_r - \alpha_z) = 0,$$

$$\alpha(\alpha_\theta - \beta - r\beta_r) + \gamma(\gamma_\theta - r\beta_z) = 0,$$

$$\alpha(\alpha_z - \gamma_r) + \frac{1}{r}\beta(r\beta_z - \gamma_\theta) = 0$$

and therefore $\mathbf{curl}\,\mathbf{B}$ is proportional to $\mathbf{B}$, say $\mathbf{curl}\,\mathbf{B} = f\mathbf{B}$. Computing ϕ, using $\phi^2 = -I + \eta \otimes \xi$ and the fact that $\eta \wedge d\eta \neq 0$, one finds that $f^2 = e^{2\sigma}$. Therefore $\mathbf{curl}\,\mathbf{B} = \pm e^\sigma\mathbf{B}$. Since we may change the sign of η without changing the problem, our study reduces to the study of the differential equation

$$\mathbf{curl}\,\mathbf{B} = |\mathbf{B}|\mathbf{B}.$$

Thus we are at the point that the existence of 3-dimensional conformally flat contact metric manifolds corresponds to finding solutions of $\mathbf{curl}\,\mathbf{B} = |\mathbf{B}|\mathbf{B}$. Had we carried out our analysis in cartesian coordinates (x, y, z) we would have been led to the same differential equation for a vector field $\mathbf{B} = \alpha\frac{\partial}{\partial x} + \beta\frac{\partial}{\partial y} + \gamma\frac{\partial}{\partial z}$. The flat case corresponds to the familiar unit vector field $\mathbf{B} = \sin z\frac{\partial}{\partial x} + \cos z\frac{\partial}{\partial y}$ that is equal to its own $\mathbf{curl}$. For the standard Sasakian structure of constant curvature $+1$ on S^3, using stereographic projection to $\mathbb{R}^3$, the corresponding vector field satisfying $\mathbf{curl}\,\mathbf{B} = |\mathbf{B}|\mathbf{B}$ is

$$\mathbf{B} = \frac{8(xz - y)}{(1 + x^2 + y^2 + z^2)^2}\frac{\partial}{\partial x} + \frac{8(x + yz)}{(1 + x^2 + y^2 + z^2)^2}\frac{\partial}{\partial y} + \frac{4(1 + z^2 - x^2 - y^2)}{(1 + x^2 + y^2 + z^2)^2}\frac{\partial}{\partial z}.$$

In neither of the two constant curvature examples above are the component functions, functions of the radial coordinate alone. If $\mathbf{curl}\,\mathbf{B} = |\mathbf{B}|\mathbf{B}$ and the component functions are functions of r alone, then $\alpha = 0$ and the other components, $\beta(r)$ and $\gamma(r)$, satisfy

$$\frac{1}{r}\beta + \beta' = \sqrt{\beta^2 + \gamma^2}\,\gamma, \quad -\gamma' = \sqrt{\beta^2 + \gamma^2}\,\beta. \tag{$*$}$$

Local existence away from $r = 0$ follows easily from the standard existence theorems for systems of ordinary differential equations.

Finally one can show that the metric $ds^2 = \frac{1}{4}e^{2\sigma}(dr^2 + r^2d\theta^2 + dz^2)$ with $e^{2\sigma} = \beta^2 + \gamma^2$ is not of constant curvature. Differentiating $e^{2\sigma} = \beta^2 + \gamma^2$ using $(*)$ we have $\sigma' = -\frac{\beta^2}{re^{2\sigma}}$. If g were of constant curvature the value of the curvature would be 0 or $+1$. Consider the unit vector field $X = 2e^{-2\sigma}\left(\frac{\gamma}{r}\frac{\partial}{\partial\theta} - \beta\frac{\partial}{\partial z}\right)$

belonging to the contact subbundle $\mathcal{D}$. Then computing the sectional curvature of the plane section spanned by ξ and X we have

$$g(R_{X\xi}\xi, X) = -4e^{-2\sigma}(\sigma'^2 + \frac{\sigma'}{r}) = \frac{4e^{-6\sigma}}{r^2}\beta^2\gamma^2.$$

If the constancy of the curvature were 0, then β or γ would vanish and from $(*)$ both would vanish making $\eta \equiv 0$, a contradiction. If the curvature is $+1$, $re^{3\sigma} = 2\beta\gamma$; differentiating this using $(*)$ and $\sigma' = -\frac{\beta^2}{re^{2\sigma}}$ we again have that $\beta = 0$.

The equation $\mathbf{curl\,B} = \alpha\mathbf{B}$ for some function α arises in solar physics and $\mathbf{curl\,B} = |\mathbf{B}|\mathbf{B}$ can be viewed as a special case. An *active region* on the sun is an area of extreme magnetic flux. These regions contain sunspots as well as the highly volatile phenomena of solar flares. For some solar phenomena, such as flares and prominences, the Lorentz force, $\mathbf{j} \times \mathbf{B}$, $\mathbf{j}$ being the current density, dominates the pressure gradient and gravitational forces and thus for a relatively slow moving plasma we have from the equations of motion the approximation $\mathbf{j} \times \mathbf{B} = 0$, the so-called "force-free" field. From Maxwell's equations we have, for an electrically neutral field, $\mathbf{curl\,B} = \mu\mathbf{j}$. Thus one is lead to $\mathbf{curl\,B} = \alpha\mathbf{B}$. There are many solutions in the literature (see e.g., Priest [1982]) which by Maxwell's equation must also satisfy $\mathbf{div\,B} = 0$. Except for the simple case $\mathbf{B} = \sin z \frac{\partial}{\partial x} + \cos z \frac{\partial}{\partial y}$, the solutions listed in the literature do not satisfy $\mathbf{curl\,B} = |\mathbf{B}|\mathbf{B}$ and we mention only a couple of them. The Lundquist solution (see e.g., Salingaros [1990] or Priest [1982, p. 138] for a generalization) is given in cylindrical coordinates by $\mathbf{B} = (0, J_1(r), J_0(r))$, J_0 and J_1 being Bessel functions, satisfies $\mathbf{curl\,B} = \mathbf{B}$ but it is not of unit length. Another common solution is $\mathbf{B} = \frac{1}{1+r^2}\left(\frac{\partial}{\partial\theta} + \frac{\partial}{\partial z}\right)$ which satisfies $\mathbf{curl\,B} = 2|\mathbf{B}|^2\mathbf{B}$ (Priest [1982, p. 262] where there are of course additional physical constants). While our vector field is another solution and one whose series expansion converges rapidly, so that its calculation for physical purposes is not prohibitive, the matter of whether the equation for a "force-free" model is even physically appropriate is questionable (see e.g., Salingaros [1986,1990]).

7.3 ϕ-sectional curvature

In this section we introduce the notion of ϕ-sectional curvature, this idea plays the role in Sasakian geometry that holomorphic sectional curvature plays in Kähler geometry. A plane section in $T_m M^{2n+1}$ is called a ϕ-*section* if there exists a vector $X \in T_m M^{2n+1}$ orthogonal to ξ such that $\{X, \phi X\}$ span the section. The sectional curvature $K(X, \phi X)$, denoted $H(X)$, is called ϕ-*sectional curvature*.

Recall that the sectional curvatures of a Riemannian manifold determine the curvature transformation $R_{XY}Z$. It is also well known that the holomorphic sectional curvatures of a Kähler manifold determine the curvature completely. We shall show that on a Sasakian manifold the ϕ-sectional curvatures determine the curvature completely (Moskal [1966]). Let $B(X,Y) = g(R_{XY}Y,X)$ and for X orthogonal to ξ, let $D(X) = B(X,\phi X)$. Also recall the tensor field P defined by

$$P(X,Y,Z,W) = d\eta(X,Z)g(Y,W) - d\eta(X,W)g(Y,Z)$$

$$-d\eta(Y,Z)g(X,W) + d\eta(Y,W)g(X,Z)$$

in Section 7.1.

Proposition 7.8 *On a Sasakian manifold, for tangent vectors X and Y orthogonal to ξ we have*

$$B(X,Y) = \frac{1}{32}(3D(X+\phi Y) + 3D(X-\phi Y) - D(X+Y) - D(X-Y)$$

$$-4D(X) - 4D(Y) - 24P(X,Y,X,\phi Y)).$$

Proof. Direct expansion and Lemma 7.1 give

$$\frac{1}{32}(3D(X+\phi Y) + 3D(X-\phi Y) - D(X+Y) - D(X-Y)$$

$$-4D(X) - 4D(Y) - 24P(X,Y,X,\phi Y))$$

$$= \frac{1}{32}(6g(R_{XY}Y,X) + 6g(R_{\phi X \phi Y}\phi Y,\phi X) + 8g(R_{X\phi X}\phi Y,Y)$$

$$+12g(R_{XY}\phi Y,\phi X) - 2g(R_{X\phi Y}\phi Y,X) - 2g(R_{\phi X Y}Y,\phi X)$$

$$+4g(R_{X\phi Y}Y,\phi X) - 24P(X,Y,X,\phi Y))$$

$$= g(R_{XY}Y,X).$$

■

Proposition 7.9 *Let M be a Sasakian manifold and $\{X,Y\}$ an orthonormal pair in T_mM with X and Y orthogonal to ξ. Set $g(X,\phi Y) = \cos\theta$, $0 \le \theta \le \pi$. Then the sectional curvature $K(X,Y)$ is given by*

$$K(X,Y) = \frac{1}{8}(3(1+\cos\theta)^2 H(X+\phi Y) + 3(1-\cos\theta)^2 H(X-\phi Y)$$

$$-H(X+Y) - H(X-Y) - H(X) - H(Y) + 6\sin^2\theta).$$

Proof. $K(X,Y) = B(X,Y)$ in the previous proposition, so we examine the terms in the expansion of $B(X,Y)$. Clearly $D(X) = g(X,X)^2 H(X)$ for any X orthogonal to ξ, and hence for the given pair $\{X,Y\}$, $g(X+\phi Y, X+\phi Y) = 2(1+\cos\theta)$, $g(X-\phi Y, X-\phi Y) = 2(1-\cos\theta)$, $g(X+Y, X+Y) = 2$ and $g(X-Y, X-Y) = 2$. Thus $D(X+\phi Y) = 4(1+\cos\theta)^2 H(X+\phi Y)$, and so on. Finally note that $P(X,Y,X,\phi Y) = -\sin^2\theta$ completing the proof.   ■

Theorem 7.13 *The ϕ-sectional curvatures of a Sasakian manifold determine the curvature completely.*

Proof. Since the sectional curvatures of a Riemannian manifold determine the curvature, it suffices to show that for an orthonormal pair $\{X,Y\}$, $K(X,Y)$ is determined uniquely by H and g. If X and Y are orthogonal to ξ, the previous proposition applies. If X or Y is ξ, $K(X,Y) = 1$. So suppose that $X = \eta(X)\xi + aZ$ and $Y = \eta(Y)\xi + bW$ where $\eta(X)$, $\eta(Y)$, $a = \sqrt{1-\eta(X)^2}$ and $b = \sqrt{1-\eta(Y)^2}$ are non-zero. Recall that on a Sasakian manifold $R_{\xi Z}\xi = -Z$ and $R_{Z\xi}Z = -\xi$ for any unit vector orthogonal to ξ. Therefore

$$K(X,Y) = g(R_{\eta(X)\xi+aZ\ \eta(Y)\xi+bW}\eta(Y)\xi + bW, \eta(X)\xi + aZ)$$

$$= b^2\eta(X)^2 - 2ab\eta(X)\eta(Y)g(Z,W) + a^2\eta(Y)^2 + a^2b^2 g(R_{ZW}W,Z).$$

Now $g(Z,W) + \dfrac{1}{ab}g(X-\eta(X)\xi, Y-\eta(Y)\xi) = -\dfrac{1}{ab}\eta(X)\eta(Y)$ and so

$$g(R_{ZW}W,Z) = (1-g(Z,W)^2)K(Z,W) = (1 - \dfrac{1}{a^2b^2}\eta(X)^2\eta(Y)^2)K(Z,W).$$

Thus

$$K(X,Y) = \eta(X)^2(1-\eta(Y)^2) + 2\eta(X)^2\eta(Y)^2 + \eta(Y)^2(1-\eta(X)^2)$$
$$+((1-\eta(X)^2)(1-\eta(Y)^2) - \eta(X)^2\eta(Y)^2))K(Z,W)$$
$$= \eta(X)^2 + \eta(Y)^2 + (1-\eta(X)^2 - \eta(Y)^2)K(Z,W)$$

and $K(Z,W)$ is given by the previous proposition completing the proof.   ■

Note that the above proof uses not only the values of the ϕ-sectional curvatures, but also the facts that on a Sasakian manifold $R_{\xi X}\xi = -X$ and $R_{X\xi}X = -\xi$ for any unit vector X orthogonal to ξ. Thus we have actually proved that any tensor field of type $(1,3)$ on a Sasakian manifold which satisfies the symmetries of the curvature tensor, the first Bianchi identity, identity a) of Lemma 7.1, $R_{\xi X}\xi = -X$ and $R_{X\xi}X = -\xi$ for any unit vector X orthogonal to ξ and which agrees with the values of the ϕ-sectional curvatures must be the curvature tensor. Therefore we can easily prove the following theorem of Ogiue [1964].

Theorem 7.14 *If the ϕ-sectional curvature at any point of a Sasakian manifold of dimension ≥ 5 is independent of the choice of ϕ-section at the point, then it is constant on the manifold and the curvature tensor is given by*

$$R_{XY}Z = \frac{c+3}{4}(g(Y,Z)X - g(X,Z)Y)$$

$$+\frac{c-1}{4}(\eta(X)\eta(Z)Y - \eta(Y)\eta(Z)X + g(X,Z)\eta(Y)\xi - g(Y,Z)\eta(X)\xi$$

$$+\Phi(Z,Y)\phi X - \Phi(Z,X)\phi Y + 2\Phi(X,Y)\phi Z)$$

where c is the constant ϕ-sectional curvature.

Proof. In view of the above remark, in order to see that the curvature tensor has the above form with c a function on the manifold, one need only check the necessary conditions and this is easily done. The Ricci tensor ρ and the scalar curvature τ are given by

$$\rho(X,Y) = \frac{n(c+3)+c-1}{2}g(X,Y) - \frac{(n+1)(c-1)}{2}\eta(X)\eta(Y)$$

and

$$\tau = \frac{1}{2}(n(2n+1)(c+3) + n(c-1)).$$

Now from the second Bianchi identity, $\nabla_\alpha \tau - 2\nabla_\beta \rho^\beta{}_\alpha = 0$ where $\rho^\beta{}_\alpha$ are the components of the Ricci tensor of type $(1,1)$ and hence

$$(n-1)dc + (\xi c)\eta = 0.$$

Applying this to ξ, we have $\xi c = 0$ and hence $dc = 0$ for $n \neq 1$ as desired. ∎

A Sasakian manifold of constant ϕ-sectional curvature c will be called a *Sasakian space form* and denoted $M(c)$. Note that a Sasakian space form has constant scalar curvature and is η-Einstein. Also if $c < 1$ we have the following pinching of the sectional curvature, similar to that in the Kähler case,

$$c \leq K(X,Y) \leq \frac{c+3}{4};$$

if $c > 1$, the inequalities are reversed.

Th. Koufogiorgos [1997a] studied (κ,μ)-manifolds of dimension ≥ 5 for which the ϕ-sectional curvature at any point is independent of the choice of ϕ-section at the point. He proved that the ϕ-sectional curvature is constant and obtained the curvature tensor explicitly.

7.4   Examples of Sasakian space forms

To begin, let (ϕ, ξ, η, g) be a contact metric structure and recall the notion of a *D-homothetic deformation*:

$$\bar{\eta} = a\eta, \quad \bar{\xi} = \frac{1}{a}\xi, \quad \bar{\phi} = \phi, \quad \bar{g} = ag + a(a-1)\eta \otimes \eta$$

where a is a positive constant. The deformed structure $(\bar{\phi}, \bar{\xi}, \bar{\eta}, \bar{g})$ is again a contact metric structure and it enjoys many of the properties of the original structure as we remarked in Section 7.2. In particular if (ϕ, ξ, η, g) is Sasakian, so is $(\bar{\phi}, \bar{\xi}, \bar{\eta}, \bar{g})$; if $M(c)$ is a Sasakian space form, then deforming the structure we obtain the Sasakian space form $M(\bar{c})$ where $\bar{c} = \frac{c+3}{a} - 3$ (see Tanno [1968], [1969] for details). We will show that there exist Sasakian space forms $M(c)$ for every value of c. Moreover we state the following theorem of Tanno and refer to [1969] for the proof.

Theorem 7.15 *Let $M(c)$ be a complete, simply connected Sasakian manifold with constant ϕ-sectional curvature c. Then $M(c)$ belongs one of the three families of examples listed below.*

7.4.1 S^{2n+1}

Let (ϕ, ξ, η, g) be the standard contact metric structure on the sphere S^{2n+1} constructed either as a hypersurface of $\mathbb{C}^{2n+2}$ (Example 4.5.2 and Section 6.3) or as a principal circle bundle over $\mathbb{C}P^n$ (Examples 4.5.4 and 6.7.2). Applying a *D*-homothetic deformation to this structure one obtains a Sasakian structure on S^{2n+1} with constant ϕ-sectional curvature $c = \frac{4}{a} - 3$. Note that from the remark on pinching in Section 7.3, these metrics on S^{2n+1} for $c > 0$ satisfy the condition of Goldberg's theorem referred to in Section 6.8.

7.4.2 $\mathbb{R}^{2n+1}$

In Examples 4.5.1 and 6.7.1 we saw that $\mathbb{R}^{2n+1}$ with coordinates (x^i, y^i, z), $i = 1, \ldots, n$ admits the Sasakian structure

$$\eta = \frac{1}{2}(dz - \sum_{i=1}^{n} y^i dx^i), \quad g = \eta \otimes \eta + \frac{1}{4}\sum_{i=1}^{n}((dx^i)^2 + (dy^i)^2).$$

With this metric, $\mathbb{R}^{2n+1}$ is a Sasakian space form with $c = -3$ (cf. Okumura [1962b]).

7.4.3 $B^n \times \mathbb{R}$

Let B^n be a simply connected bounded domain in $\mathbb{C}^n$ and (J, G) a Kähler structure with constant holomorphic sectional curvature $k < 0$. Since the fundamental 2-form Ω of the Kähler structure is closed, $\Omega = d\omega$ for some real analytic 1-form ω. Now on $B^n \times \mathbb{R}$ let π denote the projection onto B^n and t the coordinate on $\mathbb{R}$. Then $B^n \times \mathbb{R}$ with $\eta = \pi^*\omega + dt$ and $g = \pi^*G + \eta \otimes \eta$ is a Sasakian manifold. Regarding η as a connection form on $B^n \times \mathbb{R}$, let $\tilde{\pi}X$ denote the horizontal lift of a vector field X on B^n. Also we denote by $\underline{K}$ the sectional curvature of B^n. Then by direct computation (see Ogiue [1965] or use the Riemannian submersion technique of O'Neill [1966]), we have $K(\tilde{\pi}X, \tilde{\pi}Y) = \underline{K}(X, Y) - 3\eta(\nabla_{\tilde{\pi}X}\tilde{\pi}Y)^2$ where $\{X, Y\}$ is an orthonormal pair on B^n. Now $g(\nabla_{\tilde{\pi}X}\tilde{\pi}Y, \xi) = -g(\tilde{\pi}Y, \nabla_{\tilde{\pi}X}\xi) = g(\tilde{\pi}Y, \phi\tilde{\pi}X) = g(\tilde{\pi}Y, \tilde{\pi}JX)$. Therefore $B^n \times \mathbb{R}$ has constant ϕ-sectional curvature $c = k - 3$.

7.5 Locally ϕ-symmetric spaces

We saw in Theorem 7.11 that a locally symmetric K-contact manifold is of constant curvature $+1$ and Sasakian and we raised the question of the classification of locally symmetric contact metric manifolds. Certainly for Sasakian or K-contact geometry, Theorem 7.11 can be regarded as saying that the idea of being locally symmetric is too strong. For this reason T. Takahashi [1977] introduced the notion of a locally ϕ-symmetric space. A Sasakian manifold is said to be a *Sasakian locally ϕ-symmetric space* if

$$\phi^2(\nabla_V R)_{XY}Z = 0$$

for all vector fields X, Y, Z orthogonal to ξ; such spaces were called *locally $\mathcal{D}$-symmetric spaces* by Shibuya [1982]. It is easy to check that Sasakian space forms are locally ϕ-symmetric spaces. In [1987a] Blair and Vanhecke showed that a Sasakian manifold is locally ϕ-symmetric if and only if

$$g((\nabla_X R)_{X\phi X}X, \phi X) = 0$$

for all vector fields X orthogonal to ξ.

Note that on a Sasakian manifold M, or more generally on a K-contact manifold, a geodesic that is initially orthogonal to ξ remains orthogonal to ξ. We call such a geodesic a *ϕ-geodesic*. A local diffeomorphism s_m of M, $m \in M$, is a *ϕ-geodesic symmetry* if its domain contains a (possibly) smaller domain $\mathcal{U}$ such that for every ϕ-geodesic $\gamma(s)$ parametrized by arc length we have i) $\gamma(0)$ is in the intersection of $\mathcal{U}$ and the integral curve of ξ through m, and ii)

$$(s_m \circ \gamma)(s) = \gamma(-s)$$

for all s with $\gamma(\pm s) \in \mathcal{U}$. Since the points of the integral curve of ξ through m are fixed, we see that setting $S = -I + 2\eta \otimes \xi$, we have

$$s_m = \exp_m \circ S_m \circ \exp_m^{-1}.$$

Takahashi [1977] defines a Sasakian manifold to be a *Sasakian globally ϕ-symmetric space* by requiring that any ϕ-geodesic symmetry can be extended to a global automorphism of the structure and that the Killing vector field ξ generates a 1-parameter group of global transformations.

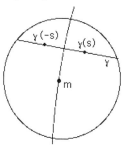

Among the main results of Takahashi [1977] are the following four theorems.

Theorem 7.16 *A Sasakian locally ϕ-symmetric space is locally isometric to a Sasakian globally ϕ-symmetric space and a complete, connected, simply-connected Sasakian locally ϕ-symmetric space is globally ϕ-symmetric.*

Theorem 7.17 *A Sasakian manifold is locally ϕ-symmetric if and only if it admits a ϕ-geodesic symmetry at every point which is a local automorphism of the structure.*

Now suppose that $\mathcal{U}$ is a neighborhood on M on which ξ is regular; then since M is Sasakian, the projection $\pi : \mathcal{U} \longrightarrow \mathcal{V} = \mathcal{U}/\xi$ gives a Kähler structure on $\mathcal{V}$. Furthermore, if $\underline{s}_{\pi(m)}$ denotes the geodesic symmetry on $\mathcal{V}$ at $\pi(m)$, then $\underline{s}_{\pi(m)} \circ \pi = \pi \circ s_m$.

Theorem 7.18 *A Sasakian manifold is locally ϕ-symmetric if and only if each Kähler manifold which is the base of a local fibering is a Hermitian locally symmetric space.*

Following Okumura [1962a] define on a Sasakian manifold a linear connection $\bar{\nabla}$ by $\bar{\nabla}_X Y = \nabla_X Y + T_X Y$ where

$$T_X Y = d\eta(X, Y)\xi - \eta(X)\phi Y + \eta(Y)\phi X$$

and it is easy to see that the structure tensors are parallel with respect to this connection. Takahashi [1977] then proves the following.

Theorem 7.19 *A Sasakian manifold is a locally ϕ-symmetric space if and only if $\bar{\nabla}\bar{R} = 0$, equivalently*

$$(\nabla_V R)_{XY}Z = -T_V R_{XY}Z + R_{T_V XY}Z + R_{X T_V Y}Z + R_{XY}T_V Z$$

for all X, Y, Z, V.

In the spirit of the fact that a Riemannian manifold is locally symmetric if and only if the local geodesic symmetries are isometries and in view of the above results of Takahashi, we state the following extension of Theorem 7.16 (sufficiency in the Sasakian case, Blair and Vanhecke [1987b]; in the K-contact case, Bueken and Vanhecke [1989]).

Theorem 7.20 *On a Sasakian locally ϕ-symmetric space, local ϕ-geodesic symmetries are isometries. Conversely if on a K-contact manifold the local ϕ-geodesic symmetries are isometries, the manifold is a Sasakian locally ϕ-symmetric space.*

The notion of a locally ϕ-symmetric space has been for the most part explored only in the Sasakian context and it is not clear what the corresponding notion should be for a general contact metric manifold. Without the K-contact property one loses the fact that a geodesic, initially orthogonal to ξ remains orthogonal to ξ. We have just seen that in the Sasakian case local ϕ-symmetry is equivalent to reflections in the integral curves of the characteristic vector field being isometries. Boeckx and Vanhecke [1997] proposed this property as the definition for local ϕ-symmetry in the contact metric case and formalized two notions in Boeckx, Bueken and Vanhecke [1999]. We adopt this formulation here. A contact metric manifold is a *weakly locally ϕ-symmetric space* if it satisfies

$$\phi^2(\nabla_V R)_{XY}Z = 0$$

for all vector fields X, Y, Z orthogonal to ξ as in the Sasakian case. A contact metric manifold is a *strongly locally ϕ-symmetric space* if reflections in the integral curves of the characteristic vector field are isometries.

From Chen and Vanhecke [1989] (see also Boeckx, Bueken and Vanhecke [1999]) one sees that on a strongly locally ϕ-symmetric space

$$g((\nabla^{2k}_{X\cdots X}R)_{XY}X, \xi) = 0,$$

$$g((\nabla^{2k+1}_{X\cdots X}R)_{XY}X, Z) = 0,$$

$$g((\nabla^{2k+1}_{X\cdots X}R)_{X\xi}X, \xi) = 0,$$

for all X, Y, Z orthogonal to ξ and all $k \in \mathbb{N}$. Conversely on an analytic Riemannian manifold these conditions are sufficient for the contact metric manifold to be a strongly locally ϕ-symmetric space. In particular taking $k = 0$

in the second condition, we note that a strongly locally ϕ-symmetric space is weakly locally ϕ-symmetric. Calvaruso, Perrone and Vanhecke [1999] determine all 3-dimensional strongly locally ϕ-symmetric spaces.

Examples of strongly locally ϕ-symmetric spaces include non-Sasakian contact metric manifolds for which ξ belongs to the (κ, μ)-nullity distribution (Boeckx [1999]). Special cases of these are the non-abelian 3-dimensional unimodular Lie groups with left-invariant contact metric structures (Boeckx, Bueken and Vanhecke [1999]). Boeckx, Bueken and Vanhecke [1999] also gave an example of a non-unimodular Lie group with a weakly locally ϕ-symmetric contact metric structure which is not strongly locally ϕ-symmetric.

To see these last examples explicitly we include the classification of simply connected homogeneous 3-dimensional contact metric manifolds as given by Perrone [1998]. Let $W = \frac{1}{8}(\tau - Ric(\xi) + 4)$ denote the Webster scalar curvature (cf. Section 10.4 below). The classification of 3-dimensional Lie groups and their left invariant metrics was given by Milnor in [1976].

Theorem 7.21 *Let (M^3, η, g) be a simply connected homogeneous contact metric manifold. Then M is a Lie group G and both g and η are left invariant. More precisely we have the following classification:*
(1) If G is unimodular, then it is one of the following Lie groups:

1. *The Heisenberg group when $W = |\mathcal{L}_\xi g| = 0$;*

2. *$SU(2)$ when $4\sqrt{2}W > |\mathcal{L}_\xi g|$;*

3. *the universal covering of the group of rigid motions of the Euclidean plane when $4\sqrt{2}W = |\mathcal{L}_\xi g| > 0$;*

4. *the universal covering of $SL(2, \mathbb{R})$ when $-|\mathcal{L}_\xi g| \neq 4\sqrt{2}W < |\mathcal{L}_\xi g|$;*

5. *the group of rigid motions of the Minkowski plane when $4\sqrt{2}W = -|\mathcal{L}_\xi g| < 0$.*

(2) If G is non-unimodular, its Lie algebra is given by

$$[e_1, e_2] = \alpha e_2 + 2\xi, \quad [e_1, \xi] = \gamma e_2, \quad [e_2, \xi] = 0,$$

where $\alpha \neq 0$, $e_1, e_2 = \phi e_1 \in \mathcal{D}$ and $4\sqrt{2}W < |\mathcal{L}_\xi g|$. Moreover, if $\gamma = 0$, the structure is Sasakian and $W = -\frac{\alpha^2}{4}$.

The structures on the unimodular Lie groups in this theorem satisfy the (κ, μ)-nullity condition and hence they are strongly locally ϕ-symmetric. The weakly locally ϕ-symmetric contact metric structure of Boeckx, Bueken and Vanhecke [1999] which is not strongly locally ϕ-symmetric is the non-unimodular case with $\gamma = 2$.

Notice also in the unimodular case the role played by the invariant $p = \frac{4\sqrt{2}W}{|\mathcal{L}_\xi g|}$. Moreover $W = \frac{(2-\mu)}{4}$ and $|\mathcal{L}_\xi g| = 2\sqrt{2}\sqrt{1-\kappa}$; thus $p = \frac{2-\mu}{2\sqrt{1-\kappa}}$ which is the invariant I_M of Boeckx discussed in Section 7.2.

J. Berndt [1997] studied the geometry of the complex Grassmannian $\mathbb{C}G_{2,m}$ of complex 2-planes in $\mathbb{C}^{m+2}$. Each point in his model of this Grassmannian represents a closed geodesic in the focal set Q^{m+1} of $\mathbb{C}P^{m+1}$ in $\mathbb{H}P^{m+1}$ and he views the Riemannian submersion $Q^{m+1} \longrightarrow \mathbb{C}G_{2,m}$ as an analogue of the Hopf fibration $S^{2m+1} \longrightarrow \mathbb{C}P^m$. Among Berndt's observations [1997, p.37] is that Q^{m+1} is a Sasakian globally ϕ-symmetric space and that Q^2 has constant ϕ-sectional curvature equal to 5.

Cho [1999] makes use of the generalized Tanaka connection $^*\nabla$ (see Section 10.4) and studies contact metric manifolds satisfying

$$(^*\nabla_{\dot\gamma} R)(\cdot, \dot\gamma)\dot\gamma = 0$$

for any unit speed $^*\nabla$-geodesic γ ($^*\nabla_{\dot\gamma}\dot\gamma = 0$). Cho shows that a contact metric manifold with ξ belonging to the (κ, μ)-nullity distribution and satisfying this condition is either Sasakian locally ϕ-symmetric, 3-dimensional with $\mu = 0$ and weakly locally ϕ-symmetric, or has $\mu = 2$ and is weakly locally ϕ-symmetric.

Returning to the Sasakian context, Vanhecke and the author [1987a] showed that complete, simply-connected globally ϕ-symmetric spaces are naturally reductive homogeneous spaces. Watanabe [1980] showed that a Sasakian locally ϕ-symmetric space is locally homogeneous and analytic. Berndt and Vanhecke [1999] proved that a simply-connected Sasakian ϕ-symmetric space is weakly symmetric, i.e. any two points can be interchanged by an isometry. Complete simply-connected globally ϕ-symmetric spaces have been classified by Jiménez and Kowalski [1993]. In dimension 3 the classification is the unit sphere, S^3, together with the universal covering of $SL(2, \mathbb{R})$, the Heisenberg group and $SU(2)$, each with a special left invariant metric (Vanhecke and the author [1987b]). In dimension 5 the classification was obtained by Kowalski and Wegrzynowski [1987]. As a corollary to this classification, Kowalski and Wegrzynowski in dimension 5 and Jiménez and Kowalski in general dimension gave a classification of the complete, simply-connected Sasakian space forms. An isospectral problem for locally ϕ-symmetric spaces was studied by Shibuya [1982].

Watanabe [1980] also showed that a 3-dimensional Sasakian manifold with constant scalar curvature is locally ϕ-symmetric. Two extensions of this are the following: A compact regular Sasakian manifold with constant scalar curvature and non-negative sectional curvature is locally ϕ-symmetric (Perrone and Vanhecke [1991]). A 3-dimensional contact metric manifold with $Q\phi = \phi Q$

is weakly locally ϕ-symmetric if and only if it is of constant scalar curvature (Koufogiorgos, Sharma and the author [1990]).

We have remarked that the product metric on $S^3 \times S^2$ is not an associated metric and referred to Tanno [1978b] for a discussion of an Einstein associated metric on this manifold. A family of Sasakian globally ϕ-symmetric structures on $S^3 \times S^2$ was constructed by Watanabe and Fujita [1988] and Perrone and Vanhecke [1991] proved that the only 5-dimensional compact, simply connected, homogeneous contact manifolds are diffeomorphic to S^5 or $S^3 \times S^2$.

On a Riemannian locally symmetric space, the local geodesic symmetries are isometries and hence volume preserving. However there exist Riemannian manifolds whose local geodesic symmetries are volume preserving but are not locally symmetric (see e.g., Kowalski and Vanhecke [1984]). In dimension 3 (Blair and Vanhecke [1987b]; see also Watanabe [1980]) and dimension 5 (Blair and Vanhecke [1987c]) Sasakian manifolds whose geodesic symmetries are volume preserving are locally ϕ-symmetric and conversely.

J. C. González-Dávila, M. C. González-Dávila and L. Vanhecke [1995] considered metrics with respect to which a given vector field ξ is a unit Killing vector field, with emphasis on the case when the dual form η is a contact form; this is again the context of the R-contact manifolds of Rukimbira [1993]. These authors study the case when local reflections with respect to the flow lines are isometries; such spaces are called *locally Killing-transversally symmetric spaces*. In particular they show that a Riemannian manifold with a unit Killing vector field ξ has a natural structure as a Sasakian locally ϕ-symmetric space if and only if it is a locally Killing-transversally symmetric space and $K(X, \xi) = 1$ for all X orthogonal to ξ.

There are a number of other geometric ideas surrounding the subjects of Sasakian space forms and locally ϕ-symmetric spaces and we will mention a few of these with a few references. One area of interest is the study of reflections and other symmetries; in addition to some of the references already given, we mention Bueken and Vanhecke [1993] which studies reflections in a submanifold. Another area is the study of Jacobi vector fields and their use in the characterization Sasakian space forms as well as in the study of symmetries. Here we will give only a few references on the use of Jacobi fields: Blair and Vanhecke [1987d], Bueken and Vanhecke [1988], and Vanhecke [1988] for a survey of the techniques involved.

8
Submanifolds of Kähler and Sasakian Manifolds

8.1 Invariant submanifolds

In this chapter we study submanifolds in both contact and Kähler geometry. These are extensive subjects in their own right and we give only a few basic results. For a submanifold M of a Riemannian manifold $(\tilde{M}, \tilde{g})$ we denote the induced metric by g. Then the Levi-Cività connection ∇ of g and the second fundamental form σ are related to the ambient Levi-Cività connection $\tilde{\nabla}$ by

$$\tilde{\nabla}_X Y = \nabla_X Y + \sigma(X, Y).$$

For a normal vector field ν we denote by A_ν the corresponding Weingarten map and we denote by $\nabla^\perp$ the connection in the normal bundle; in particular A_ν and $\nabla^\perp$ are defined by

$$\tilde{\nabla}_X \nu = -A_\nu X + \nabla^\perp_X \nu.$$

The Gauss equation is

$$\tilde{R}(X, Y, Z, W) = R(X, Y, Z, W) + \tilde{g}(\sigma(X, Z), \sigma(Y, W)) - \tilde{g}(\sigma(Y, Z), \sigma(X, W)).$$

Defining the covariant derivative of σ by $(\nabla'\sigma)(X, Y, Z) = \nabla^\perp_X \sigma(Y, Z) - \sigma(\nabla_X Y, Z) - \sigma(Y, \nabla_X Z)$ the Codazzi equation is

$$(R_{XY}Z)^\perp = (\nabla'\sigma)(X, Y, Z) - (\nabla'\sigma)(Y, X, Z).$$

Finally for normal vector fields ν and ζ the equation of Ricci–Kühn is

$$\tilde{R}(X, Y, \nu, \zeta) = R^\perp(X, Y, \nu, \zeta) - g([A_\nu, A_\zeta]X, Y).$$

122 8. Submanifolds of Kähler and Sasakian Manifolds

Let $\tilde{M}^{2n}$ be an almost Hermitian manifold with structure tensors $(\tilde{g}, \tilde{J})$. A submanifold M is said to be *invariant* if $\tilde{J}T_pM \subset T_pM$. It is well known that an invariant submanifold of a Kähler manifold is both Kähler and minimal.

For a contact metric manifold $\tilde{M}^{2n+1}$ with structure tensors $(\tilde{\phi}, \tilde{\xi}, \tilde{\eta}, \tilde{g})$ a submanifold M is said to be *invariant* if $\tilde{\phi}T_pM \subset T_pM$. Some authors also require that $\tilde{\xi}$ be tangent to M but this is a consequence. Clearly $\tilde{\xi}$ cannot be normal on any neighborhood $\mathcal{U}$ of M, for then $\mathcal{U}$ would be an integral submanifold of the contact subbundle $\mathcal{D}$ and hence as we have seen (Section 5.1) would not be invariant by $\tilde{\phi}$. Now if $\tilde{\xi} = U + \nu$ where U is tangent and ν is normal, first note that $\tilde{\phi}\nu$ is normal since $\tilde{g}(\tilde{\phi}\nu, X) = -\tilde{g}(\nu, \tilde{\phi}X) = 0$. Therefore $0 = \tilde{\phi}\tilde{\xi} = \tilde{\phi}U + \tilde{\phi}\nu$ and hence $\tilde{\phi}U = 0$ and $\tilde{\phi}\nu = 0$. As a result $U = \tilde{\eta}(U)\tilde{\xi}$ and $\nu = \tilde{\eta}(\nu)\tilde{\xi}$, but both U and ν cannot be collinear with $\tilde{\xi}$.

Clearly an invariant submanifold inherits a contact metric structure by restriction. Moreover for the induced structure (ϕ, ξ, η, g) we have $h = \tilde{h}|_M$ as well. Also for the second fundamental form we have

$$\sigma(\xi, X) = \tilde{\nabla}_X\xi - \nabla_X\xi = -\tilde{\phi}X - \tilde{\phi}\tilde{h}X - (-\phi X - \phi hX) = 0.$$

Our first result is a theorem of Chinea [1985] and independently of Endo [1985]; we give the proof of Chinea.

Theorem 8.1 *An invariant submanifold M of a contact metric manifold is minimal.*

Proof. By Lemma 7.3 we have

$$(\tilde{\nabla}_X\tilde{\phi})Y + (\tilde{\nabla}_{\tilde{\phi}X}\tilde{\phi})\tilde{\phi}Y = 2\tilde{g}(X, Y)\tilde{\xi} - \tilde{\eta}(Y)(X + \tilde{h}X + \tilde{\eta}(X)\tilde{\xi})$$

from which

$$\nabla_X\phi Y + \sigma(X, \phi Y) - \phi\nabla_X Y - \tilde{\phi}\sigma(X, Y)$$
$$+\nabla_{\phi X}(-Y + \eta(Y)\xi) - \sigma(\phi X, Y) - \phi\nabla_{\phi X}\phi Y - \tilde{\phi}\sigma(\phi X, \phi Y)$$
$$= 2\tilde{g}(X, Y)\tilde{\xi} - \tilde{\eta}(Y)(X + \tilde{h}X + \tilde{\eta}(X)\tilde{\xi}).$$

Taking the normal part we have

$$\sigma(X, \phi Y) - \sigma(\phi X, Y) - \tilde{\phi}(\sigma(X, Y) + \sigma(\phi X, \phi Y)) = 0.$$

Interchanging X and Y now yields

$$\sigma(\phi X, \phi Y) = -\sigma(X, Y)$$

and hence that M is minimal. ∎

Note also that $\sigma(X, \phi Y) = \sigma(\phi X, Y)$. Thus we have $g(A_\nu X, \phi Y) = \tilde{g}(\sigma(X, \phi Y), \nu) = \tilde{g}(\sigma(\phi X, Y), \nu) = g(A_\nu \phi X, Y)$ and hence

$$A_\nu\phi + \phi A_\nu = 0.$$

Moreover a ϕ-section on M is a $\tilde{\phi}$-section on $\tilde{M}$; the Gauss equation and $\sigma(\phi X, \phi Y) = -\sigma(X, Y)$ yield $H(X) \leq \tilde{H}(X)$ with equality holding everywhere if and only if M is totally geodesic (Endo [1986]).

Theorem 8.2 *If $\tilde{M}$ is a K-contact (resp. Sasakian) manifold and M an invariant submanifold, then M is also K-contact (resp. Sasakian).*

Proof. We have already noted that $h = \tilde{h}|_M$ and hence the K-contact result. Now again as in the last proof

$$(\tilde{\nabla}_X \tilde{\phi})Y = \nabla_X \phi Y + \sigma(X, \phi Y) - \phi \nabla_X Y - \tilde{\phi}\sigma(X, Y)$$

so if $(\tilde{\nabla}_X \tilde{\phi})Y = \tilde{g}(X, Y)\tilde{\xi} - \tilde{\eta}(Y)X$, we have $(\nabla_X \phi)Y = g(X, Y)\xi - \eta(Y)X$. ∎

In [1973a] Harada showed that invariant submanifolds of a compact regular Sasakian manifold respect the Boothby–Wang fibration. For a compact regular Sasakian manifold M, let us denote by M/ξ the base manifold of the Boothby–Wang fibration which as we have seen carries a natural Kähler structure. We can now state Harada's result.

Theorem 8.3 *Let M be a compact invariant submanifold of a compact regular Sasakian manifold $\tilde{M}$. Then M is regular and M/ξ is a Kähler (invariant) submanifold of $\tilde{M}/\xi$.*

For example consider the complex quadric Q^{n-1} in $\mathbb{C}P^n$ together with the Hopf fibration $S^{2n+1} \longrightarrow \mathbb{C}P^n$. Then the set of fibres over Q^{n-1} form a codimension 2 invariant submanifold of the Sasakian structure on S^{2n+1}; see Kenmotsu [1969], Kon [1976].

Let us now very briefly recall results of Simons [1968] and of Chern, do Carmo and Kobayashi [1970] on submanifolds of a sphere. The Simons result is that if M^n is a closed minimal submanifold of the sphere $S^{n+p}(1)$ and if $|\sigma|^2 < n/(2 - \frac{1}{p})$, then M^n is totally geodesic. Chern, do Carmo and Kobayashi proved that the only minimal submanifolds of the sphere $S^{n+p}(1)$ with $|\sigma|^2 = n/(2 - \frac{1}{p})$ are pieces of the Clifford minimal hypersurfaces and the Veronese surface. The proofs of these center around the computation of the Laplacian of $|\sigma|^2$. In [1972] Ogiue used these ideas to study complete invariant submanifolds M^{2n} in $\mathbb{C}P^{n+p}(1)$, the complex projective space with the Fubini–Study metric of constant holomorphic curvature 1. Ogiue's results of [1972] may be summarized as follows: Let M^{2n} be a complete Kähler submanifold of $\mathbb{C}P^{n+p}(1)$. If the holomorphic curvature H of M^{2n} is greater than $\frac{1}{2}$ and the scalar curvature of M^{2n} is constant or if $H > 1 - \frac{n+2}{2(n+2p)}$, then M^{2n} is totally geodesic.

In his excellent survey article [1974], Ogiue obtained other results, posed several conjectures and open problems, and continued this theme in [1976a].

For example in the context of the above results Ogiue proved in [1976a] that if the sectional curvature K of M^{2n} exceeds $\frac{n+3}{8n}$ or if $K > \frac{1}{8}$ and $H > \frac{1}{2}$, then M^{2n} is totally geodesic. Ogiue's conjectures from [1974] and [1976a] included that i) $H > \frac{1}{2}$ or ii) $K > \frac{1}{8}$ and $n \geq 2$ imply that M^{2n} is totally geodesic. In [1985a] A. Ros introduced a new technique to attack this kind of problem for M^{2n} compact, viz., for unit tangent vectors V regard $|\sigma(V,V)|^2$ as a function of the unit tangent bundle of M^{2n} and to study its behavior at its maximum point. The result of Ros in [1985a] is that Ogiue's conjecture i) is true and using the same technique Ros and Verstraelen in [1984] proved conjecture ii). Notice that the Calabi (Veronese) imbeddings of $\mathbb{C}P^n(\frac{1}{2})$ into $\mathbb{C}P^{\frac{n(n+3)}{2}}(1)$ show that these conjectures are best possible (see Section 12.7). In fact in [1985b] Ros gave a complete classification of compact Kähler submanifolds of $\mathbb{C}P^{n+p}(1)$ with $H \geq \frac{1}{2}$; there are seven interesting cases. As an illustration of the Ros technique we will give in the next section the proof of a theorem of Urbano [1985] on Lagrangian submanifolds of $\mathbb{C}P^n$.

The corresponding problem in Sasakian geometry is to study a compact invariant submanifold M^{2n+1} of a Sasakian space form $\tilde{M}^{2(n+p)+1}(\tilde{c})$ with constant ϕ-sectional curvature $\tilde{c} > -3$. This was first taken up by Harada [1973a], [1973b], Kon [1976] and later by VanLindt, Verheyen and Verstraelen [1986]. Again the later results are the stronger ones and use the Ros technique; in particular VanLindt, Verheyen and Verstraelen prove that if the ϕ-sectional curvature of M^{2n+1} exceeds $\frac{\tilde{c}+3}{2}$ or if the sectional curvature of M^{2n+1} exceeds $\frac{\tilde{c}+3}{8}$, then M^{2n+1} is totally geodesic.

8.2 Lagrangian and integral submanifolds

We have already seen in Section 1.2 that the maximum dimension of an isotropic submanifold of a symplectic manifold M^{2n} is n and that an n-dimensional isotropic submanifold is called a Lagrangian submanifold. In Section 5.1 we saw that the maximum dimension of an integral submanifold of a contact manifold M^{2n+1} is also n. In almost Hermitian geometry, isotropic submanifolds are known as *totally real submanifolds* since they are characterized by the fact that the almost complex structure maps the tangent space at any point into the normal space at the point.

In the spirit of the results of the last section we mention a few results on compact minimal Lagrangian submanifolds M^n of a complex space form $\tilde{M}^{2n}(c)$. In [1974] Yau considered a totally real minimal surface M^2 in a Kähler surface of constant holomorphic curvature c and proved the following results: 1) If M^2 has genus zero, M^2 is the standard imbedding of $\mathbb{R}P^2$ in $\mathbb{C}P^2$. 2) If M^2 is complete and non-negatively curved, it is totally geodesic or flat. 3) If M^2 is complete and non-positively curved with Gaussian curvature K and if

$\frac{c}{4} - K \geq a > 0$ for some constant a, then M^2 is totally geodesic or flat. In [1973] Houh proved that if a totally real minimal surface M^2 in $\mathbb{C}P^2$ has constant scalar normal curvature, it is totally geodesic or is non-positively curved and, combining with Yau's result, Houh showed that if in addition M^2 is complete, it is totally geodesic or flat. The flat case is realized by T^2 imbedded in $\mathbb{C}P^2$ as a flat minimal Lagrangian submanifold, Ludden, Okumura and Yano [1975].

In [1976b] Ogiue showed that if $c > 0$ and the sectional curvature K of M^n satisfies $K > \frac{(n-2)c}{4(2n-1)}$, then M^n is totally geodesic. This was extended by Chen and Houh [1979] who showed that if $c > 0$ and $K \geq \frac{(n-2)c}{4(2n-1)}$, then either M^n is totally geodesic or $n = 2$ and the surface M^2 is flat. If a minimal Lagrangian submanifold in $\tilde{M}^{2n}(c)$ is itself of constant curvature k, then it was shown by Chen and Ogiue [1974a] that the submanifold is either totally geodesic or $k \leq 0$; this was improved upon by Ejiri [1982] who showed that the submanifold is either totally geodesic or flat. Finally to give a stronger result and to illustrate the technique of A. Ros, we give the following theorem of Urbano [1985] with proof.

Theorem 8.4 *Let M be a compact Lagrangian submanifold minimally immersed in $\mathbb{C}P^n(c)$. If the sectional curvature K of M is greater than 0, then M is totally geodesic.*

Proof. We denote by $(J, \tilde{g})$ the almost Hermitian structure on $\mathbb{C}P^n(c)$ where $\tilde{g}$ is the Fubini–Study metric of constant holomorphic curvature c. For tangent vectors X and Y we have readily

$$0 = (\tilde{\nabla}_X J)Y = \tilde{\nabla}_X JY - J\tilde{\nabla}_X Y = -A_{JY}X + \nabla_X^\perp JY - J\nabla_X Y - J\sigma(X, Y)$$

giving

$$A_{JY}X = -J\sigma(X, Y), \quad \nabla_X^\perp JY = J\nabla_X Y.$$

The equations of Ricci–Kühn and Gauss give

$$R^\perp(X, Y, JZ, JW) = \tilde{g}(\tilde{R}_{XY}JZ, JW) + g([A_{JZ}, A_{JW}]X, Y)$$

$$= g(R_{XY}Z, W) - \tilde{g}(\sigma(Y, Z), \sigma(X, W)) + \tilde{g}(\sigma(X, Z), \sigma(Y, W))$$

$$+ g(A_{JW}X, A_{JZ}Y) - g(A_{JZ}X, A_{JW}Y)$$

$$= g(R_{XY}Z, W).$$

Since the ambient space is of constant holomorphic curvature, $\tilde{R}_{XY}Z = \frac{c}{4}(g(Y, Z)X - g(X, Z)Y - \Omega(Y, Z)JX + \Omega(X, Z)JY + 2\Omega(X, Y)JZ)$ and hence that $(\tilde{R}_{XY}Z)^\perp = 0$. Thus by the Codazzi equation $(\nabla'\sigma)(X, Y, Z) = \nabla_X^\perp \sigma(Y, Z) - \sigma(\nabla_X Y, Z) - \sigma(Y, \nabla_X Z)$ is symmetric. Defining $\nabla'^2\sigma$ by

$$(\nabla'^2\sigma)(X, Y, Z, W) = \nabla_X^\perp((\tilde{\nabla}\sigma)(Y, Z, W))$$

$$-(\nabla'\sigma)(\nabla_X Y, Z, W) - (\nabla'\sigma)(Y, \nabla_X Z, W) - (\nabla'\sigma)(Y, Z, \nabla_X W)$$

we have by the symmetry of $(\nabla'\sigma)(X, Y, Z)$ that

$$(\nabla'^2\sigma)(X, Y, Z, W) = (\nabla'^2\sigma)(Y, X, Z, W) + R^\perp_{XY}\sigma(Z, W)$$

$$-\sigma(R_{XY}Z, W) - \sigma(Z, R_{XY}W).$$

Define a real-valued function on the unit tangent bundle $T_1 M$ by $f(V) = \tilde{g}(\sigma(V, V), JV)$. Since $T_1 M$ is compact, f attains its maximum at a unit vector V tangent to M at a point p. For any unit tangent vector U at p, let $\gamma(t)$ be the geodesic in M with $\gamma(0) = p$ and $\gamma'(0) = U$. Let $V(t)$ be the parallel vector field along γ with $V(0) = V$. Then

$$0 = \frac{d}{dt} f(V(t))\big|_{t=0} = \tilde{g}((\nabla'\sigma)(U, V, V), JV).$$

Also

$$0 \geq \frac{d^2}{dt^2} f(V(t))\big|_{t=0} = \tilde{g}((\nabla'^2\sigma)(U, U, V, V), JV) = \tilde{g}((\nabla'^2\sigma)(U, V, V, U), JV)$$

$$= \tilde{g}((\nabla'^2\sigma)(V, U, V, U) + R^\perp_{UV}\sigma(V, U) - \sigma(R_{UV}V, U) - \sigma(R_{UV}U, V), JV)$$

$$= \tilde{g}((\nabla'^2\sigma)(V, V, U, U), JV) + g(R_{UV}V, J\sigma(U, V))$$

$$-g(A_{JU}R_{UV}V, V) - g(A_{JV}R_{UV}U, V)$$

$$= \tilde{g}((\nabla'^2\sigma)(V, V, U, U), JV) + 2g(R_{UV}V, J\sigma(U, V)) + g(R_{UV}U, J\sigma(V, V)).$$

On the other hand restricting f to the fibre of $T_1 M$ at p and taking U orthogonal to V we have

$$0 = Uf = 2\tilde{g}(\sigma(U, V), JV) + \tilde{g}(\sigma(V, V), JU)$$

$$= 2g(A_{JV}V, U) + \tilde{g}(\sigma(V, V), JU) = 3\tilde{g}(\sigma(V, V), JU).$$

Therefore since U is any unit vector orthogonal to V,

$$\sigma(V, V) = f(V)JV \quad \text{or} \quad A_{JV}V = f(V)V,$$

i.e., V is an eigenvector of A_{JV} with eigenvalue $f(V)$. Furthermore

$$0 \geq U^2 f = 6\tilde{g}(\sigma(U, V), JU) + 3\tilde{g}(\sigma(V, V), J(-V))$$

$$= 6\tilde{g}(\sigma(U, U), JV) - 3f(V)$$

$$= 6g(A_{JV}U, U) - 3f(V).$$

Thus if $\{U_1, \ldots, U_n\}$ is an orthonormal eigenvector basis of A_{JV} with $U_n = V$ and if λ_i, $i = 1, \ldots, n-1$ the other eigenvalues,

$$f(V) - 2\lambda_i \geq 0.$$

Now from the above and the minimality

$$0 \geq \sum_{i=1}^{n} \left\{ \tilde{g}((\tilde{\nabla}^2 \sigma)(V, V, U_i, U_i), JV) \right.$$

$$\left. + 2g(R_{U_i V} V, J\sigma(U_i, V)) + g(R_{U_i V} U_i, J\sigma(V, V)) \right\}$$

$$= \sum_{i=1}^{n-1} \left\{ -2\lambda_i K(V, U_i) + f(V) K(V, U_i) \right\}$$

$$= \sum_{i=1}^{n-1} K(V, U_i) \left\{ f(V) - 2\lambda_i \right\}.$$

Finally since the sectional curvature is positive, we conclude that each $\lambda_i = \frac{1}{2} f(V)$ and hence $\mathrm{tr} A_{JV} = \frac{(n+1)f(V)}{2} = 0$ giving $f(V) = 0$. Now $f(-U) = -f(U)$ and V was the maximum point of f, so $f = 0$. Thus $\sigma(V, V)$ is orthogonal to JV and to JU for any $U \perp V$ and hence $\sigma(V, V) = 0$ for any vector V and so M must be totally geodesic.   ∎

In Section 1.2 and Example 5.3.3 we remarked that there is no topological imbedding of a sphere as a Lagrangian submanifold of $\mathbb{C}^n$ and no umbilical, non-totally-geodesic, Lagrangian submanifolds of a complex space-form. The immersed Whitney sphere was the closest candidate with only one double point. Its second fundamental form σ was given by

$$\sigma(X, Y) = \frac{n}{n+2}(\tilde{g}(X, Y)\mathbf{H} + \tilde{g}(JX, \mathbf{H})JY + \tilde{g}(JY, \mathbf{H})JX)$$

where $\mathbf{H}$ denotes the mean curvature vector (Borrelli, Chen and Morvan [1995], Ros and Urbano [1998]).

More generally a Lagrangian submanifold of a Kähler manifold is said to be *Lagrangian H-umbilical* if the second fundamental form σ is of the form

$$\sigma(X, Y) = \alpha \tilde{g}(JX, \mathbf{H})\tilde{g}(JY, \mathbf{H})\mathbf{H}$$

$$+ \beta \tilde{g}(\mathbf{H}, \mathbf{H})(\tilde{g}(X, Y)\mathbf{H} + \tilde{g}(JX, \mathbf{H})JY + \tilde{g}(JY, \mathbf{H})JX)$$

for suitable functions α and β. This notion was introduced and a classification of such submanifolds in complex space forms was given by Chen in a series of papers [1997a,b],[1998].

In the contact case, integral submanifolds are often called *C-totally real submanifolds* since ϕ maps the tangent space at any point into the normal space. However, we will not adopt this term here. As a reminder, since $\eta(X) = 0$ for any vector X tangent to the integral submanifold, ξ is a normal vector field.

Examples that we might mention here are $S^n \subset S^{2n+1}$ as a totally geodesic integral submanifold as was described in Example 5.3.1 and $T^2 \subset S^5$ as a flat minimal integral submanifold as was described in Example 5.3.2.

We begin with the following lemma.

Lemma 8.1 *Let M be an integral submanifold of a K-contact manifold $\tilde{M}$. Then $A_\xi = 0$.*

Proof.

$$g(A_\xi X, Y) = \tilde{g}(\sigma(X,Y), \xi) = \tilde{g}(\tilde{\nabla}_X Y, \xi) = -\tilde{g}(Y, \tilde{\nabla}_X \xi) = -\tilde{g}(Y, \phi X) = 0.$$

∎

Now for an integral submanifold M^n of a Sasakian manifold $\tilde{M}^{2n+1}$, let $\{e_1, \dots, e_n\}$ be a local orthonormal basis on M^n. Then $\{\phi e_1, \dots, \phi e_n, \xi\}$ is an orthonormal basis of the normal space at each point of the local domain. For simplicity we write A_i for $A_{\phi e_i}$.

Lemma 8.2 *Let M be an integral submanifold of a Sasakian manifold $\tilde{M}$. Then $A_i e_j = A_j e_i$.*

Proof.

$$g(A_i e_j, e_k) = \tilde{g}(\sigma(e_j, e_k), \phi e_i) = \tilde{g}(\tilde{\nabla}_{e_k} e_j, \phi e_i)$$
$$= -\tilde{g}(e_j, (\tilde{\nabla}_{e_k}\phi)e_i + \phi\tilde{\nabla}_{e_k} e_i) = \tilde{g}(\tilde{\nabla}_{e_k} e_i, \phi e_j) = g(A_j e_i, e_k).$$

∎

Now let M^n be an integral submanifold of a Sasakian space form $\tilde{M}(c)$; the Gauss equation yields

$$g(R_{XY}Z, W) = \frac{c+3}{4}(g(X,W)g(Y,Z) - g(X,Z)g(Y,W))$$

$$+ \sum_{i=1}^{n}(g(A_i X, W)g(A_i Y, Z) - g(A_i X, Z)g(A_i Y, W))$$

by virtue of Lemma 8.1. In turn the sectional curvature $K(X,Y)$ of M^n determined by an orthonormal pair $\{X,Y\}$ is given by

$$K(X,Y) = \frac{c+3}{4} + \sum_{i=1}^{n}(g(A_i X, X)g(A_i Y, Y) - g(A_i X, Y)^2).$$

Moreover the Ricci tensor ρ and the scalar curvature τ of M^n are given by

$$\rho(X,Y) = \frac{n-1}{4}(c+3)g(X,Y) + \sum_{i=1}^{n}(\mathrm{tr}A_i)g(A_iX,Y) - \sum_{i=1}^{n}g(A_iX, A_iY),$$

$$\tau = \frac{n(n-1)}{4}(c+3) + \sum_{i=1}^{n}(\mathrm{tr}A_i)^2) - |\sigma|^2.$$

From these expressions the following proposition is not difficult and we omit the proof.

Proposition 8.1 *Let M^n be an integral submanifold of a Sasakian space form $\tilde{M}^{2n+1}(c)$ which is minimally immersed. Then the following are equivalent:*

a) M^n *is totally geodesic.*

b) M^n *is of constant curvature $\frac{c+3}{4}$.*

c) $\rho = \frac{n-1}{4}(c+3)g.$

d) $\tau = \frac{n(n-1)}{4}(c+3).$

Similar to the result of Chen and Ogiue [1974a] in the Kähler case, Yamaguchi, Kon and Ikawa [1976] proved the following result.

Theorem 8.5 *Let M^n be a minimal integral submanifold of a Sasakian space form $\tilde{M}^{2n+1}(c)$. If M^n has constant curvature k, then either M^n is totally geodesic or $k \leq 0$.*

For the standard Sasakian structure on $S^5(1)$ we give the following theorem of Yamaguchi, Kon and Miyahara [1976] and we include a proof since its techniques, though not new, are different than those presented so far in this book.

Theorem 8.6 *Let M^2 be a complete integral surface of $S^5(1)$ which is minimally immersed. If the Gaussian curvature K of M^2 is ≤ 0, then M^2 is flat.*

Proof. Choose a system of isothermal coordinates (x^1, x^2) so that the induced metric g is given by $g = E((dx^1)^2 + (dx^2)^2)$. Let $X_i = \frac{\partial}{\partial x^i}$ and $\sigma_{ij} = \sigma(X_i, X_j)$. Recall the standard formulas for the induced connection:

$$\nabla_{X_1}X_1 = -\nabla_{X_2}X_2 = \frac{X_1E}{2E}X_1 - \frac{X_2E}{2E}X_2, \quad \nabla_{X_1}X_2 = \frac{X_2E}{2E}X_1 + \frac{X_1E}{2E}X_2.$$

Using the minimality and the Codazzi equation, one readily obtains

$$\nabla^{\perp}_{X_1}\sigma_{12} - \nabla^{\perp}_{X_2}\sigma_{11} = 0, \quad \nabla^{\perp}_{X_1}\sigma_{11} + \nabla^{\perp}_{X_2}\sigma_{12} = 0.$$

Now define a complex-valued function F by

$$F = \tilde{g}(\sigma_{11}, \phi X_1) - i\tilde{g}(\sigma_{12}, \phi X_1).$$

Note that F is nowhere zero on M^2, for if $F = 0$ at some point m, then $\tilde{g}(\sigma_{11}, \phi X_1)$ and $\tilde{g}(\sigma_{12}, \phi X_1)$ vanish at m, but by minimality $\tilde{g}(\sigma_{22}, \phi X_1) = -\tilde{g}(\sigma_{11}, \phi X_1) = 0$ and by Lemma 8.2 $\tilde{g}(\sigma_{12}, \phi X_2) = \tilde{g}(\sigma_{22}, \phi X_1) = 0$ and $-\tilde{g}(\sigma_{22}, \phi X_2) = \tilde{g}(\sigma_{11}, \phi X_2) = \tilde{g}(\sigma_{12}, \phi X_1) = 0$. Thus σ vanishes at m and so by the Gauss equation, the Gaussian curvature at m is $+1$ contradicting the hypothesis $K \leq 0$.

Differentiating the real part of F with respect to X_1 we have

$$X_1 \Re F = \tilde{g}(\nabla^{\perp}_{X_1} \sigma_{11}, \phi X_1) + \tilde{g}\left(\sigma_{11}, E\xi + \frac{X_1 E}{2E} \phi X_1 - \frac{X_2 E}{2E} \phi X_2\right)$$

$$= -\tilde{g}(\nabla^{\perp}_{X_2} \sigma_{12}, \phi X_1) + \tilde{g}\left(\sigma_{11}, \frac{X_1 E}{2E} \phi X_1 - \frac{X_2 E}{2E} \phi X_2\right).$$

Differentiating with respect to X_2 and making similar calculations for the imaginary part of F, Lemma 8.2 and the minimality yield $X_1 \Re F = X_2 \Im F$ and $X_2 \Re F = -X_1 \Im F$. Thus F is analytic and therefore $\log |F|^2$ is harmonic.

Now $|F|^2 = \tilde{g}(\sigma_{11}, \phi X_1)^2 + \tilde{g}(\sigma_{12}, \phi X_1)^2$. On the other hand the Gauss equation gives the Gaussian curvature K as

$$K = 1 + \frac{1}{E^2}(\tilde{g}(\sigma_{11}, \sigma_{22})) - \tilde{g}(\sigma_{12}, \sigma_{12}) = 1 - \frac{2}{E^3}|F|^2.$$

Thus $|F|^2 = E^3\left(\frac{1-K}{2}\right)$. Note also the classical formula for the Gaussian curvature of g, namely $K = \frac{-1}{6E} \Delta \log E^3$.

Suppose now that the Gaussian curvature of M^2 is non-positive. Then

$$\Delta \log \frac{|F|^2}{E^3} = -\Delta \log E^3 = 6EK \leq 0 \tag{$*$}$$

and

$$\log \frac{|F|^2}{E^3} = \log \frac{1-K}{2} \geq \log \frac{1}{2}.$$

Thus $-\log \frac{|F|^2}{E^3}$ is a subharmonic function which is bounded above.

Now define a metric g^* on M^2 by $g^* = |F|((dx^1)^2 + (dx^2)^2)$; its Gaussian curvature is $-\frac{1}{4|F|} \Delta \log |F| = 0$. That is g^* is a flat metric on M^2 which is conformally equivalent to g and hence the universal covering surface $\tilde{M}$ of M^2 is conformally equivalent to the Euclidean plane. Thus $\tilde{M}$ is a parabolic surface; but every subharmonic function which is bounded above on a parabolic surface is a constant. Therefore $-\log \frac{|F|^2}{E^3}$, lifted to $\tilde{M}$, is a constant and hence it is constant on M^2. Equation $(*)$ now gives $K = 0$. ∎

Combining Theorems 8.5 and 8.6 we have the following corollary.

Corollary 8.1 *A complete integral surface of $S^5(1)$ with constant curvature which is minimally immersed has constant curvature 0 or +1.*

Turning to curvature conditions in the spirit of the Urbano result above, we have the following result of VanLindt, Verheyen and Verstraelen [1986].

Theorem 8.7 *Let M^n be a compact integral submanifold minimally immersed in a Sasakian space form $M^{2n+1}(c)$ with $c > -3$. If $K > 0$, then M^n is totally geodesic.*

If the sectional curvature is only ≥ 0, one can do better in dimension 7, namely, we have the following result of Dillen and Vrancken [1989].

Theorem 8.8 *Let M^3 be a compact integral submanifold of the standard Sasakian structure on $S^7(1)$ which is minimally immersed. If $K \geq 0$, then either M^3 is totally geodesic, M^3 is a covering of the 3-torus or M^3 is a covering of $S^1(\sqrt{3}) \times S^2(\frac{\sqrt{3}}{2})$.*

In dimension 5 the third case does not have an analogue and the corresponding result was given by Verstraelen and Vrancken [1988]. For the example of $S^1(\sqrt{3}) \times S^2(\frac{\sqrt{3}}{2})$ in the above theorem, the sectional curvatures satisfy $0 \leq K \leq \frac{4}{3}$ where both extremal values are attained (Dillen and Vrancken [1990]). Restricting the curvature to $0 \leq K \leq 1$, Dillen and Vrancken proved in [1990] the following theorem.

Theorem 8.9 *If M^n is a compact minimal integral submanifold of $S^{2n+1}(1)$ and if $0 \leq K \leq 1$, then K is identically 0 or 1.*

Other conditions on integral submanifolds of Sasakian space forms that have been considered include the notions of the mean curvature vector **H** and second fundamental form being C-*parallel*. That is, $\nabla_X^\perp \mathbf{H}$ is parallel to ξ for all tangent vectors X and respectively, $(\nabla'\sigma)(X,Y,Z)$ is parallel to ξ for all tangent vectors X, Y, Z. Recall also that a curve $\gamma(s)$ parametrized by arc length in a Riemannian manifold is a *Frenet curve* of osculating order r if there exist orthonormal vector fields $E_1, E_2, \dots, E_r$ along γ, such that $\dot\gamma = E_1, \nabla_{\dot\gamma} E_1 = k_1 E_2, \nabla_{\dot\gamma} E_2 = -k_1 E_1 + k_2 E_3, \dots, \nabla_{\dot\gamma} E_{r-1} = -k_{r-2}E_{r-2} + k_{r-1}E_r, \nabla_{\dot\gamma} E_r = -k_{r-1}E_{r-1}$ where $k_1, k_2, \dots, k_{r-1}$ are positive C^∞ functions of s. k_j is called the j-th curvature of γ. So, for example, a geodesic is a Frenet curve of osculating order 1, a circle is a Frenet curve of osculating order 2 with k_1 a constant; a helix of order r is a Frenet curve of osculating order r, such that $k_1, k_2, \dots, k_{r-1}$ are constants. With these ideas in mind Baikoussis and Blair [1992] proved the following theorem.

Theorem 8.10 *Let M^2 be an integral surface of a Sasakian space form $\tilde{M}^5(c)$. If the mean curvature vector $\mathbf{H}$ is C-parallel, then either M^2 is minimal or locally the Riemannian product of two curves as follows: (i) a helix of order*

4 *and a geodesic or helix of order* 3, (ii) *a helix of order* 3 *and a geodesic or helix of order* 3, *or* (iii) *a circle and a geodesic or helix of order* 3.

On the other hand Baikoussis, Blair, and Koufogiorgos [1995] obtained the following result.

Theorem 8.11 *Let M^3 be an integral submanifold of a Sasakian space form $\tilde{M}^7(c)$ with C-parallel second fundamental form. Then locally M^3 is either flat or totally geodesic or a product $\gamma \times M^2$, where γ is a curve and M^2 is a surface of constant curvature and also has C-parallel second fundamental form.*

We close this section with an example of this theorem in the case of ϕ-sectional curvature $c < -3$. In Example 7.4.3 we saw that the product $B^n \times \mathbb{R}$, where B^n is a simply connected bounded domain in $\mathbb{C}^n$, has a Sasakian structure of constant ϕ-sectional curvature $c < -3$. Now take B^n to be the unit ball in $\mathbb{C}^n$ with the metric

$$d\bar{s}^2 = 4\frac{(1 - \sum |z_i|^2)(\sum dz_i d\bar{z}_i) + (\sum \bar{z}_i dz_i)(\sum z_i d\bar{z}_i)}{(1 - \sum |z_i|^2)^2}.$$

This metric has constant holomorphic curvature -1 and the fundamental 2-form Ω of the Kähler structure is given by

$$\Omega = d\omega \;\; = \;\; -4i\frac{(1 - \sum |z_j|^2)(\sum dz_j \wedge d\bar{z}_j) + (\sum \bar{z}_j dz_j) \wedge (\sum z_j d\bar{z}_j)}{(1 - \sum |z_j|^2)^2}$$

$$= \;\; d\left(Re\frac{4i \sum \bar{z}_j dz_j}{1 - \sum |z_j|^2}\right).$$

Then $B^n \times \mathbb{R}$ with contact form $\eta = \omega + dt$ and metric $ds^2 = d\bar{s}^2 + \eta \otimes \eta$ is a Sasakian space form with constant ϕ-sectional curvature $c = -4$.

It is well known that setting the imaginary part of z_j equal to zero gives an imbedding of real hyperbolic space H^n of constant curvature $-\frac{1}{4}$ as a totally real, totally geodesic submanifold of B^n. To construct a non-trivial example of M^3 as an integral submanifold of $\tilde{M}^7(-4) = B^3 \times \mathbb{R}$ with C-parallel second fundamental form, we consider an umbilical surface in H^3 and as t varies in $\mathbb{R}$ we rotate H^3 so that the surface will trace out the desired M^3. For this purpose it will be convenient to use the polar form $r_j e^{i\theta_j}$ of z_j. Then η, ξ and ds^2 become.

$$\eta = 4\frac{\sum r_j^2 d\theta_j}{1 - \sum r_j^2} + dt, \quad \xi = \frac{\partial}{\partial t}$$

and

$$ds^2 = \frac{(1 - \sum r_j^2)(\sum (dr_j^2 + r_j^2 d\theta_j^2)) + (\sum r_j dr_j)^2 + (\sum r_j^2 d\theta_j)^2}{(1 - \sum r_j^2)^2}$$

$$+16\frac{(\sum r_j^2 d\theta_j)^2}{(1-\sum r_j^2)^2}+8\frac{\sum r_j^2 d\theta_j dt}{1-\sum r_j^2}+dt.$$

We remark that the induced metric on H^3 in $\mathbb{B}^3$ defined by $\theta_j = $ constant is the Beltrami–Klein model and not the Poincaré model of hyperbolic space. In this model an umbilical submanifold is in general an ellipsoid. With this in mind we construct our example as follows. On the Sasakian space form $\bar{M}^7(-4) = B^3 \times \mathbb{R}$ we designate the coordinates by $(x_1, \ldots, x_7) = (r_1, r_2, r_3, \theta_1, \theta_2, \theta_3, t)$. Now define $\iota : M \to \bar{M}^7(-4)$ by

$$(r_1, r_2, t) \mapsto \left(r_1, r_2, b\sqrt{1-r_1^2-r_2^2}, 0, 0, -\frac{1-b^2}{4b^2}t, t\right)$$

where b is a constant such that $b^2 < 1$. It is easy to see that $\eta(\iota_*\partial_1) = \eta(\iota_*\partial_2) = \eta(\iota_*\partial_7) = 0$. Thus M is a 3-dimensional integral submanifold of the Sasakian space form $\bar{M}^7(-4)$. Direct computation then shows that M has C-parallel second fundamental form (see Baikoussis, Blair, and Koufogiorgos [1995] for details).

8.3   Legendre curves

C. Baikoussis and Blair [1994] made a study of Legendre curves in 3-dimensional contact metric manifolds and we present some of the results from that paper here. We begin with a simple proposition that will motivate the discussion.

Proposition 8.2 *In a* 3-*dimensional Sasakian manifold, the torsion of a Legendre curve which is not a geodesic is equal to* 1.

Proof. Let γ be a Legendre curve on a 3-dimensional Sasakian manifold parametrized by arc length. Differentiating $\eta(\dot{\gamma}) = 0$ along γ we see from $\nabla_X \xi = -\phi X$ that $\nabla_{\dot{\gamma}}\dot{\gamma}$ is orthogonal to ξ and hence that

$$\nabla_{\dot{\gamma}}\dot{\gamma} = \pm\kappa\phi\dot{\gamma},$$

κ being the curvature. Thus the principal normal $E_2 = \pm\phi\dot{\gamma}$. Now by Theorem 6.3

$$\nabla_{\dot{\gamma}}E_2 = \pm(\xi + \phi\nabla_{\dot{\gamma}}\dot{\gamma}).$$

The torsion, τ, being the absolute value of the coefficient of ξ, is $\tau = +1$. ∎

It is classical that a curve in Euclidean 3-space which is not a line is a plane curve if and only if its torsion vanishes. Relative to the Euclidean metric on $\mathbb{R}^3$, the torsion of a Legendre curve is complicated, but we shall see that in the space $\mathbb{R}^3$ with its standard Sasakian structure (Examples 4.5.1, 6.7.1 and 7.4.2), denoted $\mathbb{R}^3(-3)$, a curve which is not a geodesic is a Legendre curve if and only if it starts as a Legendre curve and its torsion is equal to 1. More generally and more precisely we have the following theorem.

Theorem 8.12 *For a smooth curve γ in a 3-dimensional Sasakian manifold, set $\sigma = \eta(\dot\gamma)$. If $\tau = 1$ and at one point $\sigma = \dot\sigma = 0$, then γ is a Legendre curve.*

Remark. Without the hypothesis "at one point" it is possible to give a counterexample. The 3-dimensional sphere has three mutually orthogonal contact structures each of which together with the metric of constant curvature 1 forms a Sasakian structure (see Chapter 13). Now a curve which is a Legendre curve with respect to one of these need not be a Legendre curve with respect to the others, but the curve still has torsion 1.

Proof. Let γ be a smooth curve in a 3-dimensional Sasakian manifold other than an integral curve of ξ which would, of course, be a geodesic. In fact for our argument we may suppose that $\dot\gamma$ is never collinear with ξ. Then we may decompose $\nabla_{\dot\gamma}\dot\gamma$ as

$$\nabla_{\dot\gamma}\dot\gamma = \alpha\frac{\phi\dot\gamma}{\sqrt{1-\sigma^2}} + \beta\frac{\xi - \sigma\dot\gamma}{\sqrt{1-\sigma^2}}.$$

Thus $\kappa = \sqrt{\alpha^2 + \beta^2}$ and differentiating again we find that

$$\tau = 1 + \frac{\alpha\dot\beta - \beta\dot\alpha}{\alpha^2 + \beta^2} + \frac{\alpha\sigma}{\sqrt{1-\sigma^2}}.$$

Now if $\alpha = 0$, it is easy to see by taking the inner product of the above expression for $\nabla_{\dot\gamma}\dot\gamma$ with $\phi\dot\gamma$ that $\nabla_{\dot\gamma}\dot\gamma$ is collinear with ξ and in turn that $\dot\gamma$ is collinear with ξ, a contradiction. So we suppose $\alpha \neq 0$. Again an easy computation shows that $\beta = \dfrac{\dot\sigma}{\sqrt{1-\sigma^2}}$. Making this substitution for β and using the hypothesis $\tau = 1$ we have the equation

$$0 = \alpha\left(\ddot\sigma + \frac{2\sigma\dot\sigma^2}{1-\sigma^2}\right) + \alpha^3\sigma - \dot\sigma\dot\alpha.$$

Clearly $\sigma = 0$ is a solution and $\dot\sigma = 0$ implies $\sigma = 0$. Thus we now assume that σ is non-constant. Setting $\lambda = \frac{\dot\sigma}{\alpha}$ this equation becomes

$$\lambda\frac{d\lambda}{d\sigma} + \frac{2\sigma\lambda^2}{1-\sigma^2} + \sigma = 0.$$

Integrating gives

$$\dot\sigma^2 = \alpha^2(1-\sigma^2)(C(1-\sigma^2) - 1)$$

where C is a constant. Now suppose that at one point, $\sigma = \dot\sigma = 0$; then since $\alpha \neq 0$, $C = 1$. Finally since $\sigma^2 \leq 1$, we have that $\sigma = 0$, a contradiction. ∎

This property is characteristic of 3-dimensional Sasakian manifolds as we will now show.

Theorem 8.13 *If on a 3-dimensional contact metric manifold, the torsion of Legendre curves is equal to 1, then the manifold is Sasakian.*

Proof. First recall that $\nabla_X \xi = -\phi X - \phi h X$ (Lemma 6.2) and that on any 3-dimensional contact metric manifold

$$(\nabla_X \phi)Y = g(X + hX, Y)\xi - \eta(Y)(X + hX)$$

(Corollary 6.4). Then let $\gamma(s)$ be a Legendre curve. Differentiating $g(\dot\gamma, \xi) = 0$ along γ we see that $g(\nabla_{\dot\gamma}\dot\gamma, \xi) + g(\dot\gamma, -\phi h\dot\gamma) = 0$ and hence that

$$\nabla_{\dot\gamma}\dot\gamma = a\xi + b\phi\dot\gamma = \kappa E_2, \quad a = g(\dot\gamma, \phi h\dot\gamma).$$

Thus the principal normal $E_2 = \frac{1}{\kappa}(a\xi + b\phi\dot\gamma)$. Differentiating E_2 we have

$$\nabla_{\dot\gamma} E_2 = \left(-\frac{\dot\kappa}{\kappa^2}a + \frac{\dot a}{\kappa} + \frac{b}{\kappa}g(\dot\gamma + h\dot\gamma, \dot\gamma)\right)\xi - \frac{b^2}{\kappa}\dot\gamma + \left(-\frac{\dot\kappa}{\kappa^2}b - \frac{a}{\kappa} + \frac{\dot b}{\kappa}\right)\phi\dot\gamma - \frac{a}{\kappa}\phi h\dot\gamma$$

$$= -\kappa\dot\gamma + \tau E_3.$$

Now $h\dot\gamma$ is orthogonal to ξ and hence in terms of $\dot\gamma$ and $\phi\dot\gamma$ we have $h\dot\gamma = \delta\dot\gamma + \epsilon\phi\dot\gamma$. Applying ϕ we note that $\epsilon = -a$. Thus from our expressions for $\nabla_{\dot\gamma} E_2$ we have

$$\tau E_3 = \left(-\frac{\dot\kappa}{\kappa^2}a + \frac{\dot a}{\kappa} + \frac{b}{\kappa}(1 + \delta)\right)\xi + \left(-\frac{\dot\kappa}{\kappa^2}b + \frac{\dot b}{\kappa} - \frac{a}{\kappa}(1 + \delta)\right)\phi\dot\gamma.$$

From this and $\kappa^2 = a^2 + b^2$ we obtain

$$\tau = \left| \frac{b\dot a - a\dot b}{\kappa^2} + (1 + \delta)\right|.$$

If now γ is an eigencurve of h, then $\epsilon = -a = 0$ along the curve and we have that $\tau = 1 + \delta$. Thus if every Legendre curve has torsion 1, h has zero eigenvalues everywhere and hence $h = 0$, completing the proof. ∎

In Example 5.3.3 we saw that the projection γ^* of a closed Legendre curve γ in the space $\mathbb{R}^3(-3)$ to the xy-plane must have self-intersections and moreover that the algebraic (signed) area enclosed is zero. We now mention briefly the curvature of Legendre curves in $\mathbb{R}^3(-3)$. In the paper, Baikoussis and Blair [1994], the following result was also obtained.

Theorem 8.14 *The curvature of a Legendre curve in $\mathbb{R}^3$ with its standard Sasakian structure is equal to twice the curvature of its projection to the xy-plane with respect to the Euclidean metric.*

Now a curve is a geodesic if and only if $\kappa = 0$. Thus the Legendre curves of $\mathbb{R}^3(-3)$ which are geodesics are curves for which $x(s)$ and $y(s)$ are linear functions of s and $z(s)$ is a quadratic function of s.

9
Tangent Bundles and Tangent Sphere Bundles

9.1 Tangent bundles

In the first two sections of this chapter we discuss the geometry of the tangent bundle and the tangent sphere bundle. In Section 3 we briefly present a more general construction on vector bundles and in Section 4 specialize to the case of the normal bundle of a submanifold. The formalism for the tangent bundle and the tangent sphere bundle is of sufficient importance to warrant its own development, rather than specializing from the vector bundle case. As we saw in Chapter 1, the cotangent bundle of a manifold has a natural symplectic structure and we will see here that the same is true of the tangent bundle of a Riemannian manifold.

Let M be an (n+1)-dimensional C^∞ manifold and $\bar{\pi} : TM \longrightarrow M$ its tangent bundle. If $(x^1, \ldots, x^{n+1})$ are local coordinates on M, set $q^i = x^i \circ \bar{\pi}$; then $(q^1, \ldots, q^{n+1})$ together with the fibre coordinates $(v^1, \ldots, v^{n+1})$ form local coordinates on TM.

If X is a vector field on M, its *vertical lift* X^V on TM is the vector field defined by $X^V \omega = \omega(X) \circ \bar{\pi}$ where ω is a 1-form on M, which on the left side of this equation is regarded as a function on TM.

For an affine connection D on M, the *horizontal lift* X^H of X is defined by $X^H \omega = D_X \omega$. Since $D_X \omega$ has local expression $(X^i \frac{\partial \omega_j}{\partial x^i} - X^i \omega_k \Gamma_{ij}^k) dx^j$, if we evaluate $D_X \omega$ on a vector $t = v^l \frac{\partial}{\partial x^l}$, we have easily

$$(D_X \omega)(t) = v^j X^i \frac{\partial \omega_j}{\partial x^i} - X^i v^j \Gamma_{ij}^k \omega_k = \left(X^i \frac{\partial}{\partial q^i} - X^i v^j \Gamma_{ij}^k \frac{\partial}{\partial v^k} \right) \omega_l v^l.$$

Thus the local expression for X^H is

$$X^H = X^i \frac{\partial}{\partial q^i} - X^i v^j \Gamma_{ij}^k \frac{\partial}{\partial v^k}.$$

The span of the horizontal lifts at $t \in TM$ is called the *horizontal subspace* of $T_t TM$. The *connection map* $K : TTM \longrightarrow TM$ is defined by

$$KX^H = 0, \quad KX_t^V = X_{\bar{\pi}(t)}, \ t \in TM.$$

The connection map K may also be defined in the following way. Given $X \in T_t TM$, let γ be a smooth curve in TM with tangent vector X at $t = \gamma(0)$. Let $\alpha = \bar{\pi} \circ \gamma$ be the projection of the curve to M. Then

$$KX = D_{\dot{\alpha}} \gamma|_0$$

and the curve γ in TM is horizontal if γ, viewed as a vector field along α, is parallel. The horizontal subspace can also be defined by

$$\{(s \circ \alpha)_*(0) \in T_t TM | \alpha \text{ path in } M, s \text{ section of } TM, s(\alpha(0)) = t, D_{\dot{\alpha}(0)} s = 0\}.$$

It is immediate that $[X^V, Y^V]\omega = 0$. Furthermore

$$[X^H, Y^V] = \left[X^i \frac{\partial}{\partial q^i} - X^i v^j \Gamma_{ij}^k \frac{\partial}{\partial v^k}, Y^l \frac{\partial}{\partial v^l} \right]$$

$$= \left(X^i \frac{\partial Y^k}{\partial x^i} + X^i Y^l \Gamma_{il}^k \right) \frac{\partial}{\partial v^k} = (D_X Y)^V.$$

Similarly denoting the curvature tensor of D on M by $\mathbf{R}$ we have at the point $t \in TM$

$$[X^H, Y^H]_t = [X, Y]_t^H - (\mathbf{R}_{XY} t)^V.$$

TM admits an almost complex structure J defined by

$$JX^H = X^V, \quad JX^V = -X^H.$$

Using the above expressions for the Lie brackets in the Nijenhuis torsion of J one can easily see that J is integrable if and only if D has vanishing curvature and torsion (Hsu [1960], Dombrowski [1962]).

If now G is a Riemannian metric on M and D its Levi-Cività connection, we define a Riemannian metric $\bar{g}$ on TM called the *Sasaki metric*, Sasaki [1958], (not to be confused with a Sasakian structure), by

$$\bar{g}(X, Y) = \big(G(\bar{\pi}_* X, \bar{\pi}_* Y) + G(KX, KY) \big) \circ \bar{\pi}$$

where X and Y are vector fields on TM. Since $\bar{\pi}_* \circ J = -K$ and $K \circ J = \bar{\pi}_*$, $\bar{g}$ is Hermitian for the almost complex structure J.

On TM define a 1-form β by $\beta(X)_t = G(t, \bar{\pi}_* X)$, $t \in TM$ or equivalently by the local expression $\beta = \sum G_{ij} v^i dq^j$. Then $d\beta$ is a symplectic structure on TM and in particular $2d\beta$ is the fundamental 2-form of the almost Hermitian structure $(J, \bar{g})$. To see this first note that since $\beta(X^V) = 0$ and $[X^V, Y^V] = 0$, $2d\beta(X^V, Y^V) = 0$. Similarly

$$2d\beta(X^V, Y^H)_t = X^V(G(t, \bar{\pi}_* Y^H) \circ \bar{\pi}) = X^V(G(v^l \frac{\partial}{\partial x^l}, Y) \circ \bar{\pi})$$

$$= G(X, Y) = \bar{g}(X^V, Y^V) = \bar{g}(X^V, JY^H).$$

Now choose a vector field Z on M such that $Z_m = t$ and $(D_X Z)_m = 0$ for all X. Then

$$2d\beta(X^H, Y^H)_t = (X^H(G(Z, Y) \circ \bar{\pi}) - Y^H(G(Z, X) \circ \bar{\pi}) - G(t, [X, Y]))_m$$

$$= G(t, D_X Y)_m - G(t, D_Y X)_m - G(t, [X, Y])_m = 0.$$

Thus TM has an almost Kähler structure which is Kählerian if and only if (M, G) is flat (Tachibana and Okumura [1962]).

As before we let $\mathbf{R}$ denote the curvature tensor of D which is now the Levi-Cività connection of G. The Levi-Cività connection $\bar{\nabla}$ of $\bar{g}$ and its curvature tensor $\bar{R}$ were computed by Kowalski [1971]. These are given by the following formulas and we will give a partial proof as an illustration.

$$(\bar{\nabla}_{X^H} Y^H)_t = (D_X Y)^H_t - \frac{1}{2}(\mathbf{R}_{XY} t)^V_t,$$

$$(\bar{\nabla}_{X^H} Y^V)_t = \frac{1}{2}(\mathbf{R}_{tY} X)^H_t + (D_X Y)^V_t,$$

$$(\bar{\nabla}_{X^V} Y^H)_t = \frac{1}{2}(\mathbf{R}_{tX} Y)^H_t,$$

$$\bar{\nabla}_{X^V} Y^V = 0.$$

For example from $2\bar{g}(\bar{\nabla}_X Y, W) = X\bar{g}(Y, W) + Y\bar{g}(X, W) - W\bar{g}(X, Y) + \bar{g}([X, Y], W) + \bar{g}([W, X], Y) - \bar{g}([Y, W], X)$ we have

$$2\bar{g}(\bar{\nabla}_{X^H} Y^V, W^V) = X^H \bar{g}(Y^V, W^V) + \bar{g}((D_X Y)^V, W^V) - \bar{g}((D_X W)^V, Y^V)$$

$$= \bar{g}((D_X Y)^V, W^V) + G(D_X Y, W) \circ \bar{\pi} = 2\bar{g}((D_X Y)^V, W^V)$$

and

$$2\bar{g}(\bar{\nabla}_{X^H} Y^V, W^H)_t = Y^V \bar{g}(X^H, W^H) + \bar{g}((\mathbf{R}_{Xw} t)^V, Y^V)$$

$$= G(\mathbf{R}_{tY}X, W) \circ \bar{\pi} = \bar{g}((\mathbf{R}_{tY}X)^H, W^H)$$

giving the formula for $(\bar{\nabla}_{X^H}Y^V)_t$.

Turning to the curvature we have

$$\bar{R}_{X^V Y^V}Z^V = 0,$$

$$(\bar{R}_{X^V Y^V}Z^H)_t = \left(\mathbf{R}_{XY}Z + \frac{1}{4}\mathbf{R}_{tX}\mathbf{R}_{tY}Z - \frac{1}{4}\mathbf{R}_{tY}\mathbf{R}_{tX}Z\right)_t^H,$$

$$(\bar{R}_{X^H Y^V}Z^V)_t = -\left(\frac{1}{2}\mathbf{R}_{YZ}X + \frac{1}{4}\mathbf{R}_{tY}\mathbf{R}_{tZ}X\right)_t^H,$$

$$(\bar{R}_{X^H Y^V}Z^H)_t = \frac{1}{2}\left((D_X\mathbf{R})_{tY}Z\right)_t^H + \left(\frac{1}{2}\mathbf{R}_{XZ}Y + \frac{1}{4}\mathbf{R}_{\mathbf{R}_{tY}ZX}t\right)_t^V,$$

$$(\bar{R}_{X^H Y^H}Z^V)_t = \frac{1}{2}\left((D_X\mathbf{R})_{tZ}Y - (D_Y\mathbf{R})_{tZ}X\right)_t^H$$

$$+\left(\mathbf{R}_{XY}Z + \frac{1}{4}\mathbf{R}_{\mathbf{R}_{tZ}YX}t - \frac{1}{4}\mathbf{R}_{\mathbf{R}_{tZ}XY}t\right)_t^V,$$

$$(\bar{R}_{X^H Y^H}Z^H)_t = \left(\mathbf{R}_{XY}Z + \frac{1}{4}\mathbf{R}_{t\mathbf{R}_{ZY}t}X + \frac{1}{4}\mathbf{R}_{t\mathbf{R}_{XZ}t}Y + \frac{1}{2}\mathbf{R}_{t\mathbf{R}_{XY}t}Z\right)_t^H$$

$$+\frac{1}{2}\left((D_Z\mathbf{R})_{XY}t\right)_t^V.$$

We will prove the fourth one of these formulas. Recall that we may write the point t as $v^l\frac{\partial}{\partial x^l}$; this is important when we have to differentiate with respect to position. Also we abbreviate $\frac{\partial}{\partial x^i}$ by ∂_i.

$$\bar{R}_{X^H Y^V}Z^H = \bar{\nabla}_{X^H}\frac{1}{2}v^l(\mathbf{R}_{\partial_l Y}Z)^H$$

$$-\bar{\nabla}_{Y^V}\left((D_X Z)^H - \frac{1}{2}v^i(\mathbf{R}_{XZ}\partial_i)^V\right) - \bar{\nabla}_{(D_X Y)^V}Z^H.$$

Therefore

$$(\bar{R}_{X^H Y^V}Z^H)_t = -\frac{1}{2}X^i v^j\Gamma_{ij}^k(\mathbf{R}_{\partial_k Y}Z)^H + \frac{1}{2}v^l\left((D_X\mathbf{R}_{\partial_l Y}Z)_t^H - \frac{1}{2}(\mathbf{R}_{X\mathbf{R}_{\partial_l Y}}Zt)_t^V\right)$$

$$-\frac{1}{2}(\mathbf{R}_{tY}D_X Z)_t^H + \frac{1}{2}(\mathbf{R}_{XZ}Y)_t^V - \frac{1}{2}(\mathbf{R}_{tD_X Y}Z)_t^H$$

from which the fourth curvature formula readily follows.

The main result of Kowalski [1971] is the following theorem.

Theorem 9.1 *The tangent bundle TM with the Sasaki metric $\bar{g}$ is locally symmetric if and only if the base manifold (M, G) is flat in which case $(TM, \bar{g})$ is flat.*

Proof. Clearly if the base manifold is flat, then so is $(TM, \bar{g})$; so we have only the necessity to prove. Using the above formulas for the connection and curvature of $\bar{g}$ we have

$$\left((\bar{\nabla}_{W^H}\bar{R})_{X^H\,Y^V}Z^V\right)_t = \bar{\nabla}_{W^H}(-\frac{1}{2}\mathbf{R}_{Y\,Z}X - \frac{1}{4}\mathbf{R}_{tY}\mathbf{R}_{t\,Z}X)_t^H$$

$$-\bar{R}_{((D_WX)_t^H-\frac{1}{2}(\mathbf{R}_{Wx}t)_t^V)\,Y^V}Z^V - \bar{R}_{X^H\,(\frac{1}{2}(\mathbf{R}_{tY}W)_t^H+(D_WY)_t^V)}Z^V$$

$$-\bar{R}_{X^H\,Y^V}(\frac{1}{2}(\mathbf{R}_{tZ}W)_t^H + (D_WZ)_t^V).$$

Using the hypothesis and taking the vertical part we obtain

$$0 = \frac{1}{2}\mathbf{R}_{W\,(\frac{1}{2}\mathbf{R}_{YZ}X+\frac{1}{4}\mathbf{R}_{tY}\mathbf{R}_{tZ}X)}t$$

$$-\left(\mathbf{R}_{X\,\frac{1}{2}\mathbf{R}_{tY}W}Z + \frac{1}{4}\mathbf{R}_{\mathbf{R}_{tZ}\frac{1}{2}\mathbf{R}_{tY}W\,X}t - \frac{1}{4}\mathbf{R}_{\mathbf{R}_{tZ}X\,\frac{1}{2}\mathbf{R}_{tY}W}t\right)$$

$$-\left(\frac{1}{4}\mathbf{R}_{\mathbf{R}_{tY}\frac{1}{2}\mathbf{R}_{tZ}W\,X}t + \frac{1}{2}\mathbf{R}_{X\,\frac{1}{2}\mathbf{R}_{tZ}W}Y\right).$$

In this expression set $Y = t$ and $Z = t$ to get respectively

$$\mathbf{R}_{W\,\mathbf{R}_{tZ}X}t - \mathbf{R}_{X\,\mathbf{R}_{tZ}W}t = 0,$$

$$\mathbf{R}_{W\,\mathbf{R}_{Y_t}X}t - 2\mathbf{R}_{X\,\mathbf{R}_{tY}W}t = 0.$$

Now replace Y by Z in the second of these equations and compare with the first to obtain

$$\mathbf{R}_{X\,\mathbf{R}_{tZ}W}t = 0.$$

Setting $W = X$ and taking the inner product with Z yields $|\mathbf{R}_{tZ}X|^2 = 0$ and hence that (M, G) is flat. ∎

K. Bang [1994] obtained the corresponding result for $(TM, \bar{g})$ being conformally flat as a corollary to his result on conformally flat vector bundles, Theorem 9.11.

Theorem 9.2 *The tangent bundle TM with the Sasaki metric $\bar{g}$ is conformally flat if and only if the base manifold (M, G) is flat in which case $(TM, \bar{g})$ is flat.*

9.2 Tangent sphere bundles

We have seen that principal circle bundles over symplectic manifolds form a large class of examples of contact manifolds; they have K-contact structures which are Sasakian when the base manifolds are Kählerian. These examples together with the standard structure on $\mathbb{R}^{2n+1}$ (Examples 3.2.1, 4.5.1, 6.7.1, 7.4.2) show that Sasakian manifolds form a large and important class of contact manifolds. However, despite the example of $T_1 S^2 \cong \mathbb{R}P^3$ and more generally Theorem 9.3 below, the tangent and cotangent sphere bundles (Example 3.2.4) are not, in general, K-contact, even though they are classically an important class of contact manifolds. Boothby and Wang [1958] proved that a compact, simply connected, homogeneous contact manifold M of dimension $4r + 1$ with $r > 1$ is homeomorphic to a tangent sphere bundle only when M is the Stiefel manifold $V_{2r+2,2}$. In [1978] the author showed that the standard contact structure on the tangent sphere bundle of a compact Riemannian manifold of non-positive constant curvature cannot be regular.

We will regard the tangent sphere bundle of a Riemannian manifold as the bundle of unit tangent vectors, even though, owing to the factor $\frac{1}{2}$ in the coboundary formula for $d\eta$, a homothetic change of metric will be made. (If one adopts the convention that the $\frac{1}{2}$ does not appear in the coboundary formula, this change is not necessary. However to be consistent the odd-dimensional sphere as a standard example of a Sasakian manifold should then be taken as a sphere of radius 2 (cf. Tashiro [1969] and Sasaki and Hatakeyama [1962]).)

The tangent sphere bundle, $\pi : T_1 M \longrightarrow M$, is the hypersurface of TM defined by $\sum G_{ij} v^i v^j = 1$. The vector field $\nu = v^i \frac{\partial}{\partial v^i}$ is a unit normal as well as the position vector for a point $t \in T_1 M$. We denote by g' the Riemannian metric induced on $T_1 M$ from the Sasaki metric $\bar{g}$ on TM and by ∇ its Levi-Cività connection. We can easily find the Weingarten map of a hypersurface. For any vertical vector field U tangent to $T_1 M$,

$$\bar{\nabla}_U \nu = U v^i \frac{\partial}{\partial v^i} + v^i \bar{\nabla}_U \left(\frac{\partial}{\partial x^i}\right)^V = U.$$

For a horizontal tangent vector field X, we may suppose that X is the restriction of a horizontal lift; then

$$(\bar{\nabla}_{(\partial_j)^H} \nu)_t = ((\partial_j)^H v^i) \frac{\partial}{\partial v^i} + \frac{1}{2} v^i (\mathbf{R}_{t\,\partial_i} \partial_j)^H_t + v^i (D_{\partial_j} \partial_i)^V_t = \frac{1}{2} (\mathbf{R}_{t\,t} \partial_j)^H_t = 0$$

where we have again abbreviated $\frac{\partial}{\partial x^j}$ by ∂_j. Thus the Weingarten map A of $T_1 M$ with respect to the normal ν is given by $AU = -U$ for any vertical vector U and $AX = 0$ for any horizontal vector X.

With the simple form for the Weingarten map just obtained, many computations on $T_1 M$ can be done on TM. Yampol'skii [1985] (see also Borisenko and Yampol'skii [1987a], Tanno [1992]) computed the Levi-Cività connection and the curvature of g' but we will not need the added formalism here.

We know that as a hypersurface of the almost Kähler manifold TM, T_1M inherits an almost contact metric structure. Following the usual procedure (Example 4.5.2) we define ϕ', ξ' and η' on T_1M by

$$\xi' = -J\nu = -v^i J\left(\frac{\partial}{\partial x^i}\right)^V = v^i\left(\frac{\partial}{\partial x^i}\right)^H, \quad JX = \phi'X + \eta'(X)\nu.$$

(ϕ', ξ', η', g') is then an almost contact metric structure. Moreover η' is the form on T_1M induced from the 1-form β on TM, for

$$\eta'(X) = \bar{g}(\nu, JX) = 2d\beta(\nu, X) = 2\sum \left(d(G_{ij}v^j) \wedge dq^i\right)\left(v^k\frac{\partial}{\partial v^k}, X\right)$$

$$= \sum G_{ij}v^j dq^i(X) = \beta(X).$$

However $g'(X, \phi'Y) = 2d\eta'(X,Y)$, so strictly speaking (ϕ', ξ', η', g') is not a contact metric structure. Of course the difficulty is easily rectified and

$$\eta = \frac{1}{2}\eta', \quad \xi = 2\xi', \quad \phi = \phi', \quad g = \frac{1}{4}g'$$

is taken as the standard contact metric structure on T_1M. In local coordinates

$$\xi = 2v^i\left(\frac{\partial}{\partial x^i}\right)^H;$$

the vector field $v^i\left(\frac{\partial}{\partial x^i}\right)^H$ is the so-called *geodesic flow* on T_1M.

Before proceeding to our theorems we obtain explicitly the covariant derivatives of ξ and ϕ. For a horizontal tangent vector field we may again use a horizontal lift; then

$$(\nabla_{X^H}\xi)_t = (\bar{\nabla}_{X^H}\xi)_t = (X^H 2v^i)(\partial_i)^H_t + 2v^i(D_X\partial_i)^H_t - (\mathbf{R}_{X\,t}t)^V_t = -(\mathbf{R}_{X\,t}t)^V_t$$

and hence for any horizontal vector X we have

$$(\nabla_X\xi)_t = -(\mathbf{R}_{\pi_*X\,t}t)^V_t.$$

For a vertical vector field U we have

$$(\nabla_U\xi)_t = (\bar{\nabla}_U\xi)_t = (U2v^i)(\partial_i)^H_t - v^i(\mathbf{R}_{KU\,t}(\partial_i))^H_t = -2(\phi U)_t - (\mathbf{R}_{KU\,t}t)^H_t,$$

since $(\partial_i)^H = -J(\partial_i)^V$. Comparing with $\nabla_X\xi = -\phi X - \phi hX$, we have for X horizontal and orthogonal to ξ and for U vertical

$$hX_t = -X_t + (\mathbf{R}_{\pi_*X\,t}t)^H_t, \quad hU_t = U_t - (\mathbf{R}_{KU\,t}t)^V_t.$$

To differentiate ϕ, first note that for any tangent vector fields X and Y,

$$(\nabla_X \phi)Y = \bar{\nabla}_X JY - (\nabla_X \eta')(Y)\nu + \eta'(Y)AX$$

$$-g'(X, A\phi Y)\nu - J\bar{\nabla}_X Y - g'(X, AY)\xi'.$$

Again for X, Y horizontal we suppose that they are horizontal lifts and we let $U = U^i \frac{\partial}{\partial v^i}$ and $W = W^i \frac{\partial}{\partial v^i}$ be vertical vector fields tangent to $T_1 M$. Also denote by "tan" the component of a vector in $T_t TM$, $t \in T_1 M$ that is tangent to $T_1 M$. Then recalling that $AX = 0$ for X horizontal we have

$$((\nabla_X \phi)Y)_t = -\frac{1}{2}(\mathbf{R}_{\pi_* Y\,t}\pi_* X)_t^H + (D_{\pi_* X}\pi_* Y)_t^V - (\nabla_X \eta')(Y)\nu$$

$$-J(D_{\pi_* X}\pi_* Y)_t^H + \frac{1}{2}J(\mathbf{R}_{\pi_* X}\pi_* Y\,t)_t^V$$

$$= \frac{1}{2}(\mathbf{R}_{t\,\pi_* X}\pi_* Y)_t^H$$

where we have used the first Bianchi identity and our formula above for $\nabla_X \xi$.

$$((\nabla_X \phi)U)_t = -(XU^i)(\partial_i)_t^H - U^i(D_{\pi_* X}\partial_i)_t^H + \frac{1}{2}(\mathbf{R}_{\pi_* X\,KU}t)_t^V - (\nabla_X \eta')(U)\nu$$

$$-J(XU^i)(\partial_i)_t^V - U^i J(D_{\pi_* X}\partial_i)_t^V + \frac{1}{2}J(\mathbf{R}_{KU\,t}\pi_* X)_t^H$$

$$= \frac{1}{2}\tan(\mathbf{R}_{\pi_* X\,t}KU)_t^V$$

where we have used $(\nabla_X \eta')(U) = g'(\frac{1}{2}\nabla_X \xi, U) = -\frac{1}{2}G((\mathbf{R}_{\pi_* X}t, KU) = \frac{1}{2}\bar{g}((\mathbf{R}_{\pi_* X\,t}KU)_t^V, \nu)$. In a similar manner one obtains

$$((\nabla_U \phi)X)_t = -2\eta(X)U_t + \frac{1}{2}\tan(\mathbf{R}_{KU\,t}\pi_* X)_t^V,$$

$$((\nabla_U \phi)W)_t = 2g(U, W)\xi_t + \frac{1}{2}(\mathbf{R}_{KU\,t}KW)_t^H.$$

We now prove a theorem of Tashiro [1969] which shows that the contact metric structure on the tangent sphere bundle is almost never Sasakian.

Theorem 9.3 *The contact metric structure (ϕ, ξ, η, g) on $T_1 M$ is K-contact if and only if the base manifold (M, G) has positive constant curvature $+1$ in which case $T_1 M$ is Sasakian.*

Proof. If (ϕ, ξ, η, g) is a K-contact structure, then $\nabla_X \xi = -\phi X$ as we have seen, but for a horizontal lift we have $(\nabla_{X^H} \xi)_t = -(\mathbf{R}_{Xt}t)^V$ and hence $(\phi X^H)_t = (\mathbf{R}_{Xt}t)^V$. Now for X orthogonal to t, $\phi X^H = X^V$ and therefore $\mathbf{R}_{Xt}t = X$ for all orthogonal pairs $\{X, t\}$ on (M, G) from which we have $\mathbf{R}_{XY}Z = G(Y, Z)X - G(X, Z)Y$.

Conversely by the formulas above for the covariant derivative of ϕ, the condition $\mathbf{R}_{XY}Z = G(Y, Z)X - G(X, Z)Y$ on (M, G) implies

$$((\nabla_X \phi)Y)_t = -\frac{1}{2}(G(\pi_* Y, t)\pi_* X - G(\pi_* X, \pi_* Y)t)^H = (g(X, Y)\xi - \eta(Y)X)_t,$$

$$((\nabla_X \phi)U)_t = \frac{1}{2}\tan(G(KU, t)\pi_* X - G(\pi_* X, KU)t)^V_t = 0,$$

$$((\nabla_U \phi)X)_t = -2\eta(X)U + \frac{1}{2}\tan(G(t, \pi_* X)KU - G(KU, \pi_* X)t)^V_t = -\eta(X)U_t$$

and

$$((\nabla_U \phi)W)_t = 2g(U, W)\xi + \frac{1}{2}(G(t, KW)KU - G(KU, KW)t)^H_t = g(U, W)\xi$$

showing that $T_1 M$ is Sasakian. ∎

In particular if the base manifold is $S^3(1)$, then the standard contact metric structure on $T_1 S^3(1) \sim S^3 \times S^2$ is Sasakian. Moreover this metric is known to be Einstein but it is not a Riemannian product of constant curvature metrics. $S^3 \times S^2$ is therefore a nice example of a manifold with more than one Einstein metric, (see e.g., Tanno [1978b]). If g_0 denotes the standard metric on the sphere, $(S^3, 2g_0) \times (S^2, g_0)$ is Einstein but clearly not an associated metric for any contact structure in view of Pastore's result [1998] on 5-dimensional locally symmetric contact metric manifolds.

Recall the symmetric operator l (Section 7.2) which was defined by $lX = R_{X\xi}\xi$. For a K-contact structure, $l = I - \eta \otimes \xi$ and there exist many contact metric manifolds for which $l = 0$, as we will see in Section 4, Theorem 9.14. More strongly we have seen that $R_{XY}\xi = 0$ implies that the contact metric manifold M^{2n+1} is locally isometric to $E^{n+1} \times S^n(4)$ (Theorem 7.5) which is the tangent sphere bundle of Euclidean space. In the case of the contact metric structure on $T_1 M$ we have the following result (Blair [1977]).

Theorem 9.4 *The standard contact metric structure (ϕ, ξ, η, g) on $T_1 M$ satisfies $lU = 0$ for vertical vector fields U if and only if the base manifold (M, G) is flat in which case we have $R_{XY}\xi = 0$ for all X and Y on $T_1 M$.*

Proof. First note that for a vertical vector field U tangent to $T_1 M$ $R_{U\xi}\xi = 0$ implies $\bar{R}_{U\xi}\xi = 0$. Now recall that

$$K(\bar{R}_{X^H Y^V} Z^H)_t = \frac{1}{2}\mathbf{R}_{XZ}Y + \frac{1}{4}\mathbf{R}_{\mathbf{R}_{tY}Z X}t.$$

Since $\xi = 2v^i(\frac{\partial}{\partial x^i})^H$, combining these facts gives

$$\mathbf{R}_{\mathbf{R}_{KU t}t}t = 0$$

i.e., $\mathbf{R}_{\mathbf{R}_{YX}XX}X = 0$ for any orthonormal pair on (M, G). Taking the inner product with Y gives $|\mathbf{R}_{YX}X|^2 = 0$ from which we see that (M, G) is flat.

Conversely if (M, G) is flat, the Gauss equation for T_1M as a hypersurface of TM gives $g'(R'_{XY}\xi, Z) = 0$ and hence that $R_{XY}\xi = 0$. ∎

If the base manifold (M, G) has constant curvature -1, then the sectional curvature $K(U, \xi) = 1$ for any vertical vector U tangent to T_1M and $K(X, \xi) = -7$ for any horizontal vector tangent to T_1M.

In Section 7.2 we remarked that the classification of all locally symmetric contact metric manifolds remains open. Here we will show that the contact metric structure on T_1M is locally symmetric if and only if either the base manifold is flat or 2-dimensional and of constant curvature $+1$ (Blair [1989]). Here to do this we first give a couple of lemmas and then utilize a theorem of D. Perrone [1994].

Lemma 9.1 *If a contact metric manifold is locally symmetric, $\nabla_\xi h = 0$.*

Proof. Since the metric is locally symmetric and the integral curves of ξ are geodesics, we see that the operator l is parallel along ξ. Thus since $\nabla_\xi \phi = 0$ for any contact metric structure, differentiation of the second equation of Proposition 7.1 gives $\nabla_\xi h^2 = 0$. Then differentiating the first equation of Proposition 7.1 we have $\nabla_\xi \nabla_\xi h = 0$. Writing $\nabla_\xi h^2 = 0$ as $(\nabla_\xi h)h + h(\nabla_\xi h) = 0$ and differentiating we have $(\nabla_\xi h)^2 = 0$. Since $\nabla_\xi h$ is a symmetric operator, we have the result. ∎

For each unit tangent vector $t \in T_mM$, let $[t]^\perp$ be the subspace of T_mM orthogonal to t and define a symmetric linear transformation $L_t : [t]^\perp \longrightarrow [t]^\perp$ by $L_tX = \mathbf{R}_{Xt}t$.

Lemma 9.2 *If the contact metric structure on T_1M satisfies $\nabla_\xi h = 0$, then for any orthonormal pair $\{X, t\}$ on (M, G), $L_t^2X = L_tX$ and (M, G) is locally symmetric.*

Proof. Since $hU_t = U_t - (\mathbf{R}_{KU t}t)^V$ for a vertical vector $U \in T_tT_1M$ we have $h^2U_t = U_t - 2(L_tKU)^V + (L_t^2KU)^V$. Proceeding as in the proof of Theorem 9.4 we also have $lU_t = (L_t^2KU)^V + 2((D_t\mathbf{R})_{KU t}t)^H$. On the other hand since $\nabla_\xi h = 0$, applying ϕ to the first equation of Proposition 7.1 gives $\phi^2 + h^2 + l = 0$. Applying this to U, the vertical part gives $L_t^2 = L_t$ on (M, G). The horizontal part gives us that (M, G) is locally symmetric by a result of Cartan that a Riemannian manifold (M, G) is locally symmetric if and only if $G((D_X\mathbf{R})_{YX}Y, X) = 0$ for all orthonormal pairs $\{X, Y\}$ (English translation: Cartan [1983, pp. 257–258]). ∎

Theorem 9.5 *The standard contact metric structure (ϕ, ξ, η, g) on $T_1 M$ satisfies $\nabla_\xi h = 0$ if and only if the base manifold (M, G) is of constant curvature 0 or $+1$.*

Proof. This theorem follows from Lemma 9.2 and the purely Riemannian result (Perrone [1994]) that a Riemannian manifold (M, G) is locally symmetric and satisfies $L_t^2 = L_t$ if and only if (M, G) is of constant curvature 0 or $+1$. ∎

Theorem 9.6 *The standard contact metric structure (ϕ, ξ, η, g) on $T_1 M$ is locally symmetric if and only if either the base manifold (M, G) is flat or 2-dimensional and of constant curvature $+1$.*

Proof. By Lemma 9.1 and Theorem 9.5 we see that if the structure $T_1 M$ is locally symmetric, then (M, G) is of constant curvature 0 or 1. If (M, G) has constant curvature 1, then $T_1 M$ is Sasakian by Theorem 9.3 and therefore of constant curvature $+1$ by Theorem 7.11. That dim M is now 2 follows from Musso–Tricerri [1988] who prove that $(T_1 M, g')$ is Einstein only if dim $M = 2$ or by Theorem 9.7 below on the conformally flat case.

It remains to prove the converse, that if (M, G) is flat or of dimension 2 and constant curvature 1, then $(T_1 M, g)$ is locally symmetric. In the flat case this is virtually evident for a number of reasons. For example, $(TM, \bar{g})$ is flat and ξ is horizontal and so on $(T_1 M, g)$ we have $R_{XY} \xi = \bar{R}_{XY} \xi = 0$. Then by Theorem 7.5, $(T_1 M, g)$ is locally isometric to $E^{n+1} \times S^n(4)$ for $n > 1$ and flat for $n = 1$.

For (M, G) of dimension 2 and constant curvature 1, a direct computation shows that $(T_1 M, g')$ has constant curvature $\frac{1}{4}$ and therefore $(T_1 M, g)$ has constant curvature 1. ∎

The corresponding result for $(T_1 M, g)$ being conformally flat is stronger (Blair and Koufogiorgos [1994]).

Theorem 9.7 *The standard contact metric structure on $T_1 M$ is conformally flat if and only if the base manifold (M, G) is a surface constant Gaussian curvature 0 or $+1$.*

In Section 6.4 we saw that a contact metric structure gives rise to a strongly pseudo-convex CR-structure if and only if

$$(\nabla_X \phi) Y = g(X + hX, Y) \xi - \eta(Y)(X + hX).$$

For the tangent sphere bundle we have the following result of Mitric [1991], Tanno [1992].

Theorem 9.8 *For dim $M \geq 3$ the standard contact metric structure on $T_1 M$ gives rise to a strongly pseudo-convex CR-structure if and only if the base manifold (M, G) is of constant curvature.*

Proof. Using $hU_t = U_t - (\mathbf{R}_{KU\,t}t)^V$ evaluate the right-hand side of

$$(\nabla_X\phi)Y = g(X + hX, Y)\xi - \eta(Y)(X + hX)$$

on two vertical vectors U and W and compare with our earlier expression for the left-hand side. The right-hand side becomes $2g(U, W)\xi_t - g((\mathbf{R}_{KU\,t}t)^V_t, W)\xi_t$ while the left-hand side is $2g(U, W)\xi_t + \frac{1}{2}(\mathbf{R}_{KU\,t}KW)^H_t$. Equality implies that on the base manifold, $\mathbf{R}_{X\,t}Y$ is collinear with t for all X, Y orthogonal to t. In particular $G(\mathbf{R}_{X\,t}Y, X) = 0$ for every orthonormal triple $\{t, X, Y\}$ and hence (M, G) is of constant curvature. Conversely if $\mathbf{R}_{XY}Z = c(G(Y, Z)X - G(X, Z)Y)$, evaluate both sides of $(\nabla_X\phi)Y = g(X + hX, Y)\xi - \eta(Y)(X + hX)$ for the four cases of X, Y being both horizontal, both vertical, and each one horizontal with the other vertical and compare.   ∎

Tanno actually proves more; in [1992] (and [1991]) he introduces a gauge invariant B of type (1,3) (on the contact subbundle) whose vanishing implies the CR-condition of Theorem 6.6 and conversely under the CR-condition B reduces to the Chern–Moser–Tanaka invariant. See Section 10.5 for the notion of a gauge transformation. Tanno proves that on T_1M, $\dim M \geq 3$, B vanishes if and only if (M, G) is of constant curvature -1.

Th. Koufogiorgos studied in [1997a] the idea of constant ϕ-sectional curvature for non-Sasakian contact metric manifolds, especially for (κ, μ)-manifolds. For the tangent sphere bundle he proved the following theorem.

Theorem 9.9 *If (M, G) is of constant curvature c and dimension ≥ 3, the standard contact metric structure on T_1M has constant ϕ-sectional curvature (equal to c^2) if and only if $c = 2 \pm \sqrt{5}$.*

For (M, G) a surface of constant curvature $c \neq 1$, (T_1M, g) has constant ϕ-sectional curvature c^2 as follows readily from Theorem 7.9 and the formula $K(X, \phi X) = -(\kappa + \mu)$ on the same page.

9.3   Geometry of vector bundles

The geometric constructions on the tangent bundle described in Section 9.1 can be carried out on a general vector bundle. In this section we describe this construction and the corresponding results without proofs. We follow the treatment given by K. Bang in [1994].

Let (M^n, G) be a Riemannian manifold with Levi-Città connection D and curvature tensor $\mathbf{R}$ as before. Consider a vector bundle $\pi : \mathbf{E}^{n+k} \longrightarrow M^n$ with fibre metric $g^\perp$ and a metric connection ∇. If $(x^1, \dots, x^n)$ are local coordinates on M, set $q^i = x^i \circ \pi$; if $\{e_\alpha\}$ is a local orthonormal basis of sections of $\mathbf{E}$, writing a point $(m, U) \in \mathbf{E}$ as $U = \sum u^\alpha e_\alpha$, $(q^1, \dots, q^n)$ together with the fibre coordinates $(u^1, \dots, u^k)$ form local coordinates on $\mathbf{E}$.

For a section $\zeta = \sum \zeta^\alpha e_\alpha$ of $\mathbf{E}$, the connection ∇ is given by

$$\nabla_X \zeta = X^i \left(\frac{\partial \zeta^\alpha}{\partial x^i} + \zeta^\beta \mu^\alpha_{\beta i} \right) e_\alpha$$

where $\nabla_{\partial_i} e_\beta = \mu^\alpha_{\beta i} e_\alpha$.

If X is a tangent vector field on M and if ζ is a section of $\mathbf{E}$, the *horizontal lift* X^H of X and the *vertical lift* ζ^V of ζ are defined by

$$X^H = X^i \frac{\partial}{\partial q^i} - X^i u^\beta \mu^\alpha_{\beta i} \frac{\partial}{\partial u^\alpha}$$

and

$$\zeta^V = \zeta^\alpha \frac{\partial}{\partial u^\alpha}.$$

Then $\pi_* X^H = X$ and $\pi_* \zeta^V = 0$. Define a linear map $K : T\mathbf{E} \longrightarrow \mathbf{E}$ by

$$K X^H = 0, \quad K \zeta^V_{(m,U)} = \zeta_m, \ (m, U) \in \mathbf{E}$$

or by its local expression: If $(\tilde{X}^i, \tilde{X}^{n+\alpha})$ are the components of a vector $\tilde{X}$ tangent to $\mathbf{E}$ at (m, U) with respect to the coordinate basis, then

$$K\tilde{X} = (\tilde{X}^{n+\alpha} + \tilde{X}^i u^\beta \mu^\alpha_{\beta i}) e_\alpha.$$

We define a Riemannian metric $\bar{g}$ on $\mathbf{E}$, called the *Sasaki metric* , by

$$\bar{g}(\tilde{X}, \tilde{Y}) = G(\pi_* \tilde{X}, \pi_* \tilde{Y}) + g^\perp(K\tilde{X}, K\tilde{Y})$$

where $\tilde{X}$ and $\tilde{Y}$ are vector fields on $\mathbf{E}$. When $\mathbf{E} = TM$ and $\nabla = D$, $\bar{g}$ is the Sasaki metric on TM.

The curvature tensor R of ∇ is given by $R_{XY} \zeta = \nabla_X \nabla_Y \zeta - \nabla_Y \nabla_X \zeta - \nabla_{[X,Y]} \zeta$ and ∇ is said to be *flat* if R vanishes for all vector fields X, Y on M and all sections ζ of $\mathbf{E}$. Since $R_{XY} \zeta$ is also a section of $\mathbf{E}$, we can compute its inner product $g^\perp(R_{XY}\zeta, \psi)$ with another section ψ. We then define the adjoint $\hat{R}_{\zeta\psi} X$ by

$$G(\hat{R}_{\zeta\psi}X, Y) = g^\perp(R_{XY}\zeta, \psi).$$

With this notation in mind we give the the covariant derivatives of the Levi-Cività connection $\bar{\nabla}$ of the Sasaki metric $\bar{g}$ at a point $(m, U) \in \mathbf{E}$.

$$\bar{\nabla}_{X^H} Y^H = (D_X Y)^H - \frac{1}{2}(R_{XY}U)^V, \quad \bar{\nabla}_{X^H} \zeta^V = \frac{1}{2}(\hat{R}_{U\zeta}X)^H + (\nabla_X \zeta)^V,$$

$$\bar{\nabla}_{\zeta^V} Y^H = \frac{1}{2}(\hat{R}_{U\zeta}Y)^H, \quad \bar{\nabla}_{\zeta^V} \psi^V = 0.$$

The curvature $\bar{R}$ of the Sasaki metric at (m, U) is given as follows. $\bar{R}$ vanishes on three vertical lifts and

$$\bar{R}_{\zeta^V \psi^V} Z^H = \left(\hat{R}_{\zeta \psi} Z + \frac{1}{4} \hat{R}_{U \zeta} \hat{R}_{U \psi} Z - \frac{1}{4} \hat{R}_{U \psi} \hat{R}_{U \zeta} Z \right)^H,$$

$$\bar{R}_{X^H \zeta^V} \psi^V = -\left(\frac{1}{2} \hat{R}_{\zeta \psi} X + \frac{1}{4} \hat{R}_{U \zeta} \hat{R}_{U \psi} X \right)^H,$$

$$\bar{R}_{X^H \zeta^V} Z^H = \frac{1}{2} \left((D_X \hat{R})_{U \zeta} Z \right)^H + \left(\frac{1}{2} \mathbf{R}_{X Z} \zeta + \frac{1}{4} R_{\hat{R}_{U \zeta} Z X} U \right)^V,$$

$$\bar{R}_{X^H Y^H} \zeta^V = \frac{1}{2} \left((D_X \hat{R})_{U \zeta} Y - (D_Y \hat{R})_{U \zeta} X \right)^H$$

$$+ \left(R_{X Y} \zeta + \frac{1}{4} R_{\hat{R}_{U \zeta} Y X} U - \frac{1}{4} R_{\hat{R}_{U \zeta} X Y} U \right)^V,$$

$$\bar{R}_{X^H Y^H} Z^H = \left(\mathbf{R}_{X Y} Z + \frac{1}{4} \hat{R}_{U R_{Z Y} U} X + \frac{1}{4} \hat{R}_{U R_{X Z} U} Y + \frac{1}{2} \hat{R}_{U R_{X Y} U} Z \right)^H$$

$$+ \frac{1}{2} \left((\nabla_Z R)_{X Y} U \right)^V.$$

Concerning the questions of the vector bundle $\mathbf{E}$ with the Sasaki metric being locally symmetric or conformally flat, K. Bang proved the following theorems.

Theorem 9.10 *Let* $\pi : \mathbf{E}^{n+k} \longrightarrow M^n$ *be a vector bundle over a Riemannian manifold* (M, G) *with fibre metric* $g^{\perp}$ *and a metric connection* ∇. *Then the Sasaki metric on* $\mathbf{E}$ *is locally symmetric if and only if the connection* ∇ *is flat and* (M, G) *is locally symmetric.*

Theorem 9.11 *Let* $\pi : \mathbf{E}^{n+k} \longrightarrow M^n$ *be a vector bundle over a Riemannian manifold* (M, G) *with fibre metric* $g^{\perp}$ *and a metric connection* ∇. *Then the Sasaki metric on* $\mathbf{E}$ *is conformally flat if and only if either* (M, G) *is flat with flat connection* ∇, *or* (M, G) *has constant curvature with flat connection* ∇ *and* $k = 1$.

As a corollary K. Bang pointed out that the tangent bundle TM with the Sasaki metric $\bar{g}$ is conformally flat if and only if the base manifold is flat (Theorem 9.2).

9.4   Normal bundles

For the case of the normal bundle of a submanifold of a Riemannian manifold, the above construction of the Sasaki metric was developed by Reckziegel [1979] and Borisenko and Yampol'skii [1987b] using the normal connection, i.e.,

$\nabla = \nabla^{\perp}$. In this section we consider the special cases when the submanifold is a Lagrangian submanifold of a Kähler manifold or an integral submanifold of a Sasakian manifold. Again the results were obtained by K. Bang in his thesis [1994].

First let L be a Lagrangian submanifold of a symplectic manifold M^{2n} with associated metric $\tilde{g}$ and corresponding almost complex structure $\tilde{J}$. Recall that if X is tangent to L, then $\tilde{J}X$ is normal. Typically normal vectors will be denoted by ζ, ν, ψ. Using the ideas of horizontal and vertical lifts in the case of vector bundles from the last section, we define an almost complex structure $\bar{J}$ on the normal bundle $T^{\perp}L$ of L by

$$\bar{J}X^H = (\tilde{J}X)^V, \qquad \bar{J}\zeta^V = (\tilde{J}\zeta)^H.$$

That $\bar{J}^2 = -I$ and that the Sasaki metric $\bar{g}$ is Hermitian with respect to $\bar{J}$ are easily verified. However the symplectic nature of $(T^{\perp}L, \bar{J}, \bar{g})$ depends on the ambient manifold $(M^{2n}, \tilde{J}, \tilde{g})$ being Kähler. Let $\bar{\Omega}$ be the fundamental 2-form of the almost Hermitian structure just defined on $T^{\perp}L$.

Theorem 9.12 *Let L be a Lagrangian submanifold of a Kähler manifold $(M^{2n}, \tilde{J}, \tilde{g})$. Then the normal bundle $(T^{\perp}L, \bar{\Omega})$ is a symplectic manifold.*

Proof. Since $(\bar{J}, \bar{g})$ is an almost Hermitian structure, it is immediate that $\bar{\Omega}^n \neq 0$, so we have only to show that $\bar{\Omega}$ is a closed 2-form. First it is immediate that $\bar{\Omega}(X^H, Y^H) = 0$, $\bar{\Omega}(X^H, \zeta^V) = \tilde{g}(X, J\zeta)$, and $\bar{\Omega}(\zeta^V, \psi^V) = 0$. Now for horizontal lifts of vector fields tangent to L, we have

$$[X^H, Y^H] = [X, Y]^H - (R^{\perp}_{XY}\nu)^V$$

at the point $\nu \in T^{\perp}L$. Thus using the coboundary formula for $d\bar{\Omega}$ we have

$$3d\bar{\Omega}(X^H, Y^H, Z^H)$$

$$= X^H\bar{\Omega}(Y^H, Z^H) + Y^H\bar{\Omega}(Z^H, X^H) + Z^H\bar{\Omega}(X^H, Y^H)$$

$$-\bar{\Omega}([X^H, Y^H], Z^H) - \bar{\Omega}([Y^H, Z^H], X^H) - \bar{\Omega}([Z^H, X^H], Y^H)$$

$$= \bar{\Omega}((R^{\perp}_{XY}\nu)^V, Z^H) + \bar{\Omega}((R^{\perp}_{YZ}\nu)^V, X^H) + \bar{\Omega}((R^{\perp}_{ZX}\nu)^V, Y^H).$$

In keeping with the notation of this chapter, let G be the induced metric on L, D its Levi-Cività connection and $\mathbf{R}$ its curvature tensor. Using the Gauss–Weingarten equations and the Kähler condition, one readily obtains

$$\tilde{J}\nabla^{\perp}_X\nu = D_X\tilde{J}\nu.$$

Thus we have

$$\bar{\Omega}((R^{\perp}_{XY}\nu)^V, Z^H) = \bar{g}((R^{\perp}_{XY}\nu)^V, \bar{J}Z^H) = -\bar{g}((\tilde{J}R^{\perp}_{XY}\nu)^H, Z^H)$$

$$= -G(\mathbf{R}_{XY}\tilde{J}\nu, Z) = G(\mathbf{R}_{XY}Z, \tilde{J}\nu).$$

Substituting this and like expressions in the coboundary formula and using the Bianchi identity we have $3d\bar{\Omega}(X^H, Y^H, Z^H) = 0$. In much the same way one shows that $d\bar{\Omega}(X^H, Y^H, \zeta^V) = 0$ and $d\bar{\Omega}(X^H, \zeta^V, \psi^V) = 0$, and of course $d\bar{\Omega}$ vanishes on three vertical vectors. ∎

We now turn to the question of when the above almost Kähler structure on $T^{\perp}L$ is itself Kähler.

Theorem 9.13 *Let L be a Lagrangian submanifold of a Kähler manifold $(M^{2n}, \tilde{J}, \tilde{g})$. Then the following are equivalent: i) $(T^{\perp}L, \bar{J}, \bar{g})$ is Kähler. ii) L has flat normal connection. iii) L is flat.*

Proof. In the previous proof we noted that $\tilde{J}\nabla_X^{\perp}\nu = D_X\tilde{J}\nu$ and therefore $\mathbf{R}_{XY}\tilde{J}\nu = \tilde{J}R_{XY}^{\perp}\nu$. Thus L is flat if and only if L has flat normal connection.

To complete the proof we show that the almost complex structure $\bar{J}$ on $T^{\perp}L$ is integrable if and only if L has flat normal connection. This will be done by computing the Nijenhuis torsion of $\bar{J}$.

$$[\bar{J}, \bar{J}](X^H, \zeta^V) = -[X^H, \zeta^V] + [(\tilde{J}X)^V, (\tilde{J}\zeta)^H] - \bar{J}[(\tilde{J}X)^V, \zeta^V] - \bar{J}[X^H, (\tilde{J}\zeta)^H]$$

$$= -(\nabla_X^{\perp}\zeta)^V - (\nabla_{\tilde{J}\zeta}^{\perp}\tilde{J}X)^V - \bar{J}\big([X, \tilde{J}\zeta]^H - (R_{X\tilde{J}\zeta}^{\perp}\nu)^V\big)$$

$$= -(\nabla_X^{\perp}\zeta + \nabla_{\tilde{J}\zeta}^{\perp}\tilde{J}X + \tilde{J}[X, \tilde{J}\zeta])^V + (\tilde{J}R_{X\tilde{J}\zeta}^{\perp}\nu)^H.$$

Now using the Kähler condition and the fact that $\tilde{J}[X, \tilde{J}\zeta]$ is normal to L one can show that $\nabla_X^{\perp}\zeta + \nabla_{\tilde{J}\zeta}^{\perp}\tilde{J}X + \tilde{J}[X, \tilde{J}\zeta] = 0$. Therefore

$$[\bar{J}, \bar{J}](X^H, \zeta^V) = (\tilde{J}R_{X\tilde{J}\zeta}^{\perp}\nu)^H.$$

Similarly

$$[\bar{J}, \bar{J}](X^H, Y^H) = (R_{XY}^{\perp}\nu)^V, \quad [\bar{J}, \bar{J}](\zeta^V, \psi^V) = -(R_{\tilde{J}\zeta\tilde{J}\psi}^{\perp}\nu)^V$$

and the result follows. ∎

We remark that in Chapter 1 we proved a theorem of Weinstein that a symplectic manifold is locally the cotangent bundle of any Lagrangian submanifold. Here one may note that, since $\tilde{J}\nabla_X^{\perp}\nu = D_X\tilde{J}\nu$, $\tilde{J}$ provides a connection preserving isomorphism between the tangent bundle and the normal bundle of L.

Turning to the contact case let M^n be an integral submanifold of a contact metric manifold M^{2n+1} with structure tensors $(\tilde{\phi}, \tilde{\xi}, \tilde{\eta}, \tilde{g})$. On the normal bundle $T^{\perp}M^n$, we define an almost contact structure $(\bar{\phi}, \bar{\xi}, \bar{\eta})$ by

$$\bar{\phi}X^H = (\tilde{\phi}X)^V, \qquad \bar{\phi}\tilde{\xi}^V = 0, \qquad \bar{\phi}\zeta^V = (\tilde{\phi}\zeta)^H$$

for all tangent vectors X and normal vectors ζ orthogonal to $\tilde{\xi}$. Also let

$$\bar{\xi} = \tilde{\xi}^V, \qquad \bar{\eta}(X) = \bar{g}(X, \bar{\xi})$$

for any vector X. Then $\bar{\eta}(\bar{\xi}) = 1$ and $\bar{\phi}^2 = -I + \bar{\eta} \otimes \bar{\xi}$ follows easily.

Using the coboundary formula for $d\bar{\eta}$ we have at a point $\nu \in T^\perp M^n$,

$$2d\bar{\eta}(X^H, Y^H) = X^H \bar{\eta}(Y^H) - Y^H \bar{\eta}(X^H) - \bar{\eta}([X^H, Y^H])$$

$$= -\bar{g}([X^H, Y^H], \bar{\xi}) = \bar{g}(R^\perp_{XY}\nu, \bar{\xi}) = \tilde{g}(\tilde{R}_{XY}\nu, \tilde{\xi}) + \tilde{g}([A_\nu, A_{\tilde{\xi}}]X, Y)$$

by the equation of Ricci–Kühn. For a normal vector ζ orthogonal to $\tilde{\xi}$ we have

$$2d\bar{\eta}(X^H, \zeta^V) = -\bar{g}([X^H, \zeta^V], \bar{\xi}) = -\tilde{g}(\nabla^\perp_X \zeta, \tilde{\xi}) = -\tilde{g}(\tilde{\nabla}_X \zeta, \tilde{\xi})$$

$$= \tilde{g}(\zeta, \tilde{\nabla}_X \tilde{\xi}).$$

Similarly $d\bar{\eta}(X^H, \bar{\xi}) = 0$, $d\bar{\eta}(\zeta^V, \bar{\xi}) = 0$ and $d\bar{\eta}(\zeta^V, \psi^V) = 0$.

If now M^{2n+1} is Sasakian, by Lemma 8.1, $A_{\tilde{\xi}} = 0$ and, by Proposition 7.3, $\tilde{R}_{XY}\tilde{\xi} = \tilde{\eta}(Y)X - \tilde{\eta}(X)Y$, giving $\tilde{g}(R^\perp_{XY}\nu, \tilde{\xi}) = \tilde{g}(\tilde{R}_{XY}\nu, \tilde{\xi}) = 0$. Thus $R^\perp_{XY}\tilde{\xi} = 0$ for any tangent vectors X and Y and in turn $\hat{R}_{\tilde{\xi}\nu}X = 0$ for all $\nu \in T^\perp M^n$. Now setting

$$\phi = \bar{\phi}, \quad \xi = 2\bar{\xi}, \quad \eta = \frac{1}{2}\bar{\eta}, \quad g = \frac{1}{4}\bar{g},$$

we see that $d\eta(X, Y) = g(X, \phi Y)$ for all vector fields on $T^\perp M$ giving $T^\perp M^n$ a contact metric structure (ϕ, ξ, η, g).

In Section 7.2 we mentioned in passing that there exist many contact metric manifolds satisfying $l = 0$ ($lX = R_{X\xi}\xi$). This is seen from the following theorem of Bang [1994].

Theorem 9.14 *Let M^n be an integral submanifold of a Sasakian manifold M^{2n+1} with structure tensors $(\tilde{\phi}, \tilde{\xi}, \tilde{\eta}, \tilde{g})$. Then the normal bundle, $T^\perp M^n$, has a contact metric structure (ϕ, ξ, η, g) satisfying $l = 0$.*

Proof. We have just seen that when M^n is an integral submanifold of a Sasakian manifold M^{2n+1}, $T^\perp M^n$ has a contact metric structure (ϕ, ξ, η, g). So it remains only to show that $l = 0$. From the curvature expressions of Section 9.3 and $\hat{R}_{\tilde{\xi}\nu}X = 0$ we have

$$R_{X^H \xi}\xi = 4R_{X^H \tilde{\xi}\nu}\tilde{\xi}^V = -(\hat{R}_{\nu\tilde{\xi}}\hat{R}_{\nu\tilde{\xi}}X)^H = 0$$

and $R_{\zeta^V \xi}\xi = 0$.   ∎

In particular we see that the contact metric structure (ϕ, ξ, η, g) on $T^\perp M^n$ is never Sasakian, so we have only a partial Sasakian analogue of Theorem 9.12. This is the following result of Yano and Kon [1983, p. 50].

Theorem 9.15 *Let M^n be an integral submanifold of a Sasakian manifold M^{2n+1}. Then M^n has flat normal connection if and only if (M^n, G) has constant curvature 1.*

As an example of Theorem 9.15 recall Example 5.3.1 of S^n as a totally geodesic integral submanifold of the Sasakian manifold S^{2n+1}. In this case the normal bundle $T^{\perp}S^n$ has flat normal connection and $T^{\perp}S^n$ as a contact metric manifold with the structure (ϕ, ξ, η, g) is again the common example $E^{n+1} \times S^n(4)$ (cf. Theorem 7.5).

We also point out that for a normal vector ζ orthogonal to $\tilde{\xi}$, we have on $T^{\perp}M^n$ that $\nabla_{\zeta^V}\xi = 2\bar{\nabla}_{\zeta^V}\tilde{\xi}^V = 0$ where ∇ is the Levi-Cività connection of g. Thus from Lemma 6.2, $h\zeta^V = -\zeta^V$ and hence -1 is an eigenvalue of h with multiplicity n. Since $h\phi + \phi h = 0$, $+1$ is also an eigenvalue with multiplicity n and $hX^H = X^H$. From $[X^H, Y^H] = [X, Y]^H - (R^{\perp}_{XY}\nu)^V$ we see that the subbundle $[+1]$ is integrable if and only if M^n has flat normal connection.

We close this chapter with a continuation of Example 5.3.2. There we gave an imbedding of the 2-torus as a flat minimal integral submanifold of S^5. In keeping with the notation of Section 9.3 we let $\{x^1, x^2\}$ be the local coordinates on T^2 instead of $\{u, v\}$; $X_1 = \frac{\partial}{\partial x^1}, X_2 = \frac{\partial}{\partial x^2}$ are orthonormal in the metric G, the restriction of $\tilde{g}$ to T^2. Let $e_1 = \tilde{\phi}X_1$, $e_2 = \tilde{\phi}X_2$ and $e_3 = \tilde{\xi}$. To find $\nabla^{\perp}_{\partial_i}e_\beta = \mu^\alpha_{\beta i}e_\alpha$ explicitly, compute $\tilde{\nabla}e_\alpha$ using the Sasakian condition. Then one finds

$$\mu^1_{31} = -1, \quad \mu^2_{32} = -1, \quad \mu^3_{11} = 1, \quad \mu^3_{22} = 1$$

as the non-zero $\mu^\alpha_{\beta i}$'s. Also $\{u^1, u^2, u^3\}$ denote the fibre coordinates so that $\{q^i = x^i \circ \pi, u^\alpha\}$ are local coordinates on $T^{\perp}T^2$. Now computing the Sasaki metric $\bar{g}$ we have

$$\bar{g}\left(\frac{\partial}{\partial q^i}, \frac{\partial}{\partial q^j}\right) = \delta_{ij} + \sum \mu^\alpha_{\beta i}\mu^\alpha_{\delta j}u^\beta u^\delta,$$

$$\bar{g}\left(\frac{\partial}{\partial q^i}, \frac{\partial}{\partial u^\alpha}\right) = \mu^\alpha_{\beta i}u^\beta \quad \bar{g}\left(\frac{\partial}{\partial u^\alpha}, \frac{\partial}{\partial u^\beta}\right) = \delta_{\alpha\beta}.$$

Thus for the contact metric metric structure (η, g) on $T^{\perp}T^2$ we have

$$\eta = \frac{1}{2}(du^3 + u^1 dq^1 + u^2 dq^2)$$

and

$$g = \frac{1}{4}\begin{pmatrix} 1 + (u^1)^2 + (u^3)^2 & u^1 u^2 & -u^3 & 0 & u^1 \\ u^1 u^2 & 1 + (u^2)^2 + (u^3)^2 & 0 & -u^3 & u^2 \\ -u^3 & 0 & 1 & 0 & 0 \\ 0 & -u^3 & 0 & 1 & 0 \\ u^1 & u^2 & 0 & 0 & 1 \end{pmatrix}.$$

Compare this metric with the metric

$$g = \frac{1}{4} \begin{pmatrix} \delta_{ij} + y^i y^j + \delta_{ij} z^2 & \delta_{ij} z & -y^i \\ \delta_{ij} z & \delta_{ij} & 0 \\ -y^j & 0 & 1 \end{pmatrix}$$

associated to the Darboux form $\eta = \frac{1}{2}(dz - \sum y^i dx^i)$ on $\mathbb{R}^{2n+1}$ introduced in Section 7.2 as a formal generalization of the flat associated metric of the Darboux form on $\mathbb{R}^3$.

10

Curvature Functionals on Spaces of Associated Metrics

10.1 Introduction to critical metric problems

The study of the integral of the scalar curvature, $A(g) = \int_M \tau \, dV_g$, as a functional on the set $\mathcal{M}_1$ of all Riemannian metrics of the same total volume on a compact orientable manifold M is now classical, dating back to Hilbert [1915] (see also Nagano [1967]). A Riemannian metric g is a critical point of $A(g)$ if and only if g is an Einstein metric. Since there are so many Riemannian metrics on a manifold, one can regard, philosophically, the finding of critical metrics as an approach to searching for the best metric for the given manifold. Other functions of the curvature have been taken as integrands as well, most notably $B(g) = \int_M \tau^2 \, dV_g$, $C(g) = \int_M |\rho|^2 \, dV_g$ where ρ is the Ricci tensor, and $D(g) = \int_M |R_{kjih}|^2 \, dV_g$; the critical point conditions for these have been computed by Berger [1970]. From the critical point conditions it is easy to see that Einstein metrics are critical for $B(g)$ and $C(g)$ but not necessarily conversely. For example an η-Einstein manifold M^{2n+1} with scalar curvature equal to $2n(2n + 1)$ or $2n(2n + 3)$ is a non-Einstein critical metric of $C(g)$, Yamaguchi and Chūman [1983]. In the case of $B(g)$ Yamaguchi and Chūman showed that a Sasakian critical point is Einstein. Similarly metrics of constant curvature and Kähler metrics of constant holomorphic curvature are critical for $D(g)$, see Muto [1975]; also a Sasakian manifold of dimension m and constant ϕ-sectional curvature $3m - 1$ is critical for $D(g)$, see Yamaguchi and Chūman [1983].

Our study in this chapter is primarily motivated by two kinds of questions. 1) Given an integral functional restricted to a smaller set of metrics, what is the critical point condition; one would expect a weaker one. The smaller sets of metrics we have in mind are the sets of metrics associated to a symplectic or contact structure. 2) Given these sets of metrics, are there other natural integrands depending on the structure as well as the curvature?

To introduce the techniques for our study we will prove the classical result that a Riemannian metric is critical for $A(g)$ if and only if it is Einstein. Let M be a compact orientable manifold and $\mathcal{M}_1$ the set of all Riemannian metrics normalized by the condition of having the same total volume, usually taken to be 1, but one need not insist on the particular value in a given problem. As in Section 4.3 we often denote by the same letter a tensor field of type $(0,2)$ and its corresponding types $(1,1)$ and $(2,0)$, determined by the metric under consideration, e.g., we may write $\mathrm{tr}TD = T^i{}_j D^j{}_i = T^{ij}D_{ij}$.

Lemma 10.1 *Let T be a second order symmetric tensor field on M. Then $\int_M \mathrm{tr}TD\, dV_g = 0$ for all symmetric tensor fields D satisfying $\int_M \mathrm{tr}D\, dV_g = 0$ if and only if $T = cg$ for some constant c.*

Proof. If $T = cg$, $\mathrm{tr}TD = c\,\mathrm{tr}D$ and the sufficiency is immediate. Thus we have only to prove the necessity. Let X, Y be an orthonormal pair of vector fields on a neighborhood $\mathcal{U}$ on M and f a C^∞ function with compact support in $\mathcal{U}$. Regarding X and Y as part of a local orthonormal basis, define a tensor field D on M by $D(X, X) = f$ and $D(Y, Y) = -f$, with all other components equal to zero and $D \equiv 0$ outside $\mathcal{U}$. Then $\int_M (T(X, X) - T(Y, Y))f\, dV_g = 0$ for any C^∞ function with compact support and hence $T(X, X) = T(Y, Y)$ for every orthonormal pair X, Y. Therefore $T = cg$ for some function c and it remains to show that c is a constant. To see this let X be any vector field and $D = \pounds_X g$. Then since the integral of a divergence vanishes,

$$0 = \int_M T^{ij}(\nabla_i X_j + \nabla_j X_i)\, dV_g = -2\int_M (\nabla_i T^{ij})X_j\, dV_g,$$

but X is arbitrary so that $\nabla_i T^{ij} = 0$ (Lemma 4.7) from which we see that c must be a constant. ■

Now the approach to these critical point problems is to differentiate the functional in question along a path of metrics. The curvature functionals we study are not generally invariant under homothetic transformations; so when necessary we normalize these problems by restricting them to $\mathcal{M}_1$. Let $g(t)$ be a path of metrics in $\mathcal{M}_1$ and

$$D_{ij} = \frac{\partial g_{ij}}{\partial t}\Big|_{t=0}$$

its tangent vector at $g = g(0)$. We define two other tensor fields by

$$D_{ji}{}^h = \frac{1}{2}(\nabla_j D_i{}^h + \nabla_i D_j{}^h - \nabla^h D_{ji}),$$

$$D_{kji}{}^h = \nabla_k D_{ji}{}^h - \nabla_j D_{ki}{}^h$$

where ∇ denotes the Levi-Cività connection of $g(0)$ and we note that

$$D_{ji}{}^h = \left.\frac{\partial \Gamma_{ji}{}^h}{\partial t}\right|_{t=0}, \qquad D_{kji}{}^h = \left.\frac{\partial R_{kji}{}^h}{\partial t}\right|_{t=0}$$

where $\Gamma_{ji}{}^h$ and $R_{kji}{}^h$ denote the Christoffel symbols and curvature tensor of $g(t)$.

Theorem 10.1 *Let M be a compact orientable C^∞ manifold and $\mathcal{M}_1$ the set of all Riemannian metrics on M with unit volume. Then $g \in \mathcal{M}_1$ is a critical point of $A(g) = \int_M \tau\, dV_g$ if and only if g is Einstein.*

Proof. The proof is to compute $\frac{dA}{dt}$ at $t = 0$ for a path $g(t)$ in $\mathcal{M}_1$. First note that from $g_{ij}g^{jk} = \delta_i^k$,

$$\left.\frac{\partial g^{ij}}{\partial t}\right|_{t=0} = -D^{ij}.$$

Differentiation of the volume element gives

$$\frac{d}{dt}dV_g = \frac{d}{dt}\sqrt{\det(g(t))}dx^1 \wedge \cdots \wedge dx^n = \frac{1}{2\det(g(t))}\left(\frac{d}{dt}\det(g(t))\right)dV_g$$

$$= \frac{1}{2}g^{ij}\left(\frac{d}{dt}g_{ij}\right)dV_g = \frac{1}{2}D_i^i dV_g.$$

Now

$$\left.\frac{dA}{dt}\right|_{t=0} = \left.\frac{d}{dt}\int_M R_{kji}{}^k g^{ji}\, dV_g\right|_{t=0}$$

$$= \int_M (D_{kji}{}^k g^{ji} - \rho_{ji}D^{ji} + \frac{1}{2}\tau g^{ji}D_{ji})\, dV_g$$

$$= \int_M (-\rho^{ji} + \frac{1}{2}\tau g^{ji})D_{ji}\, dV_g$$

since the integral of a divergence vanishes. On the other hand differentiation of $\int_M dV_g = 1$ gives $\int_M D_i^i\, dV_g = 0$. Thus setting $\left.\frac{dA}{dt}\right|_{t=0} = 0$ and applying the lemma, we have

$$\rho_{ji} - \frac{1}{2}\tau g_{ji} = cg_{ji}$$

for some constant c and hence that g is Einstein. The converse is immediate. ∎

   In [1974a] Y. Muto computed the second derivative of $A(g)$ at a critical point and proved the following theorem.

Theorem 10.2 *The index of $A(g)$ and the index of $-A(g)$ are both positive at each critical point.*

Y. Muto also considered the second derivative of the functional $D(g)$ from the following point of view. Let $Diff$ denote the diffeomorphism group of M; if $f \in Diff$, then $D(f^*g) = D(g)$ and hence we have an induced mapping $\tilde{D} : \mathcal{M}_1/Diff \longrightarrow \mathbb{R}$. We say that a metric g is a critical point of $\tilde{D}$ if its orbit under $Diff$ is a critical point of $\tilde{D}$. Recall from the introduction to this section that a Riemannian metric of constant curvature is a critical point of $D(g)$; Y. Muto [1974b] proved the following result.

Theorem 10.3 *If M is diffeomorphic to a sphere and g_0 is a metric of positive constant curvature, then the index of $D(g)$ and the index of $\tilde{D}$ are both zero at g_0 and $\tilde{D}$ has a local minimum at g_0.*

We now turn to integral functionals defined on the set of metrics associated to a symplectic or contact structure. To begin we recall from Chapter 4 how the set $\mathcal{A}$ of associated metrics sits in the set $\mathcal{N}$ of all Riemannian metrics with the same volume element. In particular we saw that a symmetric tensor field D is tangent to a path in $\mathcal{A}$ at g if and only if

$$DJ + JD = 0$$

in the symplectic case and

$$D\xi = 0, \quad D\phi + \phi D = 0$$

in the contact case.

Similar to the role played by Lemma 10.1 we have the following lemma for critical point problems on $\mathcal{A}$.

Lemma 10.2 *Let T be a second order symmetric tensor field on M. Then $\int_M T^{ij} D_{ij}\, dV = 0$ for all symmetric tensor fields D satisfying $DJ + JD = 0$ in the symplectic case and $D\xi = 0$, $D\phi + \phi D = 0$ in the contact case if and only if $TJ = JT$ in the symplectic case and $\phi T - T\phi = \eta \otimes \phi T\xi - (\eta \circ T\phi) \otimes \xi$ in the contact case (i.e., ϕ and T commute when restricted to the contact subbundle).*

Proof. We give the proof in the symplectic case; the proof in the contact case being similar (see Blair and Ledger [1986]). Let $X_1, \ldots, X_{2n}$ be a local J-basis defined on a neighborhood $\mathcal{U}$ (i.e., $X_1, \ldots, X_{2n}$ is an orthonormal basis with respect to g and $X_{2i} = JX_{2i-1}$) and note that the first vector field X_1 may be any unit vector field on $\mathcal{U}$. Let f be a C^∞ function with compact support in $\mathcal{U}$ and define a path of metrics $g(t)$ as follows. Make no change in g outside $\mathcal{U}$ and within $\mathcal{U}$ change g only in the planes spanned by X_1 and X_2 by the matrix

$$\begin{pmatrix} 1 + tf + \frac{1}{2}t^2 f^2 & \frac{1}{2}t^2 f^2 \\ \frac{1}{2}t^2 f^2 & 1 - tf + \frac{1}{2}t^2 f^2 \end{pmatrix}.$$

It is easy to check that $g(t) \in \mathcal{A}$ and clearly the only non-zero components of D are $D_{11} = -D_{22} = f$. Then $\int_M T^{ij} D_{ij} \, dV = 0$ becomes

$$\int_M (T^{11} - T^{22}) f \, dV = 0.$$

Thus since X_1 was any unit vector field on $\mathcal{U}$,

$$T(X, X) = T(JX, JX)$$

for any vector field X. Since T is symmetric, linearization gives $TJ = JT$. Conversely, if T commutes with J and D anti-commutes with J, then $\mathrm{tr}TD = \mathrm{tr}TJDJ = \mathrm{tr}JTDJ = -\mathrm{tr}TD$, giving $T^{ij} D_{ij} = 0$. ∎

We end this section with our first main result (Blair and Ianus [1986]), namely we consider the functional $A(g)$ restricted to the set $\mathcal{A}$ and find the critical point condition. Since $\mathcal{A}$ is a smaller set of metrics than $\mathcal{M}_1$, we expect a weaker critical point condition than being Einstein; we will see that the condition is that the Ricci operator commute with the corresponding almost complex structure and hence is still a very natural condition.

Theorem 10.4 *Let M be a compact symplectic manifold and $\mathcal{A}$ the set of metrics associated to the symplectic form. Then $g \in \mathcal{A}$ is a critical point of $A(g) = \int_M \tau \, dV_g$ restricted to $\mathcal{A}$ if and only if the Ricci operator Q of g commutes with the almost complex structure corresponding to g.*

Proof. The proof is again to compute $\frac{dA}{dt}$ at $t = 0$ for a path $g(t)$ in $\mathcal{A}$. Since all associated metrics have the same volume element this is easier than in the Riemannian case. In particular we have,

$$\left. \frac{dA}{dt} \right|_{t=0} = \left. \frac{d}{dt} \int_M R_{kji}{}^k g^{ji} \, dV_g \right|_{t=0}$$

$$= \int_M D_{kji}{}^k g^{ji} - \rho_{ji} D^{ji} \, dV_g$$

$$= -\int_M \rho^{ji} D_{ji} \, dV_g,$$

the other terms being divergences and hence contributing nothing to the integral. Setting $\left. \frac{dA}{dt} \right|_{t=0} = 0$, the result follows from Lemma 10.2. ∎

The commutativity $QJ = JQ$ is equivalent to $\rho(JX, JY) = \rho(X, Y)$ and is often referred to as the *J-invariance of the Ricci tensor* or as the manifold having *Hermitian Ricci tensor*.

10.2   The ∗-scalar curvature

In Section 7.2 we defined the *∗-Ricci tensor* and the *∗-scalar curvature* in contact geometry. In almost Hermitian geometry these are defined similarly by

$$\rho^*_{ij} = R_{iklt} J^{kl} J_j{}^t, \quad \tau^* = \rho^{*i}_i.$$

On a Kähler manifold, $\rho^*_{ij} = \rho_{ij}$. The most important property of τ^* on an almost Kähler manifold is the analogue or forerunner of Proposition 7.7, viz.,

$$\tau - \tau^* = -\frac{1}{2}|\nabla J|^2.$$

Therefore $\tau - \tau^* \leq 0$ with equality holding if and only if the metric is Kähler. Thus for M compact, Kähler metrics are maxima of the functional

$$K(g) = \int_M \tau - \tau^* \, dV$$

on $\mathcal{A}$ and hence it is natural to ask for the critical point condition in general. This was the main question of Blair and Ianus in [1986]; the critical point condition for $K(g)$ turns out to be the same as for $A(g)$ on $\mathcal{A}$, viz., $QJ = JQ$.

Theorem 10.5 *Let M be a compact symplectic manifold and $\mathcal{A}$ the set of metrics associated to the symplectic form. Then $g \in \mathcal{A}$ is a critical point of $K(g)$ if and only if $QJ = JQ$.*

At first it may seem surprising that $A(g)$ and $K(g)$ have the same critical point condition but we will see in the course of our discussion that this is natural. The proof of Theorem 10.5 will then be an easy consequence of Theorem 10.4 and Theorem 10.6 below. It is also natural to ask whether Kähler metrics are the only critical points of $K(g)$; the answer to this is negative and a counterexample was given by Davidov and Muskarov [1990] on the twistor space of a compact Einstein, self-dual 4-manifold with negative scalar curvature.

In [1969] S. I. Goldberg showed that if $R_{XY}JZ = JR_{XY}Z$ on an almost Kähler manifold, then the metric is Kähler, and conjectured that a compact almost Kähler–Einstein manifold is Kähler. This conjecture is still open, but under the additional assumption of non-negative scalar curvature it was proved by Sekigawa [1987]. Without the assumption of compactness the conjecture is false; non-existence was shown by Armstrong [1998], Nurowski and Przanowski [1999] gave an example of a 4-dimensional non-Kähler almost Kähler which is Ricci flat, and Apostolov, Draghici and Moroianu [2001] have given non-compact counterexamples to the conjecture in dimensions ≥ 6. The scalar curvature in the twistor space example, mentioned above, of a non-Kähler, almost Kähler manifold satisfying $QJ = JQ$ is negative; thus one might ask in light of Sekigawa's result, if a compact almost Kähler manifold with $QJ = JQ$ and non-negative scalar curvature is Kähler. Draghici [1999] answered this question affirmatively in dimension 4 and negatively in general dimension. As a partial result in general dimension, Draghici proved in [1994] that if M is a compact almost Kähler manifold with Hermitian Ricci tensor and if there exists a constant $\lambda \geq 0$ such that $\lambda \leq Ric(X) \leq 2\lambda$ for any direction X, then M is Kähler.

The original proof of Theorem 10.5 was to proceed as in Theorem 10.4 and differentiate $-\tau^*$. This is very complicated and only after clever use of many identities does one see that differentiation of this term again yields a contribution of $-\rho^{ij}D_{ij}$ to the integrand and hence that one has the same critical point condition (precisely $2Q$ commuting with J). This suggests that if we consider the "total scalar curvature", $I(g) = \int_M \tau + \tau^* \, dV$, the contributions of each term to the derivative of the integrand would have canceled each other and hence every metric in $\mathcal{A}$ would have been a critical point; thus, since $\mathcal{A}$ is path connected, $I(g)$ must be constant on $\mathcal{A}$ and hence a symplectic invariant (Blair [1991a]). We now prove this using only relatively short computations. The motivation for studying the integral of the sum of τ and τ^* lies in the author's work with D. Perrone [1992] on critical point problems in the contact case and will be discussed later in this section.

Theorem 10.6 *Let M be a compact symplectic manifold. Then $\int_M \tau + \tau^* \, dV$ is a symplectic invariant and to within a constant is the cup product*

$$(c_1(M) \cup [\Omega]^{n-1})([M])$$

where $c_1(M)$ is the first Chern class of M.

Proof. Consider an almost Hermitian manifold with structure tensors (J, g). The *generalized Chern form* γ is given by

$$8\pi\gamma_{ij} = -4J^k{}_j \rho^*_{ik} - J^{kl}(\nabla_j J^h{}_k)\nabla_i J_{lh}.$$

Now on an almost Kähler manifold

$$(\nabla_k J_{ip})J_j{}^p = (\nabla_p J_{ij})J_k{}^p;$$

this is the condition for an almost Hermitian structure to be quasi-Kähler (see e.g., Gray and Hervella [1980]). Using this we have the following computation.

$$8\pi\gamma_{ij}J^{ji} = 4\tau^* - J^{kl}(\nabla^j J_k{}^h)(\nabla_i J_{hl})J_j{}^i$$

$$= 4\tau^* - J^{kl}(\nabla^j J_k{}^h)(\nabla_j J_{hi})J_l{}^i$$

$$= 4\tau^* - |\nabla J|^2,$$

but $\tau - \tau^* = -\frac{1}{2}|\nabla J|^2$ and hence

$$2(\tau + \tau^*) = 4\tau^* - |\nabla J|^2 = 8\pi \gamma_{ij} J^{ji}.$$

Thus the "total scalar curvature" of an associated metric becomes $I(g) = 4\pi \int_M \gamma_{ij} J^{ji} \, dV$. Moreover, writing Ω in terms of a J-basis, a simple computation shows that

$$\int_M \gamma_{ij} J^{ji} \, dV = \frac{1}{2^{n-1}(n-1)!} \int_M \gamma \wedge \Omega^{n-1}.$$

■

Thus one has a relatively easy proof that the "total scalar curvature" is a symplectic invariant and writing $\tau - \tau^*$ as $2\tau - (\tau + \tau^*)$ we see that $A(g)$ and $K(g)$ have the same critical point condition proving Theorem 10.5.

On a Kähler manifold the Ricci form is, up to a constant, the first Chern form γ, i.e., $2\pi \gamma_{ij} = -\rho_{ik} J^k{}_j$. On an almost Kähler manifold the Ricci tensor need not be J-invariant and hence the Ricci form is not in general defined. In [1994] Draghici defined a Ricci form on an almost Kähler manifold by decomposing the Ricci tensor ρ into its J-invariant and J-anti-invariant parts and defining the Ricci form ψ by $\psi(X,Y) = \rho^{inv}(X, JY)$. Similarly define the *-Ricci form by $\psi^*(X,Y) = \rho^*(X, JY)$; since $\rho^*(JX, JY) = \rho^*(Y, X)$, ψ^* is a well defined 2-form. Draghici then obtained a cohomological version of the Goldberg conjecture and proved that if M is a compact almost Kähler manifold whose Ricci form is cohomologous to the first Chern class, then M is a Kähler manifold.

In the contact case the results corresponding to Theorems 10.4 and 10.5 were obtained by Blair and Ledger in [1986] and the critical point conditions are different. The integral $K(g)$ for a contact manifold M^{2n+1} is taken to be $K(g) = \int_M \tau - \tau^* - 4n^2 \, dV$ (see Proposition 7.7) even though it is not necessary to include the constant $4n^2$.

Theorem 10.7 *Let M be a compact contact manifold and $\mathcal{A}$ the set of metrics associated to the contact form. Then $g \in \mathcal{A}$ is a critical point of $A(g)$ restricted to $\mathcal{A}$ if and only if Q and ϕ commute when restricted to the contact subbundle.*

Theorem 10.8 *Let M be a compact contact manifold and $\mathcal{A}$ the set of metrics associated to the contact form. Then $g \in \mathcal{A}$ is a critical point of $K(g)$ if and only if $Q - 2nh$ and ϕ commute when restricted to the contact subbundle.*

Moreover from Proposition 7.7 we see that Sasakian metrics, when they exist, are maxima of $K(g)$.

In approaching Theorem 10.6 we remarked that the study of $\int_M \tau + \tau^* \, dV$ in symplectic geometry was motivated by the corresponding study in contact geometry (Blair and Perrone [1992]). It is interesting to remark that many results in contact and Sasakian geometry were motivated by the corresponding ones in symplectic and Kähler geometry. Here we have an example of a result in contact preceding its symplectic analogue. In contact geometry the functional $I(g) = \int_M \tau + \tau^* \, dV$ is not an invariant and gives a critical point problem whose critical point condition gives the important class of K-contact metrics, i.e., associated metrics for which the characteristic vector field generates isometries.

Theorem 10.9 *Let M be a compact contact manifold and $\mathcal{A}$ the set of metrics associated to the contact form. Then $g \in \mathcal{A}$ is a critical point of $I(g) = \int_M \tau + \tau^* \, dV$ if and only if g is K-contact.*

Proof. The terms τ and τ^* are differentiated as in Blair and Ledger [1986] and the contributions of $\rho^{ij} D_{ij}$ cancel instead of doubling up as in Theorem 10.8 and the original proof of Theorem 10.5. The result is that

$$\frac{d}{dt} \int_M \tau + \tau^* \, dV \bigg|_{t=0} = -4n \int_M h^{jl} D_{jl} \, dV.$$

Thus from Lemma 10.2 and $h\xi = 0$, the critical point condition becomes $\phi h - h\phi = 0$; but $\phi h + h\phi = 0$ and hence $h = 0$. Therefore ξ is a Killing vector field (see Section 6.2). ∎

Turning to the second variation we have the following result of Blair and Perrone [1995].

Theorem 10.10 *The index of $I(g)$ and the index of $-I(g)$ are both positive at each critical point.*

Proof. Referring to Perrone and the author [1995] for details, the second derivative of $I(g)$ evaluated at a critical point is given by the following succinct formula.

$$I''(0) = 2n \int_M \operatorname{tr}(\phi D \pounds_\xi D) \, dV.$$

Now as in the proof of Lemma 10.2, let $X_1, \dots, X_{2n}, \xi$ be a local ϕ-basis defined on a neighborhood $\mathcal{U}$ and again note that the first vector field X_1 may be any unit vector field on $\mathcal{U}$ orthogonal to ξ. Let f be a C^∞ function with compact support in $\mathcal{U}$ and define a path of metrics $g(t)$ as follows. Make no change in g outside $\mathcal{U}$ and within $\mathcal{U}$ change g only in the planes spanned by X_1 and X_2 by the matrix

$$\begin{pmatrix} 1 + t^2 f^2 & tf \\ tf & 1 \end{pmatrix}.$$

It is easy to check that $g(t) \in \mathcal{A}$ and clearly the only non-zero components of D are $D_{12} = D_{21} = f$. Denoting the first vector field in the ϕ-basis by X, calculation of $I''(0)$ in the above formula yields

$$I''(0) = -4n \int_M f^2 \eta([[\xi, X], X]) \, dV$$

where X may be regarded as any unit vector field on $\mathcal{U}$ belonging to the contact subbundle. Thus the proof reduces to finding unit vector fields belonging to the contact subbundle on $\mathcal{U}$ for which $\eta([[\xi, X], X])$ has either sign and again we refer to Blair and Perrone [1995] for details.   ∎

As a application of Theorem 10.10 we prove the following result (Blair and Perrone [1995]).

Theorem 10.11 *The functional $A(g)$ restricted to $\mathcal{A}$ cannot have a local minimum at any Sasakian metric.*

Proof. Suppose that g_0 is a Sasakian metric and a local minimum of $A(g)$ in $\mathcal{A}$. Then there exists a neighborhood $\mathcal{U}$ of $g_0 \in \mathcal{A}$ on which $A(g_0) \leq A(g)$. Since all associated metrics have the same volume element, $\int_M \tau_0 \, dV \leq \int_M \tau \, dV$ for every $g \in \mathcal{U}$. From Proposition 7.7,

$$2\tau - 4n^2 \leq \tau + \tau^*$$

with equality if and only if the metric is Sasakian. Thus we have

$$I(g_0) = \int_M 2\tau_0 - 4n^2 \, dV \leq \int_M 2\tau - 4n^2 \, dV \leq I(g)$$

for every $g \in \mathcal{U}$, that is g_0 is a local minimum for $I(g)$ contradicting Theorem 10.10.   ∎

10.3   The integral of $Ric(\xi)$

The integral, $L(g) = \int_M Ric(\xi) \, dV$ was studied in general dimension by the author in [1984] and independently by Chern and Hamilton in [1985] in the 3-dimensional case. Recall (Corollary 7.1) that

$$Ric(\xi) = 2n - \mathrm{tr} h^2;$$

thus K-contact metrics, when they occur, are maxima for $L(g)$ on $\mathcal{A}$. Moreover the critical point question for $L(g)$ is the same as that for $\int_M |h|^2 \, dV$ or $\int_M |T|^2 \, dV$ where $T(X, Y) = (\mathcal{L}_\xi g)(X, Y) = 2g(X, h\phi Y)$. It is the integral $E(g) = \int_M |T|^2 \, dV$ that was studied by Chern and Hamilton [1985] for

3-dimensional contact manifolds as a functional on $\mathcal{A}$ regarded as the set of CR-structures on M (there was an error in their calculation of the critical point condition as was pointed out by Tanno [1989]). The first result concerning $L(g)$ is the following.

Theorem 10.12 *Let M be a compact regular contact manifold and $\mathcal{A}$ the set of metrics associated to the contact form. Then $g \in \mathcal{A}$ is a critical point of $L(g) = \int_M Ric(\xi)\, dV$ if and only if g is K-contact.*

Proof. As with our other critical point problems, the first step is to compute $\frac{dL}{dt}$ at $t = 0$ for a path $g(t) \in \mathcal{A}$,

$$\left.\frac{dL}{dt}\right|_{t=0} = \int_M (-h^i{}_m h^{mk} - R^k{}_{rs}{}^i \xi^r \xi^s + 2h^{ik}) D_{ik}\, dV_g.$$

Thus if $g(0)$ is a critical point, Lemma 10.2 gives

$$R_{X\xi}\xi = -\phi^2 X - h^2 X + 2hX$$

as the critical point condition. Using the first formula of Proposition 7.1 this becomes

$$(\nabla_\xi h)X = -2\phi hX.$$

From this we see that the eigenvalues of h are constant along the integral curves of ξ and that for an eigenvalue $\lambda \neq 0$ and unit eigenvector X, $g(\nabla_\xi X, \phi X) = -1$.

If now M is a regular contact manifold, then M is a principal circle bundle with ξ tangent to the fibres; locally M is $\mathcal{U} \times S^1$. Since $h\phi + \phi h = 0$, we may choose an orthonormal ϕ-basis of eigenvectors of h at some point of $\mathcal{U} \times S^1$ say, X_{2i-1}, $X_{2i} = \phi X_{2i-1}$, ξ. Since the eigenvalues are constant along the fibre, we can continue this basis along the fibre with at worst a change of orientation of some of the eigenspaces when we return to the starting point. Thus if Y is a vector field along the fibre, we may write

$$Y = \sum_i (\alpha_{2i-1} X_{2i-1} + \beta_{2i} X_{2i}) + \gamma\xi$$

where the coefficients are periodic functions.

Now suppose that the critical point g is not a K-contact metric. Since ϕ and h anti-commute, we may assume that all the $\lambda_{2i-1}, i = 1, \ldots, n$ are non-negative. Also from $(\nabla_\xi h)X = -2\phi hX$ it is easy to see that if some of the λ_{2i-1} vanish, the zero eigenspace of h is parallel along ξ and hence we may choose the corresponding X_{2i-1} and X_{2i} parallel along a fibre. Again since M is regular we may choose a vector field Y on $\mathcal{U} \times S^1$ such that at least some

$\alpha_{2i-1} \neq 0$ for some $\lambda_{2i-1} \neq 0$ and Y is horizontal and projectable, that is $[\xi, Y] = 0$. Writing $Y = \sum_i (\alpha_{2i-1} X_{2i-1} + \beta_{2i} X_{2i})$ along a fibre we have, using $\nabla_X \xi = -\phi X - \phi h X$,

$$0 = [\xi, Y] = \nabla_\xi Y - \nabla_Y \xi$$

$$= \sum_i \left((\xi \alpha_{2i-1}) X_{2i-1} + \alpha_{2i-1} \nabla_\xi X_{2i-1} + (\xi \beta_{2i}) X_{2i} + \beta_{2i} \nabla_\xi X_{2i} \right.$$

$$\left. + \alpha_{2i-1} X_{2i} + \lambda_{2i-1} \alpha_{2i-1} X_{2i} - \beta_{2i} X_{2i-1} + \lambda_{2i-1} \beta_{2i} X_{2i-1} \right).$$

Taking components we have

$$0 = \xi \alpha_{2j-1} + \sum_i \alpha_{2i-1} g(\nabla_\xi X_{2i}, X_{2j}) + \lambda_{2j-1} \beta_{2j},$$

$$0 = \xi \beta_{2j} + \sum_i \beta_{2i} g(\nabla_\xi X_{2i}, X_{2j}) + \lambda_{2j-1} \alpha_{2j-1}.$$

Multiplying the first of these by β_{2j}, the second by α_{2j-1} and summing on j we have

$$\xi \left(\sum_j \alpha_{2j-1} \beta_{2j} \right) = - \sum_j \lambda_{2j-1} (\alpha_{2j-1}^2 + \beta_{2j}^2) \leq 0.$$

Thus $\sum_j \alpha_{2j-1} \beta_{2j}$ is a non-increasing, non-constant function along the integral curve, contradicting its periodicity.   ∎

One might conjecture Theorem 10.12 without the regularity, however we have the following counterexample: The standard contact metric structure on the tangent sphere bundle of a compact contact manifold of constant curvature -1 is a critical point of L but is not K-contact (the author [1991b]). We give this in the next theorem. Recall also the result of Tashiro, Theorem 9.3, that the standard contact metric structure on the tangent sphere bundle of a Riemannian manifold is K-contact if and only if the base manifold is of constant curvature $+1$.

Theorem 10.13 *Let $T_1 M$ be the tangent sphere bundle of a compact Riemannian manifold (M, G) and $\mathcal{A}$ the set of all Riemannian metrics associated to its standard contact structure. Then the standard associated metric is a critical point of the functional $L(g)$ if and only if (M, G) is of constant curvature $+1$ or -1.*

Proof. We have seen that the critical point condition of $L(g)$ is

$$R_{X\xi} \xi = -\phi^2 X - h^2 X + 2hX$$

or $\phi^2 + h^2 + l = 2h$. Applying this to a vertical vector $U \in T_t T_1 M$ as in the proof of Lemma 9.2 we find that (M, G) is locally symmetric by the Lemma of Cartan [1983, pp. 257–258] and that $L_t^2 X = X$ for any orthonormal pair $\{X, t\}$ on (M, G). Thus the only eigenvalues of $L_t : [t]^\perp \longrightarrow [t]^\perp$ are ± 1. Now M is irreducible; for if M had a locally Riemannian product structure, then choosing t tangent to one factor and X tangent to the other, we would have $\mathbf{R}_{Xt}t = 0$, contradicting the fact that the only eigenvalues of L_t are ± 1. Now the sectional curvature of an irreducible locally symmetric space does not change sign. Thus if for some t, L_t had both $+1$ and -1 as eigenvalues, there would be sectional curvatures equal to $+1$ and -1. Consequently only one eigenvalue can occur and hence (M, G) must be a space of constant curvature $+1$ or -1.

Conversely if (M, G) has constant curvature c, our expressions for h on horizontal vectors X orthogonal to ξ and vertical vectors U, namely

$$hX_t = -X_t + (\mathbf{R}_{\pi_* X} t)^H, \quad hU_t = U_t - (\mathbf{R}_{KU} t)^V$$

become $hX = (c - 1)X$ and $hU = (1 - c)U$. Similarly $lU = c^2 U$ and $lX = (4c - 3c^2)X$. Substituting these into the critical point condition we see that it is satisfied if and only if $c = \pm 1$. ∎

In [1991] S. Deng studied the second variation of the functional $L(g)$ or equivalently, of $E(g) = \int_M |T|^2 \, dV$.

Theorem 10.14 *Let $g \in \mathcal{A}$ be a critical point of $E(g)$, then g is a minimum.*

Proof. As with Theorem 10.10 we will sketch the proof and refer to Deng [1991] for details. The first step is a lengthy calculation of $E''(0)$ yielding

$$E''(0) = 2 \int_M |\mathcal{L}_\xi D|^2 \, dV \geq 0.$$

The second step is an auxiliary result that if $\mathcal{L}_\xi D = 0$, then $|T|^2$ is constant along the geodesics $g(t) = ge^{Dt}$ in $\mathcal{A}$. Now proceed as follows. If $\mathcal{L}_\xi D = 0$ for all D, then $|T|$ is constant and hence $E(g)$ is constant. So suppose that g is not a minimum and that there exist D such that $\mathcal{L}_\xi D \neq 0$. Let V be the subspace of $T_g \mathcal{A}$ consisting of those D with $\mathcal{L}_\xi D = 0$ and $V^\perp$ its orthogonal complement. Let $\mathcal{U}$ be a neighborhood of g in $exp_g V^\perp$ (cf. Ebin [1970, p. 37]) on which E exceeds $E(g)$. Let $\mathcal{W}$ be a neighborhood of $\mathcal{U} \subset \mathcal{A}$ formed by geodesic arcs in directions belonging to V. Since g is not a minimum, there exists $\bar{g} \in \mathcal{W}$ such that $E(\bar{g}) < E(g)$. Then the geodesic γ from $\bar{g}$ in direction $\bar{D} \in V$ meets $\mathcal{U}$ at say $\hat{g}$ and E is constant along γ. Therefore $E(\hat{g}) < E(g)$ contradicting $E(\hat{g}) > E(g)$ for $\hat{g} \in \mathcal{U}$. ∎

Little has been said about the existence of critical metrics and in general this is a difficult problem. Recall the notion of an almost regular contact manifold as given by C. B. Thomas [1976] (Section 3.4). Rukimbira [1995c] proved that every almost regular contact manifold admits a critical metric for $E(g)$ (equivalently $L(g)$).

Finally in [1992b] D. Perrone considered the integral

$$F(g) = \int_M \tau + \tau^* + 2nRic(\xi)\,dV$$

or setting $\tau_1 = \tau^* + 2nRic(\xi)$, $F(g) = \int_M \tau + \tau_1\,dV$ on $\mathcal{A}$. In dimension 3, letting $K(\mathcal{D})$ denote the sectional curvature of the contact subbundle $\mathcal{D}$, $\tau = 2K(\mathcal{D}) + 2Ric(\xi)$ and $\tau^* = 2K(\mathcal{D})$, so $\tau_1 = \tau$. In higher dimensions using Proposition 7.7,

$$\tau_1 = \tau + (2n-1)(Ric(\xi) - 2n) + \frac{1}{2}(|\nabla\phi|^2 - 4n)$$

and hence if the manifold is Sasakian, $\tau_1 = \tau$. The critical point condition for $F(g)$ is $\nabla_\xi h = 0$.

10.4   The Webster scalar curvature

In Theorem 6.6 we saw that a contact metric manifold is a strongly pseudo-convex CR-manifold if and only if $(\nabla_X\phi)Y = g(X+hX,Y)\xi - \eta(Y)(X+hX)$. On a strongly pseudo-convex CR-manifold Tanaka [1976] introduced a canonical connection. In [1989] Tanno introduced the corresponding connection on a contact metric manifold called the *generalized Tanaka connection*; it agrees with the connection of Tanaka when the contact metric manifold is a strongly pseudo-convex (integrable) CR-manifold. This connection, $^*\nabla$, is defined by

$$^*\nabla_X Y = \nabla_X Y + \eta(X)\phi Y - \eta(Y)\nabla_X\xi + (\nabla_X\eta)(Y)\xi$$
$$= \nabla_X Y + \eta(X)\phi Y + \eta(Y)(\phi X + \phi hX) + d\eta(X,Y)\xi + d\eta(hX,Y)\xi.$$

The torsion of this connection $^*T$ is then

$$^*T(X,Y) = \eta(X)\phi Y - \eta(Y)\phi X - \eta(Y)\nabla_X\xi + \eta(X)\nabla_Y\xi + 2d\eta(X,Y)\xi$$
$$= \eta(Y)\phi hX - \eta(X)\phi hY + 2g(X,\phi Y)\xi.$$

Tanno [1989] then proves the following proposition.

Proposition 10.1 *The generalized Tanaka connection $^*\nabla$ on a contact metric manifold is the unique linear connection such that*

$$^*\nabla\eta = 0, \quad ^*\nabla\xi = 0, \quad ^*\nabla g = 0,$$
$$(^*\nabla_X\phi)Y = (\nabla_X\phi)Y - g(X+hX,Y)\xi + \eta(Y)(X+hX),$$
$$^*T(\xi,\phi Y) = -\phi^*T(\xi,Y),$$
$$^*T(X,Y) = 2d\eta(X,Y)\xi, \quad \text{on } \mathcal{D}.$$

Tanno also computes the curvature of $^*\nabla$ and upon contraction obtains the *generalized Tanaka–Webster scalar curvature*

$$W_1 = \tau - Ric(\xi) + 4n.$$

This is 8 times the Webster scalar curvature as defined by Chern and Hamilton [1985] on 3-dimensional contact manifolds and as used by Perrone [1998] in Theorem 7.21.

We now prove a theorem of Chern and Hamilton [1985], an alternate proof of which was given by Perrone in [1990], and a theorem of Tanno [1989]. The proofs will be given simultaneously.

Theorem 10.15 (Chern–Hamilton) *Let M be a compact 3-dimensional contact manifold and $\mathcal{A}$ the set of metrics associated to the contact form. Then $g \in \mathcal{A}$ is a critical point of $E_1(g) = \int_M W_1 \, dV$ if and only if g is K-contact.*

Theorem 10.16 (Tanno) *Let M be a compact contact manifold and $\mathcal{A}$ the set of metrics associated to the contact form. Then $g \in \mathcal{A}$ is a critical point of $E_1(g) = \int_M W_1 \, dV$ if and only if*

$$(Q\phi - \phi Q) - (l\phi - \phi l) = 4\phi h - \eta \otimes \phi Q\xi + (\eta \circ Q\phi) \otimes \xi.$$

Proofs. Clearly it is enough to consider $\int_M \tau - Ric(\xi) \, dV$ and having computed the derivatives of each term separately in our previous theorems we have

$$\frac{d}{dt} \int_M \tau - Ric(\xi) \, dV \bigg|_{t=0} = \int_M (-\rho^{ki} + h^i{}_m h^{mk} + R^k{}_{rs}{}^i \xi^r \xi^s - 2h^{ik}) D_{ik} \, dV.$$

Thus by Lemma 10.2 we see that the critical point condition is

$$(Q\phi - \phi Q) - (l\phi - \phi l) = 4\phi h - \eta \otimes \phi Q\xi + (\eta \circ Q\phi) \otimes \xi.$$

Now in dimension 3, the Ricci operator determines the full curvature tensor, i.e.,

$$R_{XY}Z = (g(Y,Z)QX - g(X,Z)QY + g(QY,Z)X - g(QX,Z)Y)$$

$$- \frac{\tau}{2}(g(Y,Z)X - g(X,Z)Y).$$

Therefore the operator l is given by

$$lX = QX - \eta(X)Q\xi + g(Q\xi,\xi)X - g(QX,\xi)\xi - \frac{\tau}{2}(X - \eta(X)\xi)$$

from which

$$(l\phi - \phi l)X = (Q\phi - \phi Q)X + \eta(X)\phi Q\xi - g(Q\phi X, \xi)\xi.$$

Combining this and the critical point condition we have $4\phi h = 0$ and hence, since $h\xi = 0$, $h = 0$.   ∎

As we saw in the closing remarks of the last section, in dimension 3 $\tau = 2K(\mathcal{D}) + 2Ric(\xi)$ and $\tau^* = 2K(\mathcal{D})$. Using this, $W_1 = \tau - Ric(\xi) + 4n$ becomes

$$W_1 = \frac{1}{2}(\tau + \tau^* + 8).$$

Thus in dimension 3 the critical point problem for $E_1(g)$ is the same as the critical point problem for $I(g)$ in Theorem 10.9, suggesting that

$$W = \frac{1}{2}(\tau + \tau^* + 4n(n+1))$$

may be the proper generalization of the Webster scalar curvature.

There are however other generalizations of the Webster scalar curvature. Again in dimension 3, we may write W_1 as $\tau^* + Ric(\xi) + 4$ and we define a second generalization of the Webster scalar curvature by

$$W_2 = \tau^* + Ric(\xi) + 4n^2.$$

In general dimension the critical point condition of $E_2(g) = \int_M W_2\, dV$ is

$$(Q\phi - \phi Q) - (l\phi - \phi l) = -4(2n-1)\phi h - \eta \otimes \phi Q\xi + (\eta \circ Q\phi) \otimes \xi$$

(see Blair and Perrone [1992]). The generalization of the Webster scalar curvature $W = \frac{1}{2}(\tau + \tau^* + 4n(n+1))$ is the average of W_1 and W_2. Th. Koufogiorgos [1997b] has considered the difference of these as we briefly mention his results.

Proposition 10.2 *Let M^{2n+1} be a contact metric manifold with a strongly pseudo-convex CR-structure. Then*

$$(Q\phi - \phi Q) = (l\phi - \phi l) + 4(n-1)h\phi - \eta \otimes \phi Q\xi + (\eta \circ Q\phi) \otimes \xi.$$

Theorem 10.17 *Let M^{2n+1} be a contact manifold with $n > 1$ and $\mathcal{A}$ the set of associated metrics. If $g \in \mathcal{A}$ gives rise to a strongly pseudo-convex CR-structure, then g is a critical point of $\int_M W_1 - W_2\, dV$.*

The idea of the proof of this theorem is to find the critical point condition of $\int_M W_1 - W_2\, dV$ which is exactly the formula of Proposition 10.2.

Finally a word is in order on the constants depending on dimension which occur in the definitions of W_1, W_2 and W. Again recall the notion of a $\mathcal{D}$-homothetic deformation, namely a change of structure tensors of the form

$$\bar{\eta} = a\eta, \quad \bar{\xi} = \frac{1}{a}\xi, \quad \bar{\phi} = \phi, \quad \bar{g} = ag + a(a-1)\eta \otimes \eta$$

where a is a positive constant. By direct computation one shows that τ, $Ric(\xi)$ and τ^* transform in the following manner.

$$\bar{\tau} = \frac{1}{a}\tau + \frac{1-a}{a^2}Ric(\xi) - 2n\left(\frac{a-1}{a}\right)^2,$$

$$\overline{Ric(\xi)} = \frac{1}{a^2}(Ric(\xi) + 2n(a^2 - 1)),$$

$$\bar{\tau}^* = \frac{1}{a}\tau^* + \frac{a-1}{a^2}Ric(\xi) + 2n\left(2n\left(\frac{1-a}{a}\right) + \frac{1-a^2}{a^2}\right).$$

From these we see that $\bar{W}_1 = \frac{1}{a}W_1$, $\bar{W}_2 = \frac{1}{a}W_2$ and $\bar{W} = \frac{1}{a}W$.

10.5   A gauge invariant

Use has often been made of the notion of a $\mathcal{D}$-homothetic deformation as we have seen. The notion of a gauge transformation of a contact metric structure has not received as much attention. No doubt this is due in part to the computational complexity; nonetheless, comparing with the notion of a contact structure in the wider sense, the notion of a gauge transformation should be fundamental and deserving of attention.

Let $(M^{2n+1}, \phi, \xi, \eta, g)$ be a contact metric manifold and consider a gauge transformation of the contact structure, $\bar{\eta} = \sigma\eta$ where σ is a positive function on M^{2n+1}. Let $\zeta = \frac{1}{2\sigma}\phi\nabla\sigma$, $\nabla\sigma$ being the g-gradient of the function σ and let z be the covariant form of ζ. Now define new structure tensors $\bar{\xi}$, $\bar{\phi}$ and $\bar{g}$ by

$$\bar{\xi} = \frac{1}{\sigma}(\xi + \zeta), \quad \bar{\phi} = \phi + \frac{1}{2\sigma}\eta \otimes (\nabla\sigma - (\xi\sigma)\xi),$$

$$\bar{g} = \sigma(g - \eta \otimes z - z \otimes \eta) + \sigma(\sigma - 1 + |\zeta|^2)\eta \otimes \eta.$$

Then $(\bar{\phi}, \bar{\xi}, \bar{\eta}, \bar{g})$ is a contact metric structure and the change from (ϕ, ξ, η, g) to $(\bar{\phi}, \bar{\xi}, \bar{\eta}, \bar{g})$ is called a *gauge transformation* of the contact metric structure. When σ is a constant this is a $\mathcal{D}$-homothetic deformation. The notion of a gauge transformation in contact metric geometry is due to Sasaki [1965, p. PTSB-1-3] (see also Tanno [1989]).

Tanno [1989] computed the change of the generalized Tanaka–Webster scalar curvature under a gauge transformation. The result of his computation is

$$\bar{W}_1 = \frac{1}{\sigma}\left(W_1 - \frac{2(n+1)}{\sigma}(\Delta\sigma - \xi\xi\sigma) - \frac{(n+1)(n-2)}{\sigma^2}(|d\sigma|^2 - (\xi\sigma)^2)\right).$$

In the same paper Tanno generalizes an invariant of Jerison and Lee [1984] for strongly pseudo-convex CR-manifolds; specifically he considers

$$\kappa_{(\eta,g)} = \inf\left\{\int_M \left(\frac{4(n+1)}{n}(|df|^2 - (\xi f)^2) + W_1 f^2\right)dV_g\right\}$$

where the infimum is taken over all non-negative functions f such that $\int_M f^p \, dV_g = 1$, $p = 2 + \frac{2}{n}$ and proves that $\kappa_{(\eta,g)}$ is a gauge invariant.

Now for a compact contact manifold (M^{2n+1}, η) let $\mathcal{F}(p)$ be the set of all non-negative functions f such that $\int_M f^p \, dV_g = 1$ and define a functional $F_\eta : \mathcal{M} \times \mathcal{F}(p) \longrightarrow \mathbb{R}$ by

$$F_\eta(g, f) = \int_M \left(\frac{4(n+1)}{n} (|df|^2 - (\xi f)^2) + W_1 f^2 \right) dV_g.$$

Also define $F_{(\eta,g)} : \mathcal{F}(p) \longrightarrow \mathbb{R}$ by $F_{(\eta,g)}(f) = F_\eta(g, f)$. In [1989] Tanno studied these functionals in detail; here we mention only the following result.

Theorem 10.18 *Let $(M^{2n+1}, \phi, \xi, \eta, g)$ be a compact contact metric manifold with constant generalized Tanaka–Webster scalar curvature W_1. Then F_η is critical at the pair (g, f) with $f = (\mathrm{vol}(M, g))^{-1/p}$, if and only if*

$$(Q\phi - \phi Q) - (l\phi - \phi l) = 4\phi h - \eta \otimes \phi Q \xi + (\eta \circ Q \phi) \otimes \xi.$$

Note that this is also the critical point condition of $E_1(g) = \int_M W_1 \, dV$ in Theorem 10.16.

10.6   The Abbena metric as a critical point

In Section 1.1 we discussed Thurston's example of a compact symplectic manifold with no Kähler structure and a natural Riemannian metric on this manifold introduced by E. Abbena [1984]. She computed the curvature and with respect to the basis $\{e_i\}$ introduced in Section 1.1 and the Ricci operator Q is given by the matrix

$$\begin{pmatrix} -\frac{1}{2} & 0 & 0 & 0 \\ 0 & -\frac{1}{2} & 0 & 0 \\ 0 & 0 & \frac{1}{2} & 0 \\ 0 & 0 & 0 & 0 \end{pmatrix}.$$

From the expression for Q it is clear that (M, g) is not Einstein nor is $QJ = JQ$. Thus this metric is not a critical point for $A(g)$ considered as a functional on $\mathcal{M}_1$ or on $\mathcal{A}$ or for $K(g)$ on $\mathcal{A}$. Also $\tau = -1/2$ and $\tau^* = +1/2$ giving zero for the "total scalar curvature".

In [1996] Park and Oh discussed a functional for which the Abbena metric on the Thurston manifold is a critical point; their results are given in the following theorem. Recall that $\mathcal{M}_1$ denotes the space of Riemannian metrics with unit volume.

Theorem 10.19 *The Abbena metric on the Thurston manifold is a critical point of the functional*

$$\int_M \left(\frac{4}{3} \mathrm{tr} Q^3 - \tau \right) dV_g$$

on $\mathcal{M}_1$. The index of this functional and its negative are both positive at the Abbena metric on the Thurston manifold.

We remark that the Abbena metric on the Thurston manifold is a critical point for $K(g)$ in a different context. C. M. Wood [1995] showed that the Abbena metric is a critical point of $K(g) = -\frac{1}{2}\int_M |\nabla J|^2 \, dV$ defined with respect to variations through almost complex structures J which preserve g. For this problem the critical point condition is

$$[J, \nabla^*\nabla J] = 0,$$

where $\nabla^*\nabla J$ is the rough Laplacian of the metric in question.

11
Negative ξ-sectional Curvature

11.1 Special Directions in the contact subbundle

The purpose of this chapter is to introduce some special directions that belong to the contact subbundle of a contact metric manifold with negative sectional curvature for plane sections containing the characteristic vector field ξ or more generally when the operator h admits an eigenvalue greater than 1; these directions were introduced by the author in [1998]. We also discuss in this chapter some questions concerning Anosov and conformally Anosov flows. For simplicity we will often refer to the sectional curvature of plane sections containing the characteristic vector field ξ as ξ-*sectional curvature*.

We may regard the equation

$$\nabla_X \xi = -\phi X - \phi h X$$

of Lemma 6.2 as indicating how ξ or, by orthogonality, the contact subbundle, rotates as one moves around on the manifold. For example when $h = 0$, as we move in a direction X orthogonal to ξ, ξ is always "turning" or "falling" toward $-\phi X$. If $hX = \lambda X$, then $\nabla_X \xi = -(1 + \lambda)\phi X$ and again ξ is turning toward $-\phi X$ if $\lambda > -1$ or toward ϕX if $\lambda < -1$. Recall that we noted above that if λ is an eigenvalue of h with eigenvector X, then $-\lambda$ is also an eigenvalue with eigenvector ϕX.

Now one can ask if there can ever be directions, say Y orthogonal to ξ, along which ξ "falls" forward or backward in the direction of Y itself.

Theorem 11.1 *Let M^{2n+1} be a contact metric manifold. If the tensor field h admits an eigenvalue $\lambda > 1$ at a point P, then there exists a vector Y*

orthogonal to ξ at P such that $\nabla_Y \xi$ is collinear with Y. In particular if M^{2n+1} has negative ξ-sectional curvature such directions Y exist.

Proof. Let λ denote a positive eigenvalue of h and X a corresponding unit eigenvector. Then

$$\nabla_X \xi = -(1+\lambda)\phi X, \quad \nabla_{\phi X} \xi = (1-\lambda)X.$$

Now let $Y = aX + b\phi X$ with $a > 0, b > 0, a^2 + b^2 = 1$ and suppose that $\nabla_Y \xi = \alpha Y$. Then

$$\alpha(aX + b\phi X) = \nabla_Y \xi = -(1+\lambda)a\phi X + (1-\lambda)bX$$

from which $\alpha a = (1-\lambda)b$, $\alpha b = -(1+\lambda)a$ and hence

$$a^2 = \frac{\lambda-1}{2\lambda}, \quad b^2 = \frac{\lambda+1}{2\lambda}, \quad \alpha = -\sqrt{\lambda^2 - 1}.$$

Thus we see that directions along which $\nabla_Y \xi$ is collinear with Y exist whenever h admits an eigenvalue greater than 1. From the fact that $Ric(\xi) = 2n - \mathrm{tr}h^2$ (Corollary 7.1) we see that if M^{2n+1} has negative ξ-sectional curvature, at least one of the eigenvalues of h must exceed 1. ∎

     Note that when there exists a direction Y along which $\nabla_Y \xi$ is collinear with Y as above, there is also a second such direction, namely $Z = aX - b\phi X$. For Z we have $\nabla_Z \xi = -\alpha Z$; thus we think of ξ as falling backward as we move in the direction Y and falling forward as we move in the direction Z.
     Next note that

$$g(Y, Z) = a^2 - b^2 = -\frac{1}{\lambda}$$

and hence that such directions Y and Z are never orthogonal. Also if λ has multiplicity $m \geq 1$, then there are m-dimensional subbundles $\mathcal{Y}$ and $\mathcal{Z}$ such that $\nabla_Y \xi = \alpha Y$ for any $Y \in \mathcal{Y}$ and $\nabla_Z \xi = -\alpha Z$ for any $Z \in \mathcal{Z}$. We refer to directions along which the covariant derivative of ξ is collinear with the direction as *special directions*.

11.2   Anosov flows

The most notable example of a contact manifold for which the characteristic vector field is Anosov is the tangent sphere bundle of a negatively curved manifold; here the characteristic vector field is (twice) the geodesic flow as we saw in Section 9.2. In the case of the tangent sphere bundle of a surface, this is closely related to the structure on $SL(2, \mathbb{R})$ from both the topological and Anosov points of view. If we set $Z_2 = \{ \left(\begin{smallmatrix} 1 & 0 \\ 0 & 1 \end{smallmatrix} \right), \left(\begin{smallmatrix} -1 & 0 \\ 0 & -1 \end{smallmatrix} \right) \}$, then $PSL(2, \mathbb{R}) =$

$SL(2,\mathbb{R})/Z_2$ is homeomorphic to the tangent sphere bundle of the hyperbolic plane. Moreover the geodesic flow on a compact surface of constant negative curvature may be realized on $PSL(2,\mathbb{R})/\Gamma$ by $\left\{\left(\begin{smallmatrix} e^t & 0 \\ 0 & e^{-t} \end{smallmatrix}\right)\right\}$, where Γ is a discrete subgroup of $SL(2,\mathbb{R})$ for which $SL(2,\mathbb{R})/\Gamma$ is compact, (see e.g., Auslander, Green and Hahn [1963, pp. 26–27]). However from the Riemannian point of view these examples are quite different as we shall see. In fact in the case of the tangent sphere bundle of a negatively curved surface, the special directions never agree with the Anosov directions (Theorem 11.3).

Classically an Anosov flow is defined as follows (Anosov [1967]). Let M be a compact differentiable manifold, ξ a non-vanishing vector field and $\{\psi_t\}$ its 1-parameter group of (C^k) diffeomorphisms. $\{\psi_t\}$ is said to be an *Anosov flow* (or ξ to be *Anosov*) if there exist subbundles E^s and E^u which are invariant along the flow and such that $TM = E^s \oplus E^u \oplus \{\xi\}$ and there exists a Riemannian metric such that

$$|\psi_{t*}Y| \le ae^{-ct}|Y| \text{ for } t \ge 0 \text{ and } Y \in E_p^s,$$

$$|\psi_{t*}Y| \le ae^{ct}|Y| \text{ for } t \le 0 \text{ and } Y \in E_p^u$$

where a and c are positive constants independent of $p \in M$ and Y in E_p^s or E_p^u. The subbundles E^s and E^u are called the *stable* and *unstable* subbundles or the *contracting* and *expanding* subbundles. The subbundles E^s and E^u are integrable with C^k integral submanifolds, but in general the subbundles themselves are only continuous.

When M is compact the notion is independent of the Riemannian metric. If M is not compact the notion is metric dependent; in fact we will give an example of a metric on $\mathbb{R}^3$ with respect to which a coordinate field is Anosov, even though a coordinate field is clearly not Anosov with respect to the Euclidean metric on $\mathbb{R}^3$. Since we are dealing with Riemannian metrics associated to a contact structure, when we speak of the *characteristic vector field being Anosov*, we will mean that it is Anosov with respect to an associated metric of the contact structure.

The following properties of Anosov flows will be of importance here. The subbundles $E^s \oplus \{\xi\}$ and $E^u \oplus \{\xi\}$ are integrable (Anosov [1967, Theorem 8]). Let μ denote the measure induced on M by the Riemannian metric. Recall that a flow is *ergodic* if for every measurable set S, $\psi_t(S) = S$ for all t implies $\mu(S)\mu(M-S) = 0$. If on a compact manifold an Anosov flow admits an integral invariant, i.e., an invariant measure which is equivalent to the measure μ, in particular if it is volume preserving, then it is ergodic (Anosov [1967, Theorem 4]) and in turn by the Ergodic Theorem almost all orbits are dense (see e.g., Walters [1975, pp. 29–30]).

As an aside we note that on a compact manifold, an Anosov flow has a countable number of periodic orbits (Anosov [1967, Theorem 2]) and if the flow admits an integral invariant, then the set of periodic orbits is dense in M (Anosov [1967, Theorem 3]). This has an immediate implication for contact geometry. In Section 3.4 we discussed the conjecture of Weinstein that on a simply connected compact contact manifold, ξ must have a closed orbit, so in particular the Weinstein conjecture holds (without the simple connectivity) for a compact contact manifold on which ξ is Anosov.

Let us now turn our attention to the case of a 3-dimensional contact metric manifold and suppose that the ξ-sectional curvature is negative and that ξ is Anosov with respect to the associated metric. We then have both the special directions of Section 11.1 and the *Anosov directions*, i.e., the 1-dimensional stable and unstable bundles. One can then ask what happens if the special directions and the Anosov directions agree.

Theorem 11.2 *Let (M^3, η, g) be a 3-dimensional contact metric manifold with negative ξ-sectional curvature. If the characteristic vector field ξ generates an Anosov flow with respect to g and the special directions agree with the Anosov directions, then the contact metric structure satisfies $\nabla_\xi h = 0$.*

Proof. Suppose that Y is a local unit vector field such that $\nabla_Y \xi = \alpha Y$, $\alpha = -\sqrt{\lambda^2 - 1}$. Since ξ is Anosov and $\mathcal{Y}$ agrees with the stable Anosov subbundle, the subbundle $\mathcal{Y} \oplus \{\xi\}$ is integrable. Thus from $[\xi, Y] = \nabla_\xi Y - \alpha Y$, $\nabla_\xi Y$ belongs to $\mathcal{Y} \oplus \{\xi\}$; but $g(\nabla_\xi Y, \xi) = 0$ and Y is unit, so $g(\nabla_\xi Y, Y) = 0$. Thus $\nabla_\xi Y = 0$. Similarly $\nabla_\xi Z = 0$. Recall the operator l defined by $lX = R_{X\xi}\xi$ for any X; clearly l is a symmetric operator. Computing $R_{Y\xi}\xi$ and $R_{Z\xi}\xi$ we have

$$lY = -(\xi\alpha + \alpha^2)Y, \quad lZ = (\xi\alpha - \alpha^2)Z;$$

but Y and Z are not orthogonal, so $\xi\alpha = 0$ and $l|_\mathcal{D} = -\alpha^2 I|_\mathcal{D}$. Now compute $\nabla_\xi h$ acting on each vector of the h-eigenvector basis, $\{X, \phi X, \xi\}$, using the first equation of Proposition 7.1; this gives

$$(\nabla_\xi h)X = \phi(X - h^2 X - lX) = \phi(X - \lambda^2 X + \alpha^2 X) = 0$$

and similarly $(\nabla_\xi h)\phi X = 0$; $(\nabla_\xi h)\xi = 0$ is immediate. Therefore $\nabla_\xi h = 0$. ∎

If M is compact in Theorem 11.2, then M is a compact quotient of $\widetilde{SL}(2, \mathbb{R})$. This follows from a result of E. Ghys [1987] that if ξ is Anosov on a compact 3-dimensional contact manifold M and the Anosov directions are smooth, then M is a compact quotient of $\widetilde{SL}(2, \mathbb{R})$. This may also be proved directly using consequences of $\nabla_\xi h = 0$, see the author's paper [1998].

We now exhibit a family of contact metric structures on the Lie group $SL(2, \mathbb{R})$, show that the characteristic vector field is Anosov and show that the special directions agree with the Anosov directions.

On a 3-dimensional unimodular Lie group we have a Lie algebra structure of the form

$$[e_2, e_3] = c_1 e_1, \quad [e_3, e_1] = c_2 e_2, \quad [e_1, e_2] = c_3 e_3.$$

In [1976] J. Milnor gave a complete classification of 3-dimensional Lie groups and their left invariant metrics. If one c_i is non-zero, the dual 1-form ω_i is a contact form and e_i is the characteristic vector field. However for the Riemannian metric defined by $g(e_i, e_j) = \delta_{ij}$ at the identity and extended by left translation to be an associated metric for ω_i, we must have $c_i = 2$. For $SL(2, \mathbb{R})$ two of the c_i's are positive and one negative in the Milnor classification, so taking ω_1 as the contact form, we write the Lie algebra as

$$[e_2, e_3] = 2e_1, \quad [e_3, e_1] = (1 - \lambda)e_2, \quad [e_1, e_2] = (1 + \lambda)e_3 \qquad (*)$$

where $\lambda > 1$. Further by way of notation, set

$$SL(2, \mathbb{R}) = \left\{ \begin{pmatrix} x & y \\ u & v \end{pmatrix} \middle| xv - yu = 1 \right\}.$$

Now consider the matrices

$$\begin{pmatrix} \frac{1}{2}\sqrt{\lambda^2 - 1} & 0 \\ 0 & -\frac{1}{2}\sqrt{\lambda^2 - 1} \end{pmatrix},$$

$$\begin{pmatrix} 0 & -\sqrt{\frac{\lambda+1}{2}} \\ \sqrt{\frac{\lambda+1}{2}} & 0 \end{pmatrix}, \quad \begin{pmatrix} 0 & -\sqrt{\frac{\lambda-1}{2}} \\ -\sqrt{\frac{\lambda-1}{2}} & 0 \end{pmatrix}$$

in the Lie algebra $\mathfrak{sl}(2, \mathbb{R})$ which we regard as the tangent space of $SL(2, \mathbb{R})$ at the identity. Applying the differential of left translation by $\begin{pmatrix} x & y \\ u & v \end{pmatrix}$ to these matrices gives the vector fields

$$\zeta_1 = \frac{1}{2}\sqrt{\lambda^2 - 1} \left(x\frac{\partial}{\partial x} - y\frac{\partial}{\partial y} + u\frac{\partial}{\partial u} - v\frac{\partial}{\partial v} \right),$$

$$\zeta_2 = \sqrt{\frac{\lambda + 1}{2}} \left(y\frac{\partial}{\partial x} - x\frac{\partial}{\partial y} + v\frac{\partial}{\partial u} - u\frac{\partial}{\partial v} \right),$$

$$\zeta_3 = -\sqrt{\frac{\lambda - 1}{2}} \left(y\frac{\partial}{\partial x} + x\frac{\partial}{\partial y} + v\frac{\partial}{\partial u} + u\frac{\partial}{\partial v} \right)$$

whose Lie brackets satisfy $(*)$. Using these matrices again, define a left invariant metric g; then $\{\zeta_1, \zeta_2, \zeta_3\}$ is an orthonormal basis. The contact form ω_1 we denote by η and it is given by

$$\eta = \frac{2}{\sqrt{\lambda^2 - 1}}(v\,dx - y\,du).$$

The characteristic vector field ξ is ζ_1. g is an associated metric and ϕ as a skew-symmetric operator is given by $\phi\xi = 0$ and $\phi\zeta_2 = \zeta_3$. The symmetric operator h is given by $h\xi = 0$, $h\zeta_2 = \lambda\zeta_2$, $h\zeta_3 = -\lambda\zeta_3$. The special directions are

$$Y = \sqrt{\frac{\lambda-1}{2\lambda}}\zeta_2 + \sqrt{\frac{\lambda+1}{2\lambda}}\zeta_3 = -\frac{\sqrt{\lambda^2-1}}{\sqrt{\lambda}}\left(x\frac{\partial}{\partial y} + u\frac{\partial}{\partial v}\right)$$

and

$$Z = \sqrt{\frac{\lambda-1}{2\lambda}}\zeta_2 - \sqrt{\frac{\lambda+1}{2\lambda}}\zeta_3 = \frac{\sqrt{\lambda^2-1}}{\sqrt{\lambda}}\left(y\frac{\partial}{\partial x} + v\frac{\partial}{\partial u}\right).$$

The 1-parameter group of $\{\psi_t\}$ of ξ is given by

$$\psi_t\begin{pmatrix} x & y \\ u & v \end{pmatrix} = \begin{pmatrix} xe^{\frac{1}{2}\sqrt{\lambda^2-1}\,t} & ye^{-\frac{1}{2}\sqrt{\lambda^2-1}\,t} \\ ue^{\frac{1}{2}\sqrt{\lambda^2-1}\,t} & ve^{-\frac{1}{2}\sqrt{\lambda^2-1}\,t} \end{pmatrix}.$$

Then

$$\psi_{t*}Y = -\frac{\sqrt{\lambda^2-1}}{\sqrt{\lambda}}e^{-\frac{1}{2}\sqrt{\lambda^2-1}\,t}\left(x\frac{\partial}{\partial y} + u\frac{\partial}{\partial v}\right) = (e^{-\frac{1}{2}\sqrt{\lambda^2-1}\,t})Y,$$

$$\psi_{t*}Z = \frac{\sqrt{\lambda^2-1}}{\sqrt{\lambda}}e^{\frac{1}{2}\sqrt{\lambda^2-1}\,t}\left(y\frac{\partial}{\partial x} + v\frac{\partial}{\partial u}\right) = (e^{\frac{1}{2}\sqrt{\lambda^2-1}\,t})Z.$$

Thus the corresponding subbundles $\mathcal{Y}$ and $\mathcal{Z}$ are invariant under the flow. Finally since $\{\zeta_1, \zeta_2, \zeta_3\}$ is orthonormal, $|Y|^2 = \frac{\lambda-1}{2\lambda} + \frac{\lambda+1}{2\lambda} = 1$ and hence

$$|\psi_{t*}Y| = (e^{-\frac{1}{2}\sqrt{\lambda^2-1}\,t})|Y|;$$

similarly

$$|\psi_{t*}Z| = (e^{\frac{1}{2}\sqrt{\lambda^2-1}\,t})|Z|.$$

Thus ξ is an Anosov vector field and the special directions Y and Z agree with the Anosov directions.

In contrast to this, consider the contact metric structure on the tangent sphere bundle. We mentioned at the beginning of this section that the tangent sphere bundle of a surface is closely related to the structure on $SL(2, \mathbb{R})$ from both the topological and Anosov points of view. Comparing the above with the following theorem we see that from the Riemannian point of view these are quite different.

Theorem 11.3 *With respect to the standard contact metric structure on the tangent sphere bundle of a negatively curved surface, the characteristic vector field is Anosov, but the special directions never agree with the stable and unstable directions.*

Proof. We noted in Section 9.2 that for the standard contact metric structure on the tangent sphere bundle of a Riemannian manifold, the characteristic vector field ξ is (twice) the geodesic flow, which is an Anosov vector field when the base manifold is negatively curved (see e.g., Anosov [1967]). By Theorem 11.2 if the special directions of the contact metric structure agree with the Anosov directions, then $\nabla_\xi h = 0$. Now by Theorem 9.5 the standard contact metric structure of the tangent sphere bundle of any Riemannian manifold satisfies $\nabla_\xi h = 0$ if and only if the base manifold is of constant curvature 0 or +1.   ∎

Perrone [2000], utilizing the special directions Y and Z belonging to the contact subbundle on a contact metric manifold of negative ξ-sectional curvature, introduced another notion. Let $\mathcal{Y}$ and $\mathcal{Z}$ denote the subbundles generated by Y and Z. The special directions are said to be *Anosov-like* if the subbundles $\mathcal{Y} \oplus \{\xi\}$ and $\mathcal{Z} \oplus \{\xi\}$ are integrable. Perrone then proved that a contact metric 3-manifold admits Anosov-like special directions if and only if it satisfies $\nabla_\xi h = 0$ and has negative Ricci curvature in the direction ξ.

Three dimensional contact metric manifolds satisfying this condition were called 3-τ-*manifolds* by F. Gouli-Andreou and Ph. Xenos [1998]. The name comes from the equivalent condition $\nabla_\xi \tau = 0$ where here $\tau = \mathcal{L}_\xi g$; in particular τ and h are related by $\tau(X,Y) = 2g(h\phi X, Y)$. A 3-dimensional contact metric manifold on which the Ricci operator Q and ϕ commute satisfies $\nabla_\xi h = 0$ but not conversely.

We close this section with an example (the author [1996]) of a contact metric manifold satisfying $\nabla_\xi h = 0$ but with $Q\phi \neq \phi Q$. We include this example here as it is also an example of a metric on $\mathbb{R}^3$ with respect to which the coordinate field $\frac{\partial}{\partial z}$ is Anosov.

Consider the standard Darboux contact form $\eta = \frac{1}{2}(dz - ydx)$ and characteristic vector field $\xi = 2\frac{\partial}{\partial z}$. Let f be a smooth function of x and y bounded below by a positive constant c. Then the metric given by

$$g = \frac{1}{4}\begin{pmatrix} \frac{e^{zf}+(1+f^2)e^{-zf}-2}{f^2}+y^2 & \frac{e^{zf}-1}{f} & -y \\ \frac{e^{zf}-1}{f} & e^{zf} & 0 \\ -y & 0 & 1 \end{pmatrix}$$

is an associated metric. The tensor fields ϕ and h are given by

$$\phi = \begin{pmatrix} \frac{e^{zf}-1}{f} & e^{zf} & 0 \\ -\left(\frac{e^{zf}+(1+f^2)e^{-zf}-2}{f^2}\right) & -\frac{e^{zf}-1}{f} & 0 \\ y\left(\frac{e^{zf}-1}{f}\right) & ye^{zf} & 0 \end{pmatrix},$$

$$h = \begin{pmatrix} e^{zf} & fe^{zf} & 0 \\ -(\frac{fe^{zf}+(1+f^2)(-f)e^{-zf}}{f^2}) & -e^{zf} & 0 \\ ye^{zf} & yfe^{zf} & 0 \end{pmatrix}.$$

By direct computation $\nabla_\xi h = 0$. Also $2\lambda^2 = \mathrm{tr}h^2 = 2(1 + f^2)$ and hence the positive eigenfunction of h is $\lambda = \sqrt{1 + f^2} > 1$. Now on a 3-dimensional contact metric manifold satisfying $Q\phi = \phi Q$, the eigenfunction λ is a constant (Blair, Koufogiorgos, and Sharma [1990]). Thus if f is not constant this structure on $\mathbb{R}^3$ satisfies $\nabla_\xi h = 0$ but not $Q\phi = \phi Q$.

For this structure the special directions discussed in Section 11.1 are given by

$$Y = f\frac{\partial}{\partial x} - \frac{\partial}{\partial y} + yf\frac{\partial}{\partial z}, \quad Z = \frac{\partial}{\partial y}.$$

To check that $\xi = 2\frac{\partial}{\partial z}$ is Anosov with respect to g, consider for simplicity just $\frac{\partial}{\partial z}$; its flow ψ_t maps a point $P_0(x, y, z)$ to the point $P(x, y, z + t)$. Now recalling that the function f was chosen to be bounded below by a positive constant c, we have for $t \leq 0$,

$$\left|\psi_{t*}\frac{\partial}{\partial y}(P_0)\right| = \left|\frac{\partial}{\partial y}(P)\right| = \frac{1}{2}e^{\frac{(z+t)f}{2}} = e^{\frac{tf}{2}}\left|\frac{\partial}{\partial y}(P_0)\right| \leq e^{\frac{ct}{2}}\left|\frac{\partial}{\partial y}(P_0)\right|.$$

Similarly for $t \geq 0$,

$$\left|\psi_{t*}\left(f\frac{\partial}{\partial x} - \frac{\partial}{\partial y} + yf\frac{\partial}{\partial z}\right)(P_0)\right| \leq e^{\frac{-ct}{2}}\left|\left(f\frac{\partial}{\partial x} - \frac{\partial}{\partial y} + yf\frac{\partial}{\partial z}\right)(P_0)\right|.$$

Thus $\frac{\partial}{\partial z}$, equivalently ξ, is Anosov with respect to this metric; Y determines the stable subbundle and Z the unstable subbundle.

11.3   Conformally Anosov flows

Recently there has been interest in a notion more general than an Anosov flow, namely, a conformally Anosov flow. We will show that if a compact contact metric 3-manifold has negative ξ-sectional curvature, then the characteristic vector field is conformally Anosov.

Mitsumatsu [1995] and Eliashberg and Thurston [1998] introduced a generalization of Anosov flows as follows. A flow ψ_t or its corresponding vector field is said to be *conformally Anosov*, Eliashberg and Thurston [1998] (*projectively Anosov* Mitsumatsu [1995]) if there is a continuous Riemannian metric and a continuous, invariant splitting, $TM = E^s \oplus E^u \oplus \{\xi\}$ as in the Anosov case such that for $Z \in E^u$ and $Y \in E^s$,

$$\frac{|\psi_{t*}Z|}{|\psi_{t*}Y|} \geq e^{ct}\frac{|Z|}{|Y|}$$

for some constant $c > 0$ and all $t \geq 0$.

Now a contact structure η on a 3-dimensional contact manifold M^3 determines an orientation on M^3. This is true in dimension 3 even for a contact structure in the wider sense, since the sign of $\eta \wedge d\eta$ is independent of the choice of local contact form η.

The main result for our purpose from Mitsumatsu [1995, p. 1418] and Eliashberg and Thurston [1998, pp. 26–27] is the following.

Theorem 11.4 *If two contact structures (in the wider sense) on a compact 3-dimensional contact manifold M^3 induce opposition orientations, then the vector field directing the intersection of the two contact subbundles is a conformally Anosov flow. Conversely given a conformally Anosov flow on M^3, there exist two contact structures giving opposite orientations on M^3 whose contact subbundles intersect tangent to the flow.*

We now show that certain curvature hypotheses on a compact contact metric 3-manifold imply that the characteristic vector field ξ is conformally Anosov; in particular negative ξ-sectional curvature is such a hypothesis (the author [2000]). A variation of this result appears in Blair and Perrone [1998].

Theorem 11.5 *Let $(M^3, \phi, \xi, \eta, g)$ be a compact 3-dimensional contact metric manifold with nowhere vanishing h and $\{e_1, e_2(= \phi e_1), \xi\}$ an orthonormal eigenvector basis of h with $he_1 = \lambda e_1$ and λ the positive eigenvalue. If $K(\xi, e_1) < (1+\lambda)^2$ and $K(\xi, e_2) < (1-\lambda)^2$, then ξ is conformally Anosov. In particular, if the ξ-sectional curvature is negative, ξ is conformally Anosov.*

Proof. Let ω^1, ω^2 be the dual 1-forms of e_1 and e_2. Since the three eigenvalues of h are everywhere distinct, the corresponding line fields are global. By the orientability, one may choose local bases directing the line fields that either agree or have two directions reversed in the overlap of coordinate neighborhoods. Thus in computing $\omega^1 \wedge d\omega^1$ and $\omega^2 \wedge d\omega^2$ on such a basis we may regard the computations as global. By straightforward computation we have

$$\omega^1 \wedge d\omega^1(e_1, e_2, \xi) = \frac{\lambda - 1}{6} - \frac{1}{6}g(\nabla_\xi e_1, e_2),$$

$$\omega^2 \wedge d\omega^2(e_1, e_2, \xi) = -\frac{\lambda + 1}{6} - \frac{1}{6}g(\nabla_\xi e_1, e_2).$$

On the other hand, applying the first equation of Proposition 7.1 to e_1 and e_2 we obtain

$$K(\xi, e_1) = 1 - \lambda^2 - 2\lambda g(\nabla_\xi e_1, e_2),$$

$$K(\xi, e_2) = 1 - \lambda^2 + 2\lambda g(\nabla_\xi e_1, e_2).$$

Combining these equations we have

$$\omega^1 \wedge d\omega^1(e_1, e_2, \xi) = \frac{1}{12\lambda}((\lambda - 1)^2 - K(\xi, e_2))$$

and
$$\omega^2 \wedge d\omega^2(e_1, e_2, \xi) = \frac{1}{12\lambda}(-(\lambda+1)^2 + K(\xi, e_1)).$$

The hypotheses now imply that $\omega^1 \wedge d\omega^1(e_1, e_2, \xi) > 0$ and $\omega^2 \wedge d\omega^2(e_1, e_2, \xi) < 0$. Therefore ξ is conformally Anosov by the result of Mitsumatsu and Eliashberg–Thurston. ∎

We remark that the curvature of the standard contact metric structure on the tangent sphere bundle of a surface of negative curvature satisfies the curvature hypotheses in the first statement of the theorem. In particular the standard contact metric structure on the tangent sphere bundle of a surface of constant curvature -1 has sectional curvature -7 for horizontal plane sections and sectional curvature $+1$ for plane sections spanned by ξ and the vertical direction; the positive eigenvalue λ is $+2$ with vertical vectors as eigenvectors (see Section 9.2).

We also remark that our argument uses two contact structures to study a third, an interesting idea in view of the result of Gonzalo [1987] (Section 3.2) that a 3-dimensional compact orientable manifold admits three independent contact structures.

It is immediate that the characteristic vector field ξ of a contact structure can never be Anosov or conformally Anosov with respect to a Sasakian metric. This is a consequence of the fact that on a Sasakian manifold ξ is a Killing vector field; thus its flow is metric preserving and therefore cannot satisfy the exponential growth behavior in the definition of a classical or conformal Anosov flow. It is possible however for a vector field to belong to the contact subbundle of a Sasakian structure and be conformally Anosov with respect to this metric. We present an example of this which has the additional feature of the example that the characteristic vector field is invariant along the flow (again see Blair and Perrone [1998]). We begin with the following lemma.

Lemma 11.1 *On a 3-dimensional contact manifold in terms of local Darboux coordinates (x, y, z) $(\eta = \frac{1}{2}(dz - y dx))$ any associated metric is of the form*

$$g = \frac{1}{4}\begin{pmatrix} a & b & -y \\ b & c & 0 \\ -y & 0 & 1 \end{pmatrix}$$

with $ac - b^2 - cy^2 = 1$; the metric is Sasakian if and only if the functions a, b and c are independent of z.

Proof. The form of the last row and column follow from the requirement that $\eta(X) = g(X, \xi)$. In dimension 3 the remaining requirements reduce to the determinant of the matrix (without the $\frac{1}{4}$) being 1. Also in dimension

3 the Sasakian condition is equivalent to the contact metric structure being K-contact. Thus evaluating the Lie derivative, $\mathcal{L}_\xi g$, on the coordinate vector fields we see that $\xi = 2\frac{\partial}{\partial z}$ is Killing if and only if the functions a, b and c are independent of z. ∎

To construct the example, consider $\mathbf{R}^3_+ = \{(x, y, z) | y > 0\}$ with the standard Darboux contact form $\eta = \frac{1}{2}(dz - ydx)$. The characteristic vector field is $2\frac{\partial}{\partial z}$ and the Riemannian metric given by the following matrix is Sasakian by Lemma 11.1

$$g = \frac{1}{4} \begin{pmatrix} e^{2y} & \sqrt{e^{2y} - y^2 - 1} & -y \\ \sqrt{e^{2y} - y^2 - 1} & 1 & 0 \\ -y & 0 & 1 \end{pmatrix}.$$

The vector field $\frac{\partial}{\partial y}$ is conformally Anosov with respect to this metric. To see this we observe that the subbundles determined by $\frac{\partial}{\partial z} = \frac{1}{2}\xi$ and $\frac{\partial}{\partial x}$ correspond to E^s and E^u respectively. The flow simply maps a point $P_0(x, y, z)$ to the point $P(x, y + t, z)$ and we have easily that

$$\frac{|\frac{\partial}{\partial x}(P)|}{|\frac{\partial}{\partial z}(P)|} = e^t \frac{|\frac{\partial}{\partial x}(P_0)|}{|\frac{\partial}{\partial z}(P_0)|}.$$

Clearly this flow satisfies $\psi_{t*}\xi = \xi$.

For a discussion of conformally Anosov flows on 3-dimensional homogeneous contact metric manifolds including an Anosov flow belonging to the contact subbundle on the Lie group $E(1, 1)$, see Blair and Perrone [1998].

12
Complex Contact Manifolds

12.1   Complex contact manifolds and associated metrics

While the study of complex contact manifolds is almost as old as the modern theory of real contact manifolds, the subject has received much less attention and as many examples are now appearing in the literature, especially twistor spaces over quaternionic Kähler manifolds (e.g., LeBrun [1991], [1995], Moroianu and Semmelmann [1994], Salamon [1982], Ye [1994]), the time is ripe for another look at the subject. As an indication of this interest we note, for example, the following result of Moroianu and Semmelmann [1994] that on a compact spin Kähler manifold M of positive scalar curvature and complex dimension $4l + 3$, the following are equivalent: (i) M is a Kähler–Einstein manifold admitting a complex contact structure, (ii) M is the twistor space of a quaternionic Kähler manifold of positive scalar curvature, (iii) M admits Kählerian Killing spinors. LeBrun [1995] proves that a complex contact manifold of positive first Chern class, i.e., a Fano contact manifold, is a twistor space if and only if it admits a Kähler–Einstein metric and conjectures that every Fano contact manifold is a twistor space.

Specifically the notion of a complex contact manifold stems from the late 1950s and early 1960s with the papers of Kobayashi [1959] and Boothby [1961], [1962]; this is just shortly after the Boothby–Wang fibration in real contact geometry. Then in [1965], J. A. Wolf studied homogeneous complex contact manifolds and their relation to quaternionic symmetric spaces. In the 1970s and early 1980s there was a development of the Riemannian theory of complex contact manifolds by Ishihara and Konishi [1979], [1980], [1982]. In this development however, the notion of normality seems too strong since it precludes the complex Heisenberg group as one of the canonical examples, although it does include complex projective spaces of odd complex dimension as one would expect. In the real case both the Heisenberg group and the odd-dimensional spheres have natural Sasakian (normal contact metric) structures. As a subject the Riemannian geometry of complex contact manifolds is just in its infancy. For brevity, in this chapter we will give more of a survey of results and refer to the references for proofs.

A *complex contact manifold* is a complex manifold of odd complex dimension $2n+1$ together with an open covering $\{\mathcal{O}_\alpha\}$ by coordinate neighborhoods such that:

1. On each $\mathcal{O}_\alpha$ there is a holomorphic 1-form θ_α such that

$$\theta_\alpha \wedge (d\theta_\alpha)^n \neq 0;$$

2. On $\mathcal{O}_\alpha \cap \mathcal{O}_\beta \neq \emptyset$ there is a non-vanishing holomorphic function $f_{\alpha\beta}$ such that $\theta_\alpha = f_{\alpha\beta}\theta_\beta$.

The subspaces $\{X \in T_m\mathcal{O}_\alpha : \theta_\alpha(X) = 0\}$ define a non-integrable holomorphic subbundle $\mathcal{H}$ of complex dimension $2n$ called the *complex contact subbundle* or *horizontal subbundle*. The quotient $L = TM/\mathcal{H}$ is a complex line bundle over M. Kobayashi [1959] proved that $c_1(M) = (n + 1)c_1(L)$ and hence for a compact complex contact manifold, a complex contact structure is given by a global 1-form if and only if its first Chern class vanishes (see also Boothby [1961], [1962]). It is for this reason that our definition of complex contact structure is analogous to that of a contact structure in the wider sense. Even for the most canonical example of a complex contact manifold, $\mathbb{C}P^{2n+1}$, the structure is not given by a global form. Since a holomorphic p-form on a compact Kähler manifold is closed (see e.g., Goldberg [1962, p. 177]), no compact Kähler manifold has a complex contact structure given by a global contact form. Moreover, Ye [1994] showed that a compact Kähler manifold with vanishing first Chern class has no complex contact structure. There are however interesting examples of complex contact manifolds with global complex contact forms as we shall see; these are called *strict complex contact manifolds* (Foreman [2000a], [2000b]).

We will not need a complex Darboux theorem in our development here, but a complex version of the real Darboux theorem is possible and such a result is discussed briefly by LeBrun [1995, p. 423].

If $(M, \{\theta_\alpha\})$ is a complex contact manifold, the transition functions $f_{\alpha\beta}$ define a holomorphic line bundle over M, viz., L^{-1}. Using local sections of this bundle we define complex-valued 1-forms $\{\pi_\alpha\}$ such that each π_α is a non-vanishing, complex-valued function multiple of θ_α and on $\mathcal{O}_\alpha \cap \mathcal{O}_\beta \neq \emptyset$,

$$\pi_\alpha = h_{\alpha\beta}\pi_\beta, \quad h_{\alpha\beta} : \mathcal{O}_\alpha \cap \mathcal{O}_\beta \longrightarrow S^1.$$

The $h_{\alpha\beta}$ are then the transition functions of a circle bundle P over M (see Ishihara and Konishi [1982]) and on $\mathcal{O}_\alpha$,

$$\pi_\alpha \wedge (d\pi_\alpha)^n \wedge \bar{\pi}_\alpha \wedge (d\bar{\pi}_\alpha)^n \neq 0.$$

Writing $\pi_\alpha = u_\alpha - iv_\alpha$, $v_\alpha = u_\alpha \circ J$ since w_α is holomorphic. Moreover u_α and v_α transform naturally with respect to S^1, namely if $h_{\alpha\beta} = a + ib$,

$$u_\beta = au_\alpha - bv_\alpha, \quad v_\beta = bu_\alpha + av_\alpha, \quad a^2 + b^2 = 1.$$

The set $\{\pi_\alpha\}$ is called a *normalized contact structure* with respect to $\{\theta_\alpha\}$, Foreman [1996].

For simplicity we will often omit the subscripts on the local tensor fields. Define a local section U of TM, i.e., a section of $T\mathcal{O}$, by $du(U, X) = 0$ for every $X \in \mathcal{H}$, $u(U) = 1$ and $v(U) = 0$. Such local sections then define a global subbundle $\mathcal{V}$ by $\mathcal{V}|_{\mathcal{O}} = \mathrm{Span}\{U, JU\}$. We now have $TM = \mathcal{H} \oplus \mathcal{V}$ and we denote the projection map to $\mathcal{H}$ by

$$p : TM \longrightarrow \mathcal{H}.$$

We assume throughout that $\mathcal{V}$ is integrable and we call $\mathcal{V}$ the *vertical subbundle* or *characteristic subbundle*.

On the other hand if M is a complex manifold with almost complex structure J, Hermitian metric g and open covering by coordinate neighborhoods $\{\mathcal{O}_\alpha\}$, M is called a *complex almost contact metric manifold* if it satisfies the following two conditions:

1) On each $\mathcal{O}_\alpha$ there exist 1-forms u_α and $v_\alpha = u_\alpha \circ J$ with orthogonal dual vector fields U_α and $V_\alpha = -JU_\alpha$ and (1,1) tensor fields G_α and $H_\alpha = G_\alpha J$ such that

$$G_\alpha^2 = H_\alpha^2 = -I + u_\alpha \otimes U_\alpha + v_\alpha \otimes V_\alpha,$$

$$G_\alpha J = -JG_\alpha, \quad G_\alpha U = 0, \quad g(X, G_\alpha Y) = -g(G_\alpha X, Y),$$

2) On $\mathcal{O}_\alpha \cap \mathcal{O}_\beta \neq \emptyset$,

$$u_\beta = au_\alpha - bv_\alpha, \quad v_\beta = bu_\alpha + av_\alpha,$$

$$G_\beta = aG_\alpha - bH_\alpha, \quad H_\beta = bG_\alpha + aH_\alpha$$

where a and b are functions with $a^2 + b^2 = 1$.

Returning to the local forms $\pi_\alpha = u_\alpha - iv_\alpha$ on a complex contact manifold, a Hermitian metric g is called an *associated metric* if there are local fields of endomorphism G_α such that the tensor fields $G_\alpha, H_\alpha = G_\alpha J, U_\alpha, V_\alpha = -JU_\alpha, u_\alpha, v_\alpha, g$ form a complex almost contact metric structure satisfying

$$g(X, G_\alpha Y) = du_\alpha(X, Y), \quad g(X, H_\alpha Y) = dv_\alpha(X, Y), \quad X, Y \in \mathcal{H}.$$

As a consequence we also have $U_\beta = aU_\alpha - bV_\alpha$, $V_\beta = bU_\alpha + aV_\alpha$ and hence $U_\alpha + iV_\alpha = h_{\alpha\beta}^{-1}(U_\beta + iV_\beta)$. In particular $\{h_{\alpha\beta}^{-1}\}$ are the transition functions of the bundle $\mathcal{V}$.

Kobayashi [1959] also showed that the structural group of a complex contact manifold is reducible to $(Sp(n) \cdot U(1)) \times U(1)$. Such a reduction is equivalent to a complex almost contact metric structure, cf. Shibuya [1978]. Shibuya's approach is the complex analogue of that of Hatakeyama [1962] which we utilized in Section 4.2. A formulation in terms of global tensor fields similar to the fundamental 4-form of a quaternionic Kähler manifold (see e.g., Ishihara [1974]) was given by Blair, Ishihara, and Ludden [1978].

Ishihara and Konishi [1982] (see also Foreman [1996]) prove that a complex contact manifold admits a complex almost contact metric structure for which the local contact form θ is of the form $u-iv$ to within a non-vanishing complex-valued function multiple and the local tensor fields G and H are related to du and dv by

$$du(X,Y) = g(X,GY)+(\sigma\wedge v)(X,Y), \quad dv(X,Y) = g(X,HY)-(\sigma\wedge u)(X,Y)$$

where $\sigma(X) = g(\nabla_X U, V)$. We refer to a complex contact manifold with a complex almost contact metric structure satisfying these conditions as a *complex contact metric manifold*. For a given normalized contact structure we denote by $\mathcal{A}$, as in the real case, the space of associated metrics. As a matter of notation we define local 2-forms $\hat{G}$ and $\hat{H}$ by $\hat{G}(X,Y) = g(X,GY)$ and $\hat{H}(X,Y) = g(X,HY)$.

Bearing in mind that the local contact form θ is $u - iv$ to within a non-vanishing complex-valued function multiple and since in the overlap $\mathcal{O}_\alpha \cap \mathcal{O}_\beta$, $u_\beta = au_\alpha - bv_\alpha$, $v_\beta = bu_\alpha + av_\alpha$, we can define an *integral submanifold* as a submanifold whose tangent spaces belong to the complex contact subbundle, i.e., $u(X) = v(X) = 0$ or equivalently $\theta(X) = 0$. If the submanifold is itself a complex submanifold, we call it a *holomorphic integral submanifold*. When the holomorphic integral submanifold has complex dimension 1, it is called a *holomorphic Legendre curve*. Recall that in real contact geometry the maximum dimension of an integral submanifold of a contact manifold of dimension $2n+1$ is only n. Similarly since U and V are normal, from the above equations we see that for any integral submanifold, GX and HX are normal for any tangent vector X. Thus an integral submanifold of a complex contact manifold of complex dimension $2n+1$ has real dimension at most $2n$.

We have seen in the earlier chapters that for a contact metric structure (ϕ, ξ, η, g), the tensor field h defined by $h = \frac{1}{2}\mathcal{L}_\xi\phi$ plays a fundamental role. Now for a complex contact metric structure we define local tensor fields by

$$h_U = \frac{1}{2}\text{sym}(\mathcal{L}_U G) \circ p, \quad h_V = \frac{1}{2}\text{sym}(\mathcal{L}_V H) \circ p$$

where sym denotes the symmetric part. h_U anti-commutes with G, h_V anti-commutes with H and

$$\nabla_X U = -GX - Gh_U X + \sigma(X)V, \quad \nabla_X V = -HX - Hh_V X - \sigma(X)U.$$

From these equations one readily sees that the integral surfaces of V are totally geodesic submanifolds. Just as the meaning of the vanishing of h in the real case is that the metric was invariant under the action of ξ, i.e., ξ is Killing, an associated metric g is projectable with respect to the foliation induced by the integrable subbundle V if and only if h_U and h_V vanish.

12.2 Examples of complex contact manifolds

12.2.1 Complex Heisenberg group

We have seen that $\mathbb{R}^3$ with the Darboux form $\eta = \frac{1}{2}(dz - ydx)$ as its contact structure and Sasakian metric $g = \frac{1}{4}(dx^2 + dy^2) + \eta \otimes \eta$ is a standard example of a contact metric manifold. Identifying $\mathbb{R}^3$ with the Heisenberg group

$$H_{\mathbb{R}} = \left\{ \begin{pmatrix} 1 & y & z \\ 0 & 1 & x \\ 0 & 0 & 1 \end{pmatrix} \middle| x, y, z \in \mathbb{R} \right\} \simeq \mathbb{R}^3,$$

left translation preserves η, and g is a left invariant metric on $H_{\mathbb{R}}$ (see Example 4.5.1).

The complex Heisenberg group is the closed subgroup $H_{\mathbb{C}}$ of $GL(3, \mathbb{C})$ given by

$$H_{\mathbb{C}} = \left\{ \begin{pmatrix} 1 & z_2 & z_3 \\ 0 & 1 & z_1 \\ 0 & 0 & 1 \end{pmatrix} \middle| z_1, z_2, z_3 \in \mathbb{C} \right\} \simeq \mathbb{C}^3.$$

If L_B denotes left translation by B, $L_B^* dz_1 = dz_1$, $L_B^* dz_2 = dz_2$, $L_B^*(dz_3 - z_2 dz_1) = dz_3 - z_2 dz_1$. The vector fields $\frac{\partial}{\partial z_1} + z_2 \frac{\partial}{\partial z_3}$, $\frac{\partial}{\partial z_2}$, $\frac{\partial}{\partial z_3}$ are dual to the 1-forms dz_1, dz_2, $dz_3 - z_2 dz_1$ and are left invariant vector fields. Moreover relative to the coordinates $(z_1, z_2, z_3, \bar{z}_1, \bar{z}_2, \bar{z}_3)$ the Hermitian metric (Jayne [1992, p. 234])

$$g = \frac{1}{8} \left(\begin{array}{c|c} O & \begin{matrix} 1+|z_2|^2 & 0 & -z_2 \\ 0 & 1 & 0 \\ -\bar{z}_2 & 0 & 1 \end{matrix} \\ \hline \begin{matrix} 1+|z_2|^2 & 0 & -\bar{z}_2 \\ 0 & 1 & 0 \\ -z_2 & 0 & 1 \end{matrix} & O \end{array} \right)$$

is a left invariant metric on $H_{\mathbb{C}}$, but it is not a Kähler metric. The form $\theta = \frac{1}{2}(dz_3 - z_2 dz_1)$ is a complex contact structure on $H_{\mathbb{C}}$ and in our view $(H_{\mathbb{C}}, \theta, g)$ plays the role in the geometry of complex contact manifolds, that $\mathbb{R}^3$ with its standard Sasakian structure does in the geometry of real contact manifolds.

As we have seen, a complex contact manifold admits a complex almost contact structure. Here $H_{\mathbb{C}} \simeq \mathbb{C}^3$ and θ is global, so the structure tensors may be taken globally. With J denoting the standard almost complex structure on $\mathbb{C}^3$, $J\frac{\partial}{\partial x_i} = \frac{\partial}{\partial y_i}$, we may give a complex almost contact structure to $H_{\mathbb{C}}$ as follows. Since θ is holomorphic, setting $\theta = u - iv$, $v = u \circ J$. Also set $4\frac{\partial}{\partial z_3} = U + iV$; then $u(X) = g(U, X)$ and $v(X) = g(V, X)$. Finally in complex coordinates G and H were given by Jayne [1992, p. 235].

$$
G = \left(
\begin{array}{c|c}
O & \begin{matrix} 0 & 1 & 0 \\ -1 & 0 & 0 \\ 0 & z_2 & 0 \end{matrix} \\
\hline
\begin{matrix} 0 & 1 & 0 \\ -1 & 0 & 0 \\ 0 & \bar{z}_2 & 0 \end{matrix} & O
\end{array}
\right),
\quad
H = \left(
\begin{array}{c|c}
O & \begin{matrix} 0 & -i & 0 \\ i & 0 & 0 \\ 0 & -iz_2 & 0 \end{matrix} \\
\hline
\begin{matrix} 0 & i & 0 \\ -i & 0 & 0 \\ 0 & i\bar{z}_2 & 0 \end{matrix} & O
\end{array}
\right).
$$

Note that the matrix gG is that of the real part $d\theta$ and the matrix gH is that of the imaginary part of $d\theta$. For this structure the 1-form σ vanishes.

Now let

$$
\Gamma = \left\{ \begin{pmatrix} 1 & \gamma_2 & \gamma_3 \\ 0 & 1 & \gamma_1 \\ 0 & 0 & 1 \end{pmatrix} \Big| \gamma_k = m_k + in_k, \ m_k, n_k \in \mathbb{Z} \right\};
$$

Γ is a subgroup of $H_{\mathbb{C}} \simeq \mathbb{C}^3$. The 1-form $dz_3 - z_2 dz_1$ is invariant under the action of Γ and hence the quotient $H_{\mathbb{C}}/\Gamma$ is a compact complex contact manifold with a global complex contact form. $H_{\mathbb{C}}/\Gamma$ is known as the *Iwasawa manifold*. The Iwasawa manifold has no Kähler structure, but it does have an indefinite Kähler structure and it has symplectic forms, see Fernández and Gray [1986].

12.2.2   Odd-dimensional complex projective space

We will need a local expression for the complex contact structure on $\mathbb{C}P^{2n+1}$; we will use homogeneous coordinates $(z_1, \ldots, z_{n+1}, w_1, \ldots, w_{n+1})$. Then the complex contact structure is given by the holomorphic 1-form

$$
\psi = \sum_{k=1}^{n+1} (z_k dw_k - w_k dz_k).
$$

To give a little more detail we remark that the complex contact structure on $\mathbb{C}P^{2n+1}$ is closely related to the Sasakian 3-structure on the sphere S^{4n+3} (see Chapter 13) and to the quaternionic Kähler structure on quaternionic projective space. $\mathbb{C}^{2n+2} \simeq \mathbb{H}^{n+1}$ has three almost complex structures I, J, K which act on the position vector $\mathbf{x}$ as

$$I\mathbf{x} = i\mathbf{x} = (iz_1, \dots, iz_{n+1}, iw_1, \dots, iw_{n+1}),$$

$$J\mathbf{x} = (i\bar{w}_1, \dots, i\bar{w}_{n+1}, -i\bar{z}_1, \dots, -i\bar{z}_{n+1}),$$

$$K\mathbf{x} = (\bar{w}_1, \dots, \bar{w}_{n+1}, -\bar{z}_1, \dots, -\bar{z}_{n+1}).$$

The vector fields on S^{4n+3} given by $\xi_1 = -I\mathbf{x}$, $\xi_2 = -J\mathbf{x}$, $\xi_3 = -K\mathbf{x}$ are the characteristic vector fields of the three contact structures η_1, η_2, η_3 on S^{4n+3}. In terms of the complex coordinates, η_1, η_2, η_3 on S^{4n+3} are the restrictions of the following forms on $\mathbb{C}^{2n+2}$ which we denote by the same letters.

$$\eta_1 = -\frac{i}{2}\sum_{k=1}^{n+1}(z_k d\bar{z}_k - \bar{z}_k dz_k + w_k d\bar{w}_k - \bar{w}_k dw_k),$$

$$\psi = \eta_3 + i\eta_2 = \sum_{k=1}^{n+1}(z_k dw_k - w_k dz_k).$$

Ishihara and Konishi [1979] proved that if one of the contact structures of a manifold $\tilde{M}^{4n+3}$ with a Sasakian 3-structure (see Chapter 13) is regular, the base manifold M of the induced fibration is a complex contact manifold. The structure is constructed as follows. If $(\phi_1, \xi_1, \eta_1, g)$, is the regular Sasakian structure, then ϕ_1 and g are projectable. Let $\tilde{\pi}$ denote the horizontal lift with respect to the principal S^1 bundle connection defined by η_1. Then J defined by $JX = \pi_*\phi_1\tilde{\pi}X$ and the projected metric form a Kähler structure on M (cf. Example 6.7.2). For a coordinate neighborhood $\mathcal{U}$ on M and a local cross section τ of $\tilde{M}^{4n+3}$ over $\mathcal{U}$, 1-forms u and v and a tensor field G defined on $\mathcal{U}$ by

$$u(X) \circ \pi = \eta_2(\tau_*X), \quad v(X) \circ \pi = \eta_3(\tau_*X),$$

$$GX = \pi_*(\phi_2\tau_*X - \eta_1(\tau_*X)\xi_3 + \eta_3(\tau_*X)\xi_1)$$

define the complex contact and complex almost contact structures on M. In the case of the Hopf fibration this is the standard Kähler structure on $\mathbb{C}P^{2n+1}$ with the Fubini–Study metric. With the Hopf fibration induced by ξ_1, $\psi = \eta_3 + i\eta_2$ is a local expression for the complex contact structure on $\mathbb{C}P^{2n+1}$.

12.2.3   Twistor spaces

Generalizing the previous example we discuss twistor spaces over quaternionic Kähler manifolds, the twistor space of quaternionic projective space being odd-dimensional complex projective space. Consider a Riemannian manifold (M^{4n}, g) whose holonomy group is contained in $Sp(n) \cdot Sp(1) = Sp(n) \times Sp(1)/\{\pm I\})$. This means that there exists a subbundle $E \subset End(TM)$ with 3-dimensional fibres such that locally there exists a basis of E consisting of almost complex structures $\{\mathcal{I}, \mathcal{J}, \mathcal{K}\}$ satisfying, $\mathcal{I}\mathcal{J} = -\mathcal{J}\mathcal{I} = \mathcal{K}$ and $\nabla_X \mathcal{I}, \nabla_X \mathcal{J}, \nabla_X \mathcal{K}$ belong to the span of $\{\mathcal{I}, \mathcal{J}, \mathcal{K}\}$ for any vector field X on M. In dimension 4 ($n = 1$) $Sp(1) \cdot Sp(1) = SO(4)$ so the holonomy group condition is not a restriction. For $n > 1$, (M^{4n}, g) is called a *quaternionic Kähler manifold*. A well-known result of Alekseevskii [1968] (see also Berger [1966], Ishihara [1974]) is that in this case, the metric g is Einstein.

Since $Sp(1) \cdot Sp(1) = SO(4)$, a stronger definition of quaternionic Kähler manifold is needed in dimension 4. A 4-dimensional manifold is said to be a *quaternionic Kähler manifold* if it is Einstein and self-dual with non-zero scalar curvature (see e.g., LeBrun [1991]). A 4-dimensional self-dual manifold is sometimes said to be *half-conformally flat*.

Returning to the subbundle $\bar{\pi} : E \longrightarrow M$, regardless of dimension, we define the *horizontal* subspace $\bar{\mathcal{H}}_p$ at $p \in E$ as follows. Let α denote a curve in M such that $\alpha(0) = \bar{\pi}(p)$ and let s denote a section of E such that $s(\alpha(0)) = p$. Then set

$$\bar{\mathcal{H}}_p = \{(s \circ \alpha)_*(0) \in T_p E \mid \nabla_{\dot{\alpha}(0)} s = 0\}.$$

Now induce a bundle metric on E by making each quaternionic Kähler frame $\{\mathcal{I}, \mathcal{J}, \mathcal{K}\}$ an orthonormal basis. Let Z be the space of all unit elements of E. Locally

$$Z = \{x\mathcal{I} + y\mathcal{J} + z\mathcal{K} \in E \mid x^2 + y^2 + z^2 = 1\}.$$

$\pi : Z \longrightarrow M$ is called the *twistor space* of M and each element $j \in Z$ is an almost complex structure on the tangent space of M at $\pi(j)$.

Let $\mathcal{V} = \ker \pi_*$. Since each fibre of Z is a unit sphere, then at $j = x\mathcal{I} + y\mathcal{J} + z\mathcal{K}$, $x^2 + y^2 + z^2 = 1$, we may make the identification

$$\mathcal{V}_j = \{X \in E_{\pi(j)} \mid X \perp j\} = \{a\mathcal{I} + b\mathcal{J} + c\mathcal{K} \mid ax + by + cz = 0\}.$$

Setting $\mathcal{H} = \bar{\mathcal{H}}|_Z$, we have the splitting $TM \cong \mathcal{V} \oplus \mathcal{H}$. Define an almost complex structure J on Z as follows. For a vertical vector $V \in \mathcal{V}_j$, set $JV = j \times V$ where $\times$ denotes the usual vector product in Euclidean 3-space; in particular this is the usual almost complex structure on S^2. To define the action of J on $X \in \mathcal{H}_j$, first let $\tilde{}$ denote the horizontal lift with respect to the connection on E determined by $\bar{\mathcal{H}}$. Then since j is an almost complex structure on the tangent space of M at $\pi(j)$, set $JX = \widetilde{j\pi_* X}$. Now local $\mathcal{V}$-valued forms

θ defining $\mathcal{H}$ then give Z a complex contact structure, see Salamon [1982] or for verification of the integrability Besse [1987, pp. 413–415] and for $\theta \wedge (d\theta)^n \neq 0$, Besse [1987, p. 416].

Now let $\{\mathcal{U}_\alpha\}$ be an open covering of the quaternionic Kähler manifold (M^{4n}, g) and corresponding to a neighborhood $\mathcal{U}_\alpha$ set

$$\mathcal{O}_\alpha = \{x\mathcal{I} + y\mathcal{J} + z\mathcal{K} \in Z \mid z \neq 1\}, \quad \mathcal{O}'_\alpha = \{x\mathcal{I} + y\mathcal{J} + z\mathcal{K} \in Z \mid z \neq -1\}.$$

The collection of pairs $\{\mathcal{O}_\alpha, \mathcal{O}'_\alpha\}$ is then an atlas on Z. On $\mathcal{O}_\alpha$ define vector fields $\hat{U}$ and $\hat{V}$ by

$$\hat{U} = \frac{1 - z - x^2}{1 - z}\mathcal{I} - \frac{xy}{1 - z}\mathcal{J} + x\mathcal{K}$$

and $\hat{V} = -J\hat{U}$. Then $\{\hat{U}, \hat{V}\}$ forms a basis of $\mathcal{V}$ on $\mathcal{O}$. As elements of E, $\hat{U}$ and $\hat{V}$ are unit and there exist real 1-forms $\hat{u}$ and $\hat{v}$ such that $\theta = \hat{u} \otimes \hat{U} + \hat{v} \otimes \hat{V}$ and $\hat{u}(\hat{U}) = \hat{v}(\hat{V}) = 1$, $\hat{u}(\hat{V}) = \hat{v}(\hat{U}) = 0$. On $\mathcal{O}'_\alpha$ define vector fields $\hat{U}'$ and $\hat{V}'$ by

$$\hat{U}' = \frac{1 + z - x^2}{1 + z}\mathcal{I} - \frac{xy}{1 + z}\mathcal{J} + x\mathcal{K}$$

and $\hat{V}' = -J\hat{U}'$. Again we have the corresponding 1-forms $\hat{u}'$, $\hat{v}'$ and

$$\hat{U} + i\hat{V} = h(\hat{U}' + i\hat{V}'), \quad \hat{u} - i\hat{v} = h(\hat{u}' - i\hat{v}')$$

where $h : \mathcal{O}_\alpha \cap \mathcal{O}'_\alpha \longrightarrow S^1$. Thus $\{\hat{u} - i\hat{v}\}$ is a normalized contact structure on Z and $\hat{u} \otimes \hat{u} + \hat{v} \otimes \hat{v}$ is a global tensor field on Z.

Let $c > 0$ and define a Riemannian metric g_c on Z by

$$g_c = c^2(\hat{u} \otimes \hat{u} + \hat{v} \otimes \hat{v}) + \pi^* g.$$

$g_c|_{\mathcal{V}} = c^2 <,>$ where $<,>$ is the Euclidean metric on fibres of E restricted to S^2. g_c is called the *Salamon–Bérard–Bergery metric with vertical coefficient c.* Setting $u_c = c\hat{u}$, $v_c = c\hat{v}$, $\{u_c - iv_c\}$ is a normalized contact structure on Z and

$$g_c = u_c \otimes u_c + v_c \otimes v_c + \pi^* g.$$

In [1996] Foreman proved the following theorem.

Theorem 12.1 *Let (M^{4n}, g) be a quaternionic Kähler manifold and τ the scalar curvature of g. Consider the twistor space Z with the Salamon–Bérard–Bergery metric g_c, $c > 0$ and normalized contact structure $\{u_c - iv_c\}$. Then we have the following:*

1. *g_c is Hermitian with respect to the complex structure J on Z.*
2. *g_c is an associated metric if and only if $c|\tau| = 8n(n + 2)$.*
3. *g_c is Kähler if and only if $c^2\tau = 4n(n + 2)$.*
4. *g_c is associated and Kähler if and only if $c = \frac{1}{2}$ and $\tau = 16n(n + 2)$.*

Recall that in Section 1 we noted conditions for an associated metric g to be projectable with respect to the foliation induced by the integrable subbundle $\mathcal{V}$, namely that h_U and h_V vanish. Foreman [1996] also proves the following result.

Theorem 12.2 *Let Z be the twistor space over a quaternionic Kähler manifold (M^{4n}, g) of constant scalar curvature of τ and let c be such that $c|\tau| = 8n(n+2)$. Then in the space $\mathcal{A}$ of all associated metrics with respect to the normalized contact structure $\{u_c - iv_c\}$, the Salamon-Bérard-Bergery metric g_c is the only projectable associated metric in $\mathcal{A}$.*

Again it seems worthwhile to mention the result of LeBrun [1995].

Theorem 12.3 *A complex contact manifold of positive first Chern class is a twistor space if and only if it admits a Kähler–Einstein metric.*

Foreman [2000b] also gives curvature conditions for a complex contact manifold to be the twistor space of a quaternionic Kähler manifold; these conditions are conditions on the curvatures $R_{XY}U$ and $R_{XY}V$.

12.2.4   The complex Boothby–Wang fibration

In [2000a] and again in [2000b] Foreman constructed complex contact manifolds with global complex contact forms and fibrations with vertical fibres $S^1 \times S^1$ and hence these examples are quite different from the twistor space examples. Here we discuss the complex Boothby–Wang fibration of Foreman [2000a]. Let (M, Ω) be a *complex symplectic manifold* of complex dimension $2n$ and complex structure J_0, i.e., M is a complex manifold together with a closed holomorphic 2-form Ω such that $\Omega^n \neq 0$. Writing $\Omega = \Omega_1 + i\Omega_2$ we see that Ω_1 and Ω_2 are closed 2-forms. On the other hand Ω may be written in terms of a local basis of holomorphic 1-forms as $\theta_1 \wedge \theta_{n+1} + \cdots + \theta_n \wedge \theta_{2n}$. Then taking real and imaginary parts of $\theta_k = \alpha_k + i\beta_k$ we have

$$\Omega_1 = \alpha_1 \wedge \alpha_{n+1} - \beta_1 \wedge \beta_{n+1} + \cdots + \alpha_n \wedge \alpha_{2n} - \beta_n \wedge \beta_{2n},$$

$$\Omega_2 = \alpha_1 \wedge \beta_{n+1} + \beta_1 \wedge \alpha_{n+1} + \cdots + \alpha_n \wedge \beta_{2n} - \beta_n \wedge \alpha_{2n}$$

from which we see that $\Omega_1^{2n} \neq 0$ and $\Omega_2^{2n} \neq 0$. Thus we have two distinct symplectic structures on M. If each of these is of integral class, we have two principal circle bundles P_1 and P_2 with contact (connection) forms η_1 and η_2 as in Sections 3.3, 4.5.4 and 6.7.2 and each $d\eta_k = \Omega_k$. Finally let ξ_1 and ξ_2 be the characteristic vector fields of these contact structures.

Define a principal $S^1 \times S^1$-bundle P over M by $P = P_1 \oplus P_2$ and let p denote the projection map. For $z \in P$, set $\mathcal{V}_z P = \ker p_*|_z$. This defines a vector bundle $\mathcal{V}$ over P and by extending each ξ_k to be trivial on the other factor, we may regard ξ_1 and ξ_2 as vector fields on P. Moreover the pair (η_1, η_2) defines a connection on P (see Foreman [2000b] for details). We denote the horizontal subspace determined by the connection at z by $\mathcal{H}_z P$ and the horizontal lift by $\tilde{p}$ or $\tilde{X}$.

The bundle space P carries an almost complex structure J defined as follows. On $\mathcal{V}$ define J by $J\xi_1 = \xi_2$ and $J\xi_2 = -\xi_1$, and for $X \in \mathcal{H}_z P$, set $JX = \tilde{p} J_0 p_* X$. Since the Lie algebra $\mathfrak{s}^1 \oplus \mathfrak{s}^1$ is abelian $[\xi_1, \xi_2] = 0$ and hence $[J, J](\xi_1, \xi_2) = 0$. Also it is easy to check that $[\xi_k, \tilde{X}] = 0$ and hence that $[J, J](\xi_k, \tilde{X}) = 0$. Finally utilizing the integrability of J_0 on M one can readily show that $p_*[J, J](\tilde{X}, \tilde{Y}) = 0$. Therefore P is a complex manifold.

We can now define a complex contact structure on the complex manifold P. First note that $\eta_2 = -\eta_1 \circ J$ where again by extending each η_k to be trivial on the other factor we regard η_1 and η_2 as 1-forms on P. Set $\theta = \eta_1 + i\eta_2$; then $\theta(X + iJX) = 0$ so that θ is a holomorphic 1-form. Moreover since $d\eta_k$ is the pullback of Ω_k, $d\theta = p^*\Omega$ and a straightforward computation shows that $\theta \wedge (d\theta)^n \neq 0$. Thus P becomes a complex contact manifold with a global complex contact form.

We summarize the above construction in the following theorem.

Theorem 12.4 *Let M be a complex symplectic manifold with a complex symplectic form $\Omega = \Omega_1 + i\Omega_2$ such that both Ω_1 and Ω_2 determine integral classes. Then the $S^1 \times S^1$-bundle defined by $([\Omega_1], [\Omega_2]) \in H^2(M, \mathbb{Z}) \oplus H^2(M, \mathbb{Z})$ has an integrable complex structure and a complex contact structure given by a holomorphic connection form whose curvature form is Ω.*

Foreman now proves a converse to this theorem as a complex Boothby–Wang fibration; we state the result here and refer to Foreman [2000b] for the proof. For a global complex contact form θ we write $\theta = u - iv$ where u and v are real forms with $v = u \circ J$; the vertical bundle $\mathcal{V}$ is then spanned by global vector fields U and $V = -JU$ where

$$u(U) = 1, \quad v(U) = 0, \quad \iota(U)du = 0,$$

$$u(V) = 0, \quad v(V) = 1, \quad \iota(V)dv = 0.$$

Theorem 12.5 *Let P be a $(2n + 1)$-dimensional compact complex contact manifold with a global form $\theta = u - iv$ such that the corresponding vertical vector fields U and JU are regular. Then θ generates a free $S^1 \times S^1$-action on P and $p : P \longrightarrow M$ is a principal $S^1 \times S^1$-bundle over a complex symplectic manifold M such that θ is a connection form for this fibration and the complex symplectic form Ω on M is given by $p^*\Omega = d\theta$.*

12.2.5   3-dimensional homogeneous examples

In the previous example B. Foreman gave a complex Boothby–Wang fibration for (regular) complex contact manifold with a global contact form. In [1999] Foreman studied 3-dimensional complex homogeneous complex contact manifolds with a global complex contact form and obtained the following classification.

Theorem 12.6 *If M is a 3-dimensional complex homogeneous complex contact manifold with global complex contact form, then M is of the form $M = G/\Gamma$ where G is a simply connected 3-dimensional complex Lie group and $\Gamma \subset G$ is a discrete subgroup.*

1. *Suppose G is unimodular. Then G is one of the following:*

   (a) $SL(2, \mathbb{C})$, if $\mathrm{rk}(ad(\mathcal{V})) = 2$.

   (b) The universal cover of the group of rigid motions of the complex Euclidean plane, if $\mathrm{rk}(ad(\mathcal{V})) = 1$.

   (c) $H_{\mathbb{C}}$, if $\mathrm{rk}(ad(\mathcal{V})) = 0$.

2. *Suppose G is not unimodular. Then G is solvable; $\mathrm{rk}(ad(\mathcal{V})) = 1$; and G is one of the following complex Lie groups:*

   (a) The semi-direct product $G_\alpha = \mathbb{C} \times_{\tau_\alpha} \mathbb{C}^2$, for any $\alpha \in \mathbb{C}^* \backslash 1$, where τ_α is a certain representation of $\mathbb{C}$ in $GL(2, \mathbb{C})$.

   (b) $G = \left\{ \begin{pmatrix} e^t & te^t & u \\ 0 & e^t & v \\ 0 & 0 & 1 \end{pmatrix} \,\middle|\, t, u, v \in \mathbb{C} \right\}$.

12.2.6 $\mathbb{C}^{n+1} \times \mathbb{C}P^n$(16)

Korkmaz [1998] gives a complex analogue of the real contact metric manifold $E^{n+1} \times S^n$(4) together with a curvature characterization analogous to Theorem 7.5. We describe the example and state the theorem.

Let $(t_0, \dots, t_n)$ be homogeneous coordinates on $\mathbb{C}P^n$ and $\mathcal{U}_i$ the neighborhood defined by $t_i \neq 0$. On $\mathcal{U}_i$ introduce non-homogeneous coordinates by $w^j = \frac{t_j}{t_i}$, $j = 0, \dots, n$, $j \neq i$. Let $\mathcal{O}_i = \mathbb{C}^{n+1} \times \mathcal{U}_i$ and let $(z^0, \dots, z^n)$ be coordinates on $\mathbb{C}^{n+1}$. Define a holomorphic 1-form θ_i on $\mathcal{O}_i$ by $\theta_i = \dfrac{1}{t_i} \displaystyle\sum_{k=0}^{n} t_k dz^k$.

Then $\theta_i \wedge (d\theta_i)^n \neq 0$ on $\mathcal{O}_i$ and $\theta_j = \frac{t_i}{t_j}\theta_i$ on $\mathcal{O}_i \cap \mathcal{O}_j$. Thus $\{\theta_i\}_{i=0}^n$ is a complex contact structure on $\mathbb{C}^{n+1} \times \mathbb{C}P^n$.

For convenience we continue our work on $\mathcal{O}_0$ with $\theta_0 = dz^0 + \sum_{k=0}^n w^k dz^k$. The product metric on $\mathbb{C}^{n+1} \times \mathbb{C}P^n(16)$ is given by the matrix

$$
g = \frac{1}{8}
\left(
\begin{array}{cc|cc}
 & O & I_{n+1} & 0 \\
 & & 0 & \mathbf{g} \\
\hline
I_{n+1} & 0 & & \\
0 & \mathbf{g}^T & & O
\end{array}
\right)
$$

where $\frac{1}{8}\mathbf{g}$ is the metric on $\mathbb{C}P^n(16)$, i.e.,

$$
\mathbf{g}_{ij} = \frac{(1+\sum_{k=1}^n |w^k|^2)\delta_{ij} - \bar{w}^i w^j}{(1+\sum_{k=1}^n |w^k|^2)^2}.
$$

Let $f_0 = 1 + \sum_{k=1}^n |w^k|^2$ and define real 1-forms u_0 and v_0 by

$$
u_0 = \frac{1}{4\sqrt{f_0}}\left(dz^0 + d\bar{z}^0 + \sum_{k=1}^n (w^k dz^k + \bar{w}^k d\bar{z}^k)\right),
$$

$$
v_0 = \frac{i}{4\sqrt{f_0}}\left(dz^0 - d\bar{z}^0 + \sum_{k=1}^n (w^k dz^k - \bar{w}^k d\bar{z}^k)\right).
$$

Similarly define vector fields U_0 and V_0 by

$$
U_0 = \frac{2}{\sqrt{f_0}}\left(\frac{\partial}{\partial z^0} + \frac{\partial}{\partial \bar{z}^0} + \sum_{k=1}^n (\bar{w}^k \frac{\partial}{\partial z^k} + w^k \frac{\partial}{\partial \bar{z}^k})\right),
$$

$$
V_0 = \frac{-2i}{\sqrt{f_0}}\left(\frac{\partial}{\partial z^0} - \frac{\partial}{\partial \bar{z}^0} + \sum_{k=1}^n (\bar{w}^k \frac{\partial}{\partial z^k} - w^k \frac{\partial}{\partial \bar{z}^k})\right).
$$

Then $\theta_0 = 2\sqrt{f_0}(u_0 - iv_0)$, $du_0(U_0, X) = 0$ for all $X \in \mathcal{H}$, $u_0(U_0) = 1$, $v_0(U_0) = 0$ and $g(U_0, X) = u_0(X)$ for all X.
    Let

$$
G_0 =
\left(
\begin{array}{cc|cc}
 & O & 0 & A \\
 & & B & 0 \\
\hline
0 & \bar{A} & & \\
\bar{B} & 0 & & O
\end{array}
\right)
$$

where

$$
A = \begin{pmatrix} w^1 & w^2 & \cdots & w^n \\ |w^1|^2 - f_0 & \bar{w}^1 w^2 & \cdots & \bar{w}^1 w^n \\ w^1 \bar{w}^2 & |w^2|^2 - f_0 & \cdots & \bar{w}^2 w^n \\ \vdots & \vdots & \ddots & \vdots \\ w^1 \bar{w}^n & w^2 \bar{w}^n & \cdots & |w^n|^2 - f_0 \end{pmatrix}, \quad B = f_0^2 \begin{pmatrix} -w^1 \\ -w^2 \\ \vdots & & I_n \\ -w^n \end{pmatrix}.
$$

Then $G_0^2 = -I + u_0 \otimes U_0 + v_0 \otimes V_0$, $G_0 J + J G_0 = 0$, etc. On $\mathcal{O}_1$ set $f_1 = 1 + |w^0|^2 + \sum_{k=2}^{n} |w^k|^2$. Then $\frac{f_0}{f_1} = \frac{|t_1|^2}{|t_0|^2}$. Setting $a - bi = \sqrt{\frac{f_0}{f_1} \frac{t_0}{t_1}}$ on $\mathcal{O}_0 \cap \mathcal{O}_1$ we have $a^2 + b^2 = 1$ and

$$
u_1 = a u_0 - b v_0, \quad v_1 = b u_0 + a v_0, \quad G_1 = a G_0 - b H_0, \quad H_1 = b G_0 + a H_0
$$

where u_1, v_1, G_1, H_1 are the structure tensors on $\mathcal{O}_1$. Therefore

$$
\{(u_k, v_k, U_k, V_k, G_k, H_k, g)\} \text{ on } \{\mathcal{O}_k\}_{k=0}^{n}
$$

is a complex contact metric structure on $\mathbb{C}^{n+1} \times \mathbb{C}P^n(16)$.

It can be shown that for this structure $h_U = h_V$, see Korkmaz [1998]. We now state a characterization of this example, due to Korkmaz [1998], and analogous to Theorem 7.5, that a contact metric manifold on which $R_{XY}\xi = 0$ is locally isometric to $E^{n+1} \times S^n(4)$.

Theorem 12.7 *Let M be a complex contact metric manifold with $h_U = h_V$. If $R_{XY}\mathcal{V} = 0$, then M is locally isometric to $\mathbb{C}^{n+1} \times \mathbb{C}P^n(16)$.*

12.3   Normality of complex contact manifolds

As we have seen in real contact geometry, the product $M \times \mathbb{R}$ of an almost contact manifold and the real line carries a natural almost complex structure J and the almost contact structure is said to be *normal*, if J is integrable. Recall also that a Sasakian manifold is a normal contact metric manifold.

Ishihara and Konishi [1979], [1980] introduced a notion of normality for complex contact structures. Their notion is the vanishing of the two tensor fields S and T given by

$$
S(X,Y) = [G,G](X,Y) + 2\hat{G}(X,Y)U - 2\hat{H}(X,Y)V + 2(v(Y)HX - v(X)HY)
$$

$$
+ \sigma(GY)HX - \sigma(GX)HY + \sigma(X)GHY - \sigma(Y)GHX,
$$

$$
T(X,Y) = [H,H](X,Y) - 2\hat{G}(X,Y)U + 2\hat{H}(X,Y)V + 2(u(Y)GX - u(X)GY)
$$

$$
+ \sigma(HX)GY - \sigma(HY)GX + \sigma(X)GHY - \sigma(Y)GHX.
$$

However this notion seems to be too strong; among its implications is that the underlying Hermitian manifold (M, g) is Kähler. Thus while indeed one of the canonical examples of a complex contact manifold, the odd-dimensional complex projective space, is normal in this sense, the complex Heisenberg group, is not. B. Korkmaz [2000] generalized the notion of normality and we adopt her definition here. A complex contact metric structure is *normal* if

$$S(X, Y) = T(X, Y) = 0, \text{ for every } X, Y \in \mathcal{H},$$

$$S(U, X) = T(V, X) = 0, \text{ for every } X.$$

Even though the definition appears to depend on the special nature of U and V, it respects the change in overlaps, $\mathcal{O}_\alpha \cap \mathcal{O}_\beta$, and is a global notion (Korkmaz [2000]). With this notion of normality both odd-dimensional complex projective space and the complex Heisenberg group with their standard complex contact metric structures are normal. We remark also that for Ishihara and Konishi's notion of normality, the holonomy group is the full unitary group $U(2m + 1)$, Houh [1976].

One consequence of normality is that $h_U = 0$ for every $U \in \mathcal{V}$; this is analogous to the fact that every Sasakian structure is K-contact. Another is that the sectional curvature of a plane section spanned by a vector in $\mathcal{V}$ and a vector in $\mathcal{H}$ is equal to $+1$ (cf. Korkmaz [2000], Foreman [2000b]).

We now give expressions for the covariant derivatives of G and H on a normal complex contact metric manifold; for proofs see Korkmaz [2000]. Setting $\Omega = d\sigma$ we have that a complex contact metric manifold is normal if and only if the covariant derivatives of G and H have the following forms.

$$g((\nabla_X G)Y, Z) = 2\sigma(X)g(HY, Z) + 2v(X)\Omega(GZ, GY) - 2v(X)g(HGY, Z)$$

$$-u(Y)g(X, Z) - v(Y)g(JX, Z) + u(Z)g(X, Y) + v(Z)g(JX, Y).$$

$$g((\nabla_X H)Y, Z) = -2\sigma(X)g(GY, Z) - 2u(X)\Omega(HZ, HY) - 2u(X)g(GHY, Z)$$

$$+u(Y)g(JX, Z) - v(Y)g(X, Z) + u(Z)g(X, JY) + v(Z)g(X, Y).$$

In these formulas the first two terms on the right vanish for the complex Heisenberg group (Example 12.2.1) and the second and third terms cancel on $P\mathbb{C}^{2n+1}$ (Example 12.2.2). Also one has

$$\nabla_X U = -GX, \quad \nabla_X V = -HX$$

on $H_\mathbb{C}$ and

$$\nabla_X U = -GX + \sigma(X)V, \quad \nabla_X V = -HX - \sigma(X)U$$

on $\mathbb{C}P^{2n+1}$. Finally on a normal complex contact manifold

$$g((\nabla_X J)Y, Z)$$

$$= 2u(X)(\Omega(Z, GY) - g(HY, Z)) + 2v(X)(\Omega(Z, HY) + g(GY, Z)).$$

When the complex contact structure is strict, i.e., given by a global complex contact form, the situation is more restrictive. In particular $\sigma = 0$ and therefore some of the above formulas simplify. Foreman [2000b] defines a *complex Sasakian manifold* to be a normal complex contact metric manifold whose complex contact structure is given by a global complex contact form. His paper gives a number of basic properties and examples including local projectivity to a hyper-Kähler manifold.

Finally we mention that in [toap] Korkmaz develops a theory of complex (κ, μ)-spaces analogous to that described in Section 7.2.

12.4   GH-sectional curvature

Corresponding to the ideas of holomorphic curvature in complex geometry and ϕ-sectional curvature in real contact geometry, B. Korkmaz [2000] defined the notion of GH- sectional curvature for a complex contact metric manifold. For a unit vector $X \in \mathcal{H}_m$, the plane in $T_m M$ spanned by X and $Y = aGX + bHX, a, b \in \mathbb{R}, a^2 + b^2 = 1$ is called a *GH-plane section* and its sectional curvature, $K(X, Y)$, the *GH-sectional curvature* of the plane section. For a given vector X, $K(X, Y)$ is independent of the vector Y in the plane of GX and HX if and only if $K(X, GX) = K(X, HX)$ and $g(R_{XGX} HX, X) = 0$. Let M be a normal complex contact metric manifold; if the GH-sectional curvature is independent of the choice of GH-section at each point, it is constant on the manifold and we say that M is a *complex contact space form*. The curvature tensor and the following theorems were obtained by Korkmaz [2000]; explicitly the curvature tensor is

$$R_{XY} Z = \frac{c+3}{4} \Big(g(Y, Z)X - g(X, Z)Y$$

$$+ g(Z, JY)JX - g(Z, JX)JY + 2g(X, JY)JZ \Big)$$

$$+ \frac{c-1}{4} \Big(- \big(u(Y)u(Z) + v(Y)v(Z)\big)X + \big(u(X)u(Z) + v(X)v(Z)\big)Y$$

$$+ 2u \wedge v(Z, Y)JX - 2u \wedge v(Z, X)JY + 4u \wedge v(X, Y)JZ$$

$$+ g(Z, GY)GX - g(Z, GX)GY + 2g(X, GY)GZ$$

$$+ g(Z, HY)HX - g(Z, HX)HY + 2g(X, HY)HZ$$

$$+\big(-u(X)g(Y,Z)+u(Y)g(X,Z)$$
$$+v(X)g(JY,Z)-v(Y)g(JX,Z)+2v(Z)g(X,JY)\big)U$$
$$+\big(-v(X)g(Y,Z)+v(Y)g(X,Z)$$
$$-u(X)g(JY,Z)+u(Y)g(JX,Z)-2u(Z)g(X,JY)\big)V\Big)$$

$$-\frac{4}{3}(\Omega(U,V)+c+1)\Big(\big(v(X)u\wedge v(Z,Y)-v(Y)u\wedge v(Z,X)+2v(Z)u\wedge v(X,Y)\big)U$$

$$-\big(u(X)u\wedge v(Z,Y)-u(Y)u\wedge v(Z,X)+2u(Z)u\wedge v(X,Y)\big)V\Big).$$

Odd-dimensional complex projective space with the Fubini–Study metric of constant holomorphic curvature 4 is of constant *GH*-sectional curvature 1. The complex Heisenberg group has holomorphic curvature 0 for horizontal and vertical holomorphic sections and constant *GH*-sectional curvature -3.

Theorem 12.8 *Let M be a normal complex contact metric manifold. Then M has constant GH-sectional curvature c, if and only if for X horizontal, the holomorphic sectional curvature of the plane spanned by X and JX is $c+3$.*

Theorem 12.9 *Let M be a normal complex contact metric manifold of constant GH-sectional curvature $+1$ and satisfying $d\sigma(V,U)=2$, then M has constant holomorphic curvature 4. If, in addition, M is complete and simply connected, then M is isometric to $\mathbf{C}P^{2n+1}$ with the Fubini–Study metric of constant holomorphic curvature 4.*

Korkmaz [2000] then introduced the idea of an $\mathcal{H}$-homothetic deformation of a complex contact metric structure. Let α be a positive constant and consider the local structure tensors (G,H,U,V,u,v,g). Then define new structure tensors by

$$\tilde{u}=\alpha u, \quad \tilde{v}=\alpha v, \quad \tilde{U}=\frac{1}{\alpha}U, \quad \tilde{V}=\frac{1}{\alpha}V, \quad \tilde{G}=G, \quad \tilde{H}=H,$$

$$\tilde{g}=\alpha g+\alpha(\alpha-1)(u\otimes u+v\otimes v).$$

This change of structure is called an $\mathcal{H}$-*homothetic deformation*. The new structure then respects the transitions on the overlaps of coordinate neighborhoods and hence gives a new complex contact metric structure on M. Moreover $\tilde{S}(X,Y)=S(X,Y)$ on $\mathcal{H}$, $\tilde{S}(X,\tilde{U})=\frac{1}{\alpha}S(X,U)$, etc., so if the given structure is normal, so is the new structure. Korkmaz computed the curvature and showed that if on a normal complex contact metric manifold the original structure has constant *GH*-sectional curvature c, then the new structure has constant *GH*-sectional curvature $\tilde{c}=\dfrac{c+3}{\alpha}-3$; in particular she proved the following results.

Theorem 12.10 *Complex projective space* $\mathbf{C}P^{2n+1}$ *carries a normal complex contact metric structure with constant GH-section curvature* $\dfrac{4}{\alpha} - 3$ *for every* $\alpha > 0$.

Theorem 12.11 *A normal complex contact metric manifold with metric $\tilde{g}$ of constant GH-sectional curvature $\tilde{c} > -3$ is $\mathcal{H}$-homothetic to a normal complex contact metric manifold with metric g of constant GH-section curvature $c = 1$. Moreover, if $d\sigma(\tilde{V}, \tilde{U}) = \dfrac{(\tilde{c} + 3)^2}{8}$, the metric g is Kähler and has constant holomorphic curvature 4.*

12.5　The set of associated metrics and integral functionals

In [1996] B. Foreman began the study of the set of all associated metrics on a complex contact manifold. In Chapter 4 we studied how the set of associated metrics $\mathcal{A}$ for a symplectic or contact form sits in the set of Riemannian metrics with the same volume element and in particular we studied the tangent space to $\mathcal{A}$ at an associated metric g. As before we will typically use the same letter for a symmetric tensor field as a tangent vector and for the corresponding tensor field of type $(1,1)$ determined by the metric.

As already mentioned, and in keeping with the notation in the real case, we will denote by $\mathcal{A}$ the set of all associated metrics for a complex contact structure as defined in Section 12.1. As in the real case, $\mathcal{A}$ is infinite dimensional; all associated metrics to a given complex contact structure have the same volume element; $\mathcal{A}$ is totally geodesic in $\mathcal{N}$ in the sense that if $D \in T_g\mathcal{A}$, the geodesic $g_t = ge^{tD}$ is a path in $\mathcal{A}$; and two metrics in $\mathcal{A}$ may be joined by a geodesic. We now give a characterization of $T_g\mathcal{A}$; for details of this and the preceeding statements see Foreman [1996].

Lemma 12.1 *Let g be a metric in $\mathcal{A}$. Then a symmetric tensor field D of type $(0,2)$ is in $T_g\mathcal{A}$ if and only if D, as a tensor field of type $(1,1)$, satisfies*

$$DJ = JD, \quad D|_\mathcal{V} = 0, \quad DG = -GD$$

for any local tensor field G as above.

As before, to study integral functionals defined on $\mathcal{A}$ we will be differentiating such functionals along paths of metrics and for this we need the following fundamental lemma (Foreman [1996]). For an endomorphism T of a complex vector space, set

$$T^s = \frac{1}{2}(T - JTJ), \quad T^d = \frac{1}{2}(T + JTJ),$$

i.e., T^s is the part of T that commutes with J and T^d the part of T that anti-commutes with J.

Lemma 12.2 *For $g \in \mathcal{A}$ and T a symmetric $(1,1)$ tensor field,*

$$\int_M \mathrm{tr} TD \, dV = 0$$

for every $D \in T_g\mathcal{A}$ if and only if

$$p(TJ + JT)p = HTG - GTH$$

or equivalently

$$pT^s p = -GT^s G.$$

In Section 10.3 we studied the integral of the Ricci curvature in the direction of the characteristic vector field ξ, i.e., $L(g) = \int_M Ric(\xi) \, dV$, as a functional on the set of associated metrics. In [1996] Foreman gave two analogues of $Ric(\xi)$ for complex contact manifolds and proved the following results. For a unit vertical vector field U, $Ric(U) + Ric(JU)$ is global, i.e., respects transition on $\mathcal{O}_\alpha \cap \mathcal{O}_\beta$, and we denote it by $Ric(\mathcal{V})$. For the second analogue, Foreman utilizes the $*$-Ricci tensor and it is easy to check that for a unit vertical vector field U, $Ric^*(U) = R^*_{ij}U^iU^j$ is independent of the unit vertical vector field U and globally defined; we denote it by $Ric^*(\mathcal{V})$. Define functionals L and L^* on $\mathcal{A}$ by

$$L(g) = \int_M Ric(\mathcal{V}) \, dV, \qquad L^*(g) = \int_M Ric^*(\mathcal{V}) \, dV.$$

Theorem 12.12 *Let M be a compact complex contact manifold and $\mathcal{A}$ the set of associated metrics. Then $g \in \mathcal{A}$ is a critical point of $L(g)$ if and only if*

$$(\nabla_U h_U)^s + (\nabla_V h_V)^s = \sigma(U)h_V^s - \sigma(V)h_U^s + 4k_U h_U^d$$

for any local unit vertical vector field U.

Clearly any projectable associated metric is a critical point of $L(g)$. In fact Foreman showed that $Ric(\mathcal{V}) = 8n - 2K(\mathcal{V}) - \mathrm{tr} h_U^2 - \mathrm{tr} h_V^2$ where $K(\mathcal{V})$ is the sectional curvature of a vertical plane section and moreover that $\int_M K(\mathcal{V}) \, dV$ is independent of the associated metric. Thus projectable metrics are maxima of $L(g)$. If g is Kähler, $h_U^s = 0$ for any vertical vector field U and moreover $(\nabla_W h_U^d)^s = 0$ for any vertical vector field W. Thus we also have the following corollary.

Corollary 12.1 *If $g \in \mathcal{A}$ is Kähler and a critical point of $L(g)$, then it is projectable.*

Theorem 12.13 *Let M be a compact complex contact manifold and $\mathcal{A}$ the set of associated metrics. Then $g \in \mathcal{A}$ is a critical point of $L^*(g)$ if and only if*

$$h_U^d(\nabla_U J) = 0$$

for any local unit vertical vector field U.

In particular, if g is projectable or Kähler, it is critical. In complex dimension 3, this theorem may be improved to g being critical if and only if locally either $h_U^d = 0$ or J is parallel along $\mathcal{V}$, i.e., $\nabla_U J = 0$ for any local unit vertical vector field U. For the twistor spaces as described in Example 12.2.3 this theorem may be improved as follows. If Z is the twistor space of a compact quaternionic Kähler manifold (M^{4n}, g), $4n \geq 8$, with normalized contact structure $\{u_c - iv_c\}$ and if the Salamon–Bérard–Bergery metric g_c is an associated metric, then an associated metric g is critical for L^* if and only if $h_U^d = 0$ for every vertical vector U.

B. Foreman also obtained some important results on the constancy of L^*, i.e., when L^* is independent of the associated metric.

Theorem 12.14 *Let M be a complex contact manifold with corresponding almost complex structure J. Then L^* is constant on $\mathcal{A}$ if and only if J is projectable.*

Theorem 12.15 *Suppose M is a compact complex contact manifold with a global complex contact structure and $\mathcal{A}$ the set of associated metrics. Then L^* is constant on $\mathcal{A}$.*

Turning now to the scalar curvatures, in real contact geometry the "total scalar curvature" defined by $\int_M \tau + \tau^* \, dV$ is a functional on $\mathcal{A}$ whose critical points are precisely the K-contact metrics as we have seen (Theorem 10.9). B. Foreman (unpublished) has also defined a $**$-*scalar curvature* by first contracting the curvature with the local tensor fields G and H as in the $*$-scalar curvature giving $*$-scalar curvatures, τ_G^* and τ_H^*; $\tau^{**} = \tau_G^* + \tau_H^*$ is then globally defined and one defines the "total scalar curvature" $I(g)$ by

$$I(g) = \int_M \tau + \tau^* + \tau^{**} \, dV.$$ Recall that an almost Hermitian structure (J, g) is *semi-Kähler* if the fundamental 2-form Ω is co-closed. Foreman computed the critical point condition for this functional and, his computation, though complicated, yields the following result.

Theorem 12.16 *A projectable semi-Kähler metric is a critical point of the functional $I(g)$.*

12.6   Holomorphic Legendre curves

In Section 8.3 we saw that in a 3-dimensional Sasakian manifold, Legendre curves are characterized by their torsion being equal to 1 and initial conditions at one point and that without the initial conditions, a counterexample can be given. Moreover, in $\mathbb{R}^3$ with its standard Sasakian structure the curvature of a Legendre curve is twice the curvature of its projection to the xy-plane with respect to the Euclidean metric. Our first purpose here is to study the complex Heisenberg group $H_{\mathbb{C}}$ as a complex contact manifold and to prove similar results for this space. One important difference from the real case is that while holomorphic Legendre curves in $H_{\mathbb{C}}$ have torsion equal to 1, this is the only possible non-zero constant. In fact the first main result here is that a holomorphic Frenet curve in $H_{\mathbb{C}}$ whose torsion is not identically zero and satisfies an initial condition at one point is a holomorphic Legendre curve.

Let $\tilde{M}$ be a Hermitian manifold of complex dimension n with complex structure J and corresponding Riemannian metric g. Following Chern, Cowen and Vitter [1974], S. Dolbeault [1977] we describe holomorphic curves and Frenet frames. A holomorphic curve in $\tilde{M}$ is a non-constant holomorphic map $\iota: M \to \tilde{M}$, where M is a Riemann surface. If $z_i = z_i(w)$ is a local representation of M in a neighborhood of $w = 0$, then its holomorphic tangent vector at $\iota(0)$ is given by $\sum z_i'(0)\frac{\partial}{\partial z_i} = w^p V$ where p is a non-negative integer and V a non-zero vector. The isolated points where $p > 0$ are called *stationary points of order* 0. A unitary frame $\{f_1, \ldots, f_n\}$ is called a *Frenet frame* if $f_1 = \frac{V}{|V|}$ and

$$\tilde{\nabla}_X f_i = \omega_{ii-1}(X)f_{i-1} + \omega_{ii}(X)f_i + \omega_{ii+1}(X)f_{i+1}$$

for $i = 1$ to $n - 1$ where ω_{ii+1} is a holomorphic 1-form and $\omega_{i+1i}(X) = -\omega_{ii+1}(X)$. Points where ω_{ii+1} vanish are called *stationary points of order* i. In general a unitary frame is not holomorphic but we will be interested in the case where the f_i's are holomorphic vector fields and we then speak of a *holomorphic Frenet curve* and a *holomorphic Frenet frame*.

Not every holomorphic curve in a Hermitian manifold $\tilde{M}$ has a Frenet frame. Chern, Cowen and Vitter [1974] (or see Dolbeault [1977]) give curvature conditions on $\tilde{M}$ under which every holomorphic curve has a Frenet frame; in particular a Kähler manifold of complex dimension ≥ 3 has a Frenet frame along every holomorphic curve if and only if it has constant holomorphic curvature.

M being a holomorphic Frenet curve has implications on M as a submanifold. In the following lemma we give three such implications, the first two of which are immediate in any Kähler manifold. For our purpose we give the lemma only for Hermitian manifolds of complex dimension 3 (see Baikoussis, Blair, and Gouli-Andreou [1998] for the proof). Recall that the span of the second fundamental form α is called the *first normal space* and will be denoted by ν_1. Let 'proj' denote projection to the orthogonal complement of $TM \oplus \nu_1$. Define $\beta(X, Y, Z)$ by

$$\beta(X, Y, Z) = \text{proj}\, \tilde{\nabla}_X \tilde{\nabla}_Y Z (= \text{proj}\, \nabla_X^{\perp} \alpha(Y, Z)).$$

The span of β is called the *second normal space* and will be denoted by ν_2. Successively the higher normal spaces may be defined in this manner taking higher order derivatives.

Lemma 12.3 *Let M be a holomorphic Frenet curve in a 3-dimensional Hermitian manifold $(\tilde{M}, J, g)$ with second fundamental form α. Let $\tilde{R}$ denote the curvature tensor of $\tilde{M}$. Then*

1) $\alpha(X, JY) = J\alpha(X, Y)$ for any tangent vectors X, Y and hence the real dimension of the first normal space is 2,

2) $\tilde{\nabla}_X J|_{TM \oplus T^{\perp}M} = 0$,

3) for any tangent vectors X, Y, Z, we have that $\tilde{R}(X, Y)Z$ is orthogonal to the second normal space.
Conversely if M is a holomorphic curve satisfying 1), 2) and 3), then it has a holomorphic Frenet frame.

Related to the idea of a Frenet frame are the curvatures themselves. These Frenet curvatures for a holomorphic curve date back to Calabi [1953a] who defined $(n-1)$ real-valued curvature functions (see also Lawson [1970]). When the ambient space is a complex space form these curvatures are actually intrinsic (Lawson [1970]). We follow the development as presented by Lawson [1970].

Let M be a holomorphic curve in a 3-dimensional Hermitian manifold $\tilde{M}$ for which the properties of the lemma hold. From $\alpha(X, JY) = J\alpha(X, Y)$ we have easily that for all unit tangent vectors X, Y (Y not necessarily distinct from X), $|\alpha(X, Y)|^2$ is a function of position alone, say $\kappa_1(p), p \in M$; κ_1 is called the *curvature* or *first curvature* of M. Now with $\beta(X, Y, Z)$ defined as above, from property 2) of the lemma we have $\beta(X, Y, JZ) = J\beta(X, Y, Z)$. From the second expression for β in the definition we see that β is symmetric in the second and third variables giving $\beta(X, JY, Z) = J\beta(X, Y, Z)$. Let ν be a vector in the second normal space, then by property 3), $g(\tilde{\nabla}_X \tilde{\nabla}_Y Z, \nu) = g(\tilde{\nabla}_Y \tilde{\nabla}_X Z, \nu)$, giving that β is symmetric in the first and second variables and therefore $\beta(JX, Y, Z) = J\beta(X, Y, Z)$. From these properties of β we have that

$$\kappa_2(p) = \frac{|\beta(X, Y, Z)|^2}{\kappa_1(p)}$$

is well defined, X, Y, Z being any unit tangent vectors. We call κ_2 the *torsion* or *second curvature* and also denote it by τ.

It is interesting to compare the curvature and torsion with the derivatives of the Frenet frames and the holomorphic connection forms ω_{ii+1}. In particular writing f_1 as $\frac{e_1 - iJe_1}{\sqrt{2}}$, $\kappa = |\omega_{12}(e_1)|^2$ and $\tau = |\omega_{23}(e_1)|^2$.

Now analogous to the starting point in the real case, Proposition 8.2, we begin with the following proposition from Baikoussis, Blair, and Gouli-Andreou [1998].

Proposition 12.1 *Let M be a real surface in $(H_{\mathbb{C}}, \theta, g)$ such that $\theta(X) = 0$ for any tangent vector X. Then M is a holomorphic Legendre curve as well as a holomorphic Frenet curve with torsion $\tau \equiv 1$.*

Our first result, due to Baikoussis, Blair, and Gouli-Andreou [1998], is the complex analogue of Theorem 8.12.

Theorem 12.17 *If the torsion of a holomorphic Frenet curve in the complex Heisenberg group is not identically zero and at one point the complex contact form annihilates the tangent space, then the curve is Legendre.*

Corresponding to Theorem 8.14 we also have the following result (Baikoussis, Blair, and Gouli-Andreou [1998]).

Theorem 12.18 *Let M be a holomorphic Legendre curve in $(H_{\mathbb{C}}, \theta, g)$ and N its projection to $\mathbb{C}^2 = \{(z_1, z_2)\}$ with its standard complex structure and Kähler (Euclidean) metric. Then the Gaussian curvature of M is 8 times that of N.*

Turning to $\mathbb{C}P^3$ we give a description of holomorphic Legendre curves due to Bryant [1982].

Theorem 12.19 *For any compact Riemann surface there exists a holomorphic Legendre imbedding into $\mathbb{C}P^3$.*

These holomorphic Legendre curves in $\mathbb{C}P^3$ can be described as follows. Let (z_1, z_2, w_1, w_2) be homogeneous coordinates on $\mathbb{C}P^3$ and $\psi = z_1 dw_1 - w_1 dz_1 + z_2 dw_2 - w_2 dz_2$ as in Example 12.2.2. Now for a connected Riemann surface M, let f and g be meromorphic functions on M with g non-constant. Define $\iota : M \longrightarrow \mathbb{C}P^3$ by

$$\zeta \longrightarrow (1, g, f - g\frac{f'}{2g'}, \frac{f'}{2g'}),$$

then it is easy to check that $\psi(\iota_* \frac{\partial}{\partial \zeta}) = 0$. Conversely Bryant shows that any holomorphic Legendre curve $\iota : M \longrightarrow \mathbb{C}P^3$ is either of this form or has its image in some $\mathbb{C}P^1 \subset \mathbb{C}P^3$. As an application of his result Bryant uses the fibration of $\mathbb{C}P^3$ over S^4 to show that given a compact Riemann surface, there exists a conformal superminimal generically 1-1 immersion into S^4 whose image in S^4 is an algebraic surface.

12.7   The Calabi (Veronese) imbeddings as integral submanifolds of $\mathbb{C}P^{2n+1}$

In [1953b] Calabi showed that up to holomorphic congruence there is a unique holomorphic imbedding of $\mathbb{C}P^n(\frac{4}{\nu})$ into $\mathbb{C}P^N(4)$, $N = \binom{n+\nu}{\nu} - 1$ which does not lie in any totally geodesic complex projective space of lower dimension. Nakagawa and Ogiue [1976] showed that the only full isometric immersions of positively curved complex space forms into positively curved complex space forms are local versions of these imbeddings. These imbeddings are known as the *Calabi imbeddings* or as the *Veronese imbeddings*, especially in the case $\mathbb{C}P^2(2) \longrightarrow \mathbb{C}P^5(4)$. For $n = 1$ these imbeddings are called *Calabi curves*. Classically the Calabi imbeddings are given as follows. Let $\zeta_1, \dots, \zeta_{n+1}$ be homogeneous coordinates for $\mathbb{C}P^n(\frac{4}{\nu})$. The Calabi imbedding of $\mathbb{C}P^n(\frac{4}{\nu})$ into $\mathbb{C}P^N(4)$, in terms of homogeneous coordinates for $\mathbb{C}P^N(4)$ is given by

$$(\zeta_1, \dots, \zeta_{n+1}) \longrightarrow (\zeta_1^\nu, \sqrt{\nu}\zeta_1^{\nu-1}\zeta_2, \dots, \sqrt{\frac{\nu!}{a_1! \cdots a_{n+1}!}}\zeta_1^{a_1} \cdots \zeta_{n+1}^{a_{n+1}}, \dots, \zeta_{n+1}^\nu)$$

where $\sum_{i=1}^{n+1} a_i = \nu$, the a_i's being non-negative integers. The meaning of the integer ν is that there are $\nu - 1$ normal spaces for these imbeddings.

The question to be addressed in this section is the following. For which of these imbeddings is there a holomorphic congruence of $\mathbb{C}P^{2n+1}$ which positions the submanifold as a holomorphic integral submanifold of the complex contact structure?

For the Calabi curves we have the following positive answer due to Blair, Dillen, Verstraelen, and Vrancken [1996].

Theorem 12.20 *There exists a holomorphic congruence of $\mathbb{C}P^{2n+1}(4)$ which positions the Calabi curve $\mathbb{C}P^1(\frac{4}{2n+1})$ as a holomorphic Legendre curve in $\mathbb{C}P^{2n+1}(4)$.*

The imbedding is given explicitly as follows. For simplicity set $A_k = \sqrt{\binom{2n+1}{k-1}}$; we position $\mathbb{C}P^1(\frac{4}{2n+1})$ by a holomorphic congruence of (1) as

$$(\zeta_1, \zeta_2) \longrightarrow$$

$$(\zeta_1^{2n+1}, \dots, A_k\zeta_1^{2n+2-k}\zeta_2^{k-1}, \dots, A_{n+1}\zeta_1^{n+1}\zeta_2^n,$$
$$\zeta_2^{2n+1}, \dots, (-1)^{k-1}A_k\zeta_2^{2n+2-k}\zeta_1^{k-1}, \dots, (-1)^n A_{n+1}\zeta_2^{n+1}\zeta_1^n).$$

Then

$$\frac{\partial}{\partial \zeta_1} = \sum_{k=1}^{m+1} \left(A_k(2n+2-k)\zeta_1^{2n+1-k}\zeta_2^{k-1}\frac{\partial}{\partial z_k} - (-1)^k A_k(k-1)\zeta_2^{2n+2-k}\zeta_1^{k-2}\frac{\partial}{\partial w_k} \right),$$

$$\frac{\partial}{\partial \zeta_2} = \sum_{k=1}^{m+1} \left(A_k(k-1)\zeta_1^{2n+2-k}\zeta_2^{k-2}\frac{\partial}{\partial z_k} - (-1)^k A_k(2n+2-k)\zeta_2^{2n+1-k}\zeta_1^{k-1}\frac{\partial}{\partial w_k} \right)$$

represent the tangent space and the proof is to show that $\psi(\frac{\partial}{\partial \zeta_1}) = \psi(\frac{\partial}{\partial \zeta_2}) = 0$ where $\psi = \sum_{k=1}^{n+1}(z_k dw_k - w_k dz_k)$ as in Example 12.2.2.

On the other hand there is no holomorphic congruence of $\mathbb{C}P^5(4)$ that brings the Veronese surface $\mathbb{C}P^2(2)$ into position as a Legendre submanifold of the complex contact structure on $\mathbb{C}P^5(4)$ even though the codimension is large enough. This has to do with the fact that the integer ν is equal to 2.

Theorem 12.21 *Assume* $N = \binom{n+2}{2} - 1$ *is odd. There is no holomorphic congruence of* $\mathbb{C}P^N(4)$ *that brings the Calabi imbedding of* $\mathbb{C}P^n(2)$ *into position as an integral submanifold of the complex contact structure on* $\mathbb{C}P^N(4)$.

Proof. Recall that the meaning of the condition $\nu = 2$ is that the first normal space is the whole normal space. Suppose now that $\mathbb{C}P^n(2) \longrightarrow \mathbb{C}P^N(4)$ is an integral submanifold of the complex contact structure with second fundamental form α. We have already noted that the vector fields U and V are normal and that for any tangent vector X, GX is normal. Using $\tilde{\nabla}_X U = -GX + \sigma(X)V$ we have

$$0 = Xg(Y,U) = g(\tilde{\nabla}_X Y, U) - g(Y, GX) = g(\alpha(X,Y), U)$$

and similarly $g(\alpha(X,Y), V) = 0$. Thus U and V are orthogonal to both the tangent space and the first normal space, but the first normal space is the whole normal space, giving a contradiction.   ∎

For $\nu = 3$ we first give a non-existence result for $n = 2$ and then a positive result for n odd; these results are due to Blair, Korkmaz, and Vrancken [2000].

Theorem 12.22 *There is no holomorphic congruence of* $\mathbb{C}P^9(4)$ *that brings the Calabi imbedding of* $\mathbb{C}P^2(\frac{4}{3})$ *into position as an integral submanifold of the complex contact structure on* $\mathbb{C}P^9(4)$.

Theorem 12.23 *When* $\nu = 3$ *and* n *is odd,* $N = \binom{n+3}{3} - 1$ *is odd and there exists a holomorphic congruence of* $\mathbb{C}P^N(4)$ *that brings the Calabi imbedding of* $\mathbb{C}P^n(\frac{4}{3})$ *into position as an integral submanifold of the complex contact structure on* $\mathbb{C}P^N(4)$.

Finally we give an example with $\nu = 5$. In terms of the homogeneous coordinates $(z_1, \ldots, z_{28}, w_1, \ldots, w_{28})$, $\mathbb{C}P^3(\frac{4}{5})$ may be realized as an integral submanifold of the complex contact structure on $\mathbb{C}P^{55}(4)$ in the following way:

$$z_1 = \zeta_1^5 \qquad\qquad\qquad w_1 = -\zeta_3^5$$
$$z_2 = \zeta_2^5 \qquad\qquad\qquad w_2 = -\zeta_4^5$$
$$z_3 = \sqrt{5}\zeta_1^4\zeta_2 \qquad\qquad w_3 = -\sqrt{5}\zeta_3^4\zeta_4$$
$$z_4 = \sqrt{5}\zeta_1^4\zeta_3 \qquad\qquad w_4 = \sqrt{5}\zeta_3^4\zeta_1$$
$$z_5 = \sqrt{5}\zeta_1^4\zeta_4 \qquad\qquad w_5 = \sqrt{5}\zeta_3^4\zeta_2$$
$$z_6 = \sqrt{5}\zeta_2^4\zeta_1 \qquad\qquad w_6 = -\sqrt{5}\zeta_4^4\zeta_3$$
$$z_7 = \sqrt{5}\zeta_2^4\zeta_3 \qquad\qquad w_7 = \sqrt{5}\zeta_4^4\zeta_1$$
$$z_8 = \sqrt{5}\zeta_2^4\zeta_4 \qquad\qquad w_8 = \sqrt{5}\zeta_4^4\zeta_2$$
$$z_9 = \sqrt{10}\zeta_1^3\zeta_2^2 \qquad\qquad w_9 = -\sqrt{10}\zeta_3^3\zeta_4^2$$
$$z_{10} = \sqrt{10}\zeta_1^3\zeta_3^2 \qquad\qquad w_{10} = -\sqrt{10}\zeta_3^3\zeta_1^2$$
$$z_{11} = \sqrt{10}\zeta_1^3\zeta_4^2 \qquad\qquad w_{11} = -\sqrt{10}\zeta_3^3\zeta_2^2$$
$$z_{12} = \sqrt{10}\zeta_2^3\zeta_1^2 \qquad\qquad w_{12} = -\sqrt{10}\zeta_4^3\zeta_3^2$$
$$z_{13} = \sqrt{10}\zeta_2^3\zeta_3^2 \qquad\qquad w_{13} = -\sqrt{10}\zeta_4^3\zeta_1^2$$
$$z_{14} = \sqrt{10}\zeta_2^3\zeta_4^2 \qquad\qquad w_{14} = -\sqrt{10}\zeta_4^3\zeta_2^2$$
$$z_{15} = \sqrt{20}\zeta_1^3\zeta_2\zeta_3 \qquad\quad w_{15} = \sqrt{20}\zeta_3^3\zeta_1\zeta_4$$
$$z_{16} = \sqrt{20}\zeta_1^3\zeta_2\zeta_4 \qquad\quad w_{16} = \sqrt{20}\zeta_3^3\zeta_2\zeta_4$$
$$z_{17} = \sqrt{20}\zeta_1^3\zeta_3\zeta_4 \qquad\quad w_{17} = -\sqrt{20}\zeta_3^3\zeta_1\zeta_2$$
$$z_{18} = \sqrt{20}\zeta_2^3\zeta_1\zeta_3 \qquad\quad w_{18} = \sqrt{20}\zeta_4^3\zeta_1\zeta_3$$
$$z_{19} = \sqrt{20}\zeta_2^3\zeta_1\zeta_4 \qquad\quad w_{19} = \sqrt{20}\zeta_4^3\zeta_2\zeta_3$$
$$z_{20} = \sqrt{20}\zeta_2^3\zeta_3\zeta_4 \qquad\quad w_{20} = -\sqrt{20}\zeta_4^3\zeta_1\zeta_2$$
$$z_{21} = \sqrt{30}\zeta_1^2\zeta_2^2\zeta_3 \qquad\quad w_{21} = \sqrt{30}\zeta_3^2\zeta_4^2\zeta_1$$
$$z_{22} = \sqrt{30}\zeta_1^2\zeta_2^2\zeta_4 \qquad\quad w_{22} = \sqrt{30}\zeta_3^2\zeta_4^2\zeta_2$$
$$z_{23} = \sqrt{30}\zeta_1^2\zeta_3^2\zeta_2 \qquad\quad w_{23} = -\sqrt{30}\zeta_3^2\zeta_1^2\zeta_4$$
$$z_{24} = \sqrt{30}\zeta_1^2\zeta_4^2\zeta_2 \qquad\quad w_{24} = -\sqrt{30}\zeta_3^2\zeta_2^2\zeta_4$$
$$z_{25} = \sqrt{30}\zeta_1^2\zeta_4^2\zeta_3 \qquad\quad w_{25} = \sqrt{30}\zeta_3^2\zeta_2^2\zeta_1$$
$$z_{26} = \sqrt{30}\zeta_2^2\zeta_4^2\zeta_1 \qquad\quad w_{26} = -\sqrt{30}\zeta_2^2\zeta_4^2\zeta_3$$
$$z_{27} = \sqrt{60}\zeta_1^2\zeta_2\zeta_3\zeta_4 \qquad\quad w_{27} = -\sqrt{60}\zeta_3^2\zeta_1\zeta_2\zeta_4$$
$$z_{28} = \sqrt{60}\zeta_2^2\zeta_1\zeta_3\zeta_4 \qquad\quad w_{28} = -\sqrt{60}\zeta_4^2\zeta_1\zeta_2\zeta_3$$

13
3-Sasakian Manifolds

13.1 3-Sasakian manifolds

As with the last chapter we will give more of a survey and only a few proofs. Another survey of both history and recent work on 3-Sasakian manifolds is Boyer and Galicki [1999].

If a manifold M^{2m+1} admits three almost contact structures $(\phi_i, \xi_i, \eta_i), i = 1, 2, 3$ satisfying the following for an even permutation (i, j, k) of $(1, 2, 3)$,

$$\phi_k = \phi_i \phi_j - \eta_j \otimes \xi_i = -\phi_j \phi_i + \eta_i \otimes \xi_j,$$

$$\xi_k = \phi_i \xi_j = -\phi_j \xi_i, \quad \eta_k = \eta_i \circ \phi_j = -\eta_j \circ \phi_i,$$

then the manifold is said to have an *almost contact 3-structure*. This notion was introduced by Kuo [1970] and independently under the name *almost coquaternion structure* by Udriste [1969]. Some authors follow different sign conventions, taking $\phi_k = -\phi_i \phi_j + \eta_j \otimes \xi_i$, etc. (see e.g., the latter part of Section 13.2 and Baikoussis and Blair [1995]). Note that given two almost contact metric structures, satisfying

$$\phi_1 \phi_2 - \eta_2 \otimes \xi_1 = -\phi_2 \phi_1 + \eta_1 \otimes \xi_2,$$

$$\phi_1 \xi_2 = -\phi_2 \xi_1, \quad \eta_1 \circ \phi_2 = -\eta_2 \circ \phi_1, \quad \eta_2(\xi_1) = \eta_1(\xi_2) = 0,$$

there exists a third almost contact structure defined by

$$\phi_3 = \phi_1 \phi_2 - \eta_2 \otimes \xi_1, \quad \xi_3 = \phi_1 \xi_2, \quad \eta_3 = -\eta_2 \circ \phi_1$$

giving an almost contact 3-structure.

Now given an almost contact 3-structure (ϕ_i, ξ_i, η_i), define on $M^{2m+1} \times \mathbb{R}$ three almost complex structures J_i using each of the almost contact structures as in Section 6.1. It is then easy to check that $J_k = J_i J_j = -J_j J_i$. Therefore $M^{2m+1} \times \mathbb{R}$ has an almost quaternionic structure and hence its dimension is a multiple of 4. Thus the dimension of a manifold with an almost contact 3-structure is of the form $4n + 3$. Tachibana and Yu [1970] used this idea to show that there cannot be a fourth almost contact structure (ϕ_4, ξ_4, η_4) with $\eta_i(\xi_4) = \eta_4(\xi_i) = 0, i = 1, 2, 3$, and satisfying the anti-commutativity conditions with the first three structures. To see this let J_4 be the almost complex structure on $M^{2m+1} \times \mathbb{R}$ constructed using (ϕ_4, ξ_4, η_4). Then pairing J_4 with each of J_1, J_2, J_3 giving $J_4 J_i = -J_i J_4, i = 1, 2, 3$. This contradicts $J_3 J_4 = J_1 J_2 J_4 = -J_1 J_4 J_2 = J_4 J_1 J_2 = J_4 J_3$.

The normality of these almost contact structures was discussed by Yano, Ishihara and Konishi [1973]. In particular if two of the almost contact structures are normal, then so is the third.

Kuo [1970] proved that given an almost contact 3-structure, there exists a Riemannian metric compatible with each of then and hence we can speak of an *almost contact metric 3-structure* $(\phi_i, \xi_i, \eta_i, g), i = 1, 2, 3$. He also showed that the structural group of the tangent bundle is reducible to $Sp(n) \times I_3$. Moreover the vector fields $\{\xi_1, \xi_2, \xi_3\}$ are orthonormal with respect to the compatible metric.

If the three structures $(\phi_i, \xi_i, \eta_i, g)$ are contact metric structures we say that M^{4n+3} has a *contact metric 3-structure*. If the three structures are Sasakian, we say that M^{4n+3} has 3-*Sasakian structure*, also called a *Sasakian 3-structure* and M^{4n+3} is a 3-*Sasakian manifold*. Similarly we could define, and will occasionally use, the term *K-contact 3-structure*; however Kashiwada [toap] recently proved the very remarkable result that every contact metric 3-structure is 3-Sasakian and we now give the proof in detail. The crux of Kashiwada's proof is a generalization of a lemma of Hitchin [1987] to the effect that given three almost complex structures J_i, $i = 1, 2, 3$, satisfying $J_k = J_i J_j = -J_j J_i$ and a metric g that is compatible with each of them, then if each of the three fundamental 2-forms is closed, each J_i is integrable. Kashiwada [1998] generalized this lemma as follows.

Lemma 13.1 *Let M^{4n} be a differentiable manifold admitting three almost complex structures J_i, $i = 1, 2, 3$, satisfying $J_k = J_i J_j = -J_j J_i$ and a metric g that is compatible with each of them, and let Ω_i, $i = 1, 2, 3$, be the corresponding fundamental 2-forms. If $d\Omega_i = 2\omega \wedge \Omega_i$, $i = 1, 2, 3$, for some 1-form ω, then each J_i is integrable.*

Proof. First observe that

$$J_1(\nabla_Z J_2)J_1 = -J_1 \nabla_Z J_3 - J_3 \nabla_Z J_1 = -J_1 \nabla_Z J_3 - \nabla_Z J_2 + (\nabla_Z J_3)J_1$$

and hence that

$$(\nabla_Z \Omega_2)(J_1 X, J_1 Y) = -g(X, J_1(\nabla_Z J_2)J_1 Y)$$

$$= (\nabla_Z \Omega_2)(X, Y) - (\nabla_Z \Omega_3)(J_1 X, Y) - (\nabla_Z \Omega_3)(X, J_1 Y). \qquad (*)$$

Now recall that the Nijenhuis tensor of J_1 can be written

$$[J_1, J_1](X, Y) = (J_1 \nabla_Y J_1 - \nabla_{J_1 Y} J_1)X - (J_1 \nabla_X J_1 - \nabla_{J_1 X} J_1)Y.$$

Then

$$J_2[J_1, J_1](X, Y) = -J_3(\nabla_Y J_1)X - J_2(\nabla_{J_1 Y} J_1)X + J_3(\nabla_X J_1)Y + J_2(\nabla_{J_1 X} J_1)Y$$

and by straightforward computation

$$g(J_2[J_1, J_1](X, Y), Z) = -3d\Omega_2(X, Y, Z) + (\nabla_Z \Omega_2)(X, Y)$$

$$+3d\Omega_3(X, J_1 Y, Z) - (\nabla_Z \Omega_3)(X, J_1 Y)$$

$$+3d\Omega_3(J_1 X, Y, Z) - (\nabla_Z \Omega_3)(J_1 X, Y)$$

$$+3d\Omega_2(J_1 X, J_1 Y, Z) - (\nabla_Z \Omega_2)(J_1 X, J_1 Y).$$

Equation $(*)$ and $d\Omega_i = 2\omega \wedge \Omega_i$ then yield $[J_1, J_1] = 0$. Similarly one proves that J_2 and J_3 are integrable. ∎

We now prove Kashiwada's theorem.

Theorem 13.1 *Every contact metric 3-structure is 3-Sasakian.*

Proof. Let M^{4n+3} be manifold with contact metric 3-structure $(\phi_i, \xi_i, \eta_i, g)$, $i = 1, 2, 3$. On $M^{4n+3} \times \mathbb{R}^1$ define three almost Hermitian structures (J_i, G) as in Chapter 6, i.e.,

$$J_i(X, f\frac{d}{dt}) = (\phi_i X - f\xi_i, \eta_i(X)\frac{d}{dt}), \quad G = g + dt^2.$$

Then $J_k = J_i J_j = -J_j J_i$ and the fundamental 2-forms Ω_i satisfy $d\Omega_i = 2\omega \wedge \Omega_i$ where $\omega = -dt$. Now by Lemma 13.1 each J_i is integrable and hence each contact metric structure $(\phi_i, \xi_i, \eta_i, g)$ is Sasakian. ∎

As we remarked at the end of Section 6.5, Boyer and Galicki [1999] define a 3-Sasakian manifold by considering the cone $\mathbb{R}_+ \times M$. A Riemannian manifold (M^m, g) is a 3-Sasakian manifold if and only if the holonomy group of the cone $(\mathbb{R}_+ \times M, dr^2 + r^2 g)$ reduces to a subgroup of $Sp(\frac{m+1}{4})$. Again we see that $m = 4n + 3$, $n \geq 1$. Moreover $(\mathbb{R}_+ \times M, dr^2 + r^2 g)$ is hyper-Kähler, i.e., it has a quaternionic structure consisting of three global almost complex structures which are Kähler with respect to the metric $dr^2 + r^2 g$. For a proof of the equivalence of these definitions see Boyer, Galicki, Mann [1994].

Again if M^{4n+3} has two Sasakian structures $(\phi_i, \xi_i, \eta_i, g), i = 1, 2$ with ξ_1 and ξ_2 orthogonal then the third structure defined by $\xi_3 = \phi_1 \xi_2$ and $\phi_3 X = -\nabla_X \xi_3$ gives a Sasakian 3-structure. Exploring this idea further, Tachibana and Yu [1970] prove the following theorem.

Theorem 13.2 *Let M be a complete simply connected Riemannian manifold admitting Sasakian structures $(\phi_i, \xi_i, \eta_i, g), i = 1, 2$ such that $g(\xi_1, \xi_2)$ is a non-constant function on M; then M is isometric to a unit sphere.*

The proof of this theorem is to differentiate the function $f = g(\xi_1, \xi_2)$ twice and use the Sasakian conditions to show that f satisfies $\nabla_l \nabla_k f_j + 2 f_l g_{kj} + f_k g_{lj} + f_j g_{lk} = 0$ where $f_j = \nabla_j f$. Then by a well-known theorem of Obata [1965] (see also Tanno [1978a]) M is isometric to a unit sphere.

Using $\nabla_X \xi_i = -\phi_i X$ one readily obtains on a manifold with a 3-Sasakian structure that $[\xi_i, \xi_j] = 2\xi_k$. Thus the subbundle spanned by $\{\xi_1, \xi_2, \xi_3\}$ is integrable with totally geodesic leaves which are easily seen to be of constant curvature $+1$. Notice also that from $\phi_i \xi_j = \xi_k$, etc., each leaf of the foliation is itself a 3-Sasakian manifold.

The canonical example of a manifold with a Sasakian 3-structure is the sphere S^{4n+3}. Its structure is readily obtained by taking S^{4n+3} as hypersurface in $\mathbb{H}^{n+1}$. Each of the three almost complex structures forming the quaternionic structure of $\mathbb{H}^{n+1}$ applied to the outer normal of the sphere gives a vector field $\xi_i, i = 1, 2, 3$ on S^{4n+3}. These three vector fields are orthogonal and give rise to the standard Sasakian 3-structure on S^{4n+3}. This Sasakian 3-structure on S^{4n+3} also projects under the Hopf fibration to the quaternionic structure on quaternionic projective space $\mathbb{H}P^n$. We will briefly discuss generalizations of this fibration below.

We first however prove an early result of Kashiwada [1971] that a 3-Sasakian manifold is Einstein; this can be regarded as analogous to the well-known fact that a quaternionic Kähler manifold is Einstein.

Theorem 13.3 *Let M^{4n+3} be a manifold with a Sasakian 3-structure*

$$(\phi_i, \xi_i, \eta_i, g), \quad i = 1, 2, 3.$$

Then M^{4n+3} is an Einstein space with positive scalar curvature.

Proof. First taking an orthonormal basis $\{e_A\}$, observe that from equation a) of Lemma 7.1, we have, summing on A,

$$\sum g(R_{e_A Y} Z, \phi e_A) + \rho(Y, \phi Z) = -\sum P(e_A, Y, Z, e_A) = (\dim M - 2)d\eta(Y, Z)$$

for any Sasakian structure (ϕ, ξ, η, g). Now consider equation a) of Lemma 7.1 for the first Sasakian structure and set $Z = e_A$ and $W = \phi_2 e_A$, i.e.,

$$g(R_{XY}e_A, \phi_1\phi_2 e_A) + g(R_{XY}\phi_1 e_A, \phi_2 e_A) = -P(X, Y, e_A, \phi_2 e_A).$$

In the first term replace $\phi_1\phi_2 e_A$ by $\phi_3 e_A + \eta_2(e_A)\xi_1$. Now the idea is to sum on A, but in the second term assume that ξ_1 is part of the basis and replace the basis $\{e_A\}$ by $\{\phi_1 e_A, \xi_1\}$. The result of the sum is $2\sum g(R_{XY}e_A, \phi_3 e_A) = 4g(X, \phi_3 Y)$. Canceling the factor of 2 and applying the Bianchi identity we have

$$-\sum g(R_{e_A X} Y, \phi_3 e_A) + \sum g(R_{e_A Y} X, \phi_3 e_A) = 2g(X, \phi_3 Y).$$

Using our first observation applied to the third Sasakian structure yields

$$\rho(X, \phi_3 Y) = (4n + 2)g(X, \phi_3 Y)$$

which with the fact that $Q\xi = (\dim M - 1)\xi$ for any Sasakian structure gives the result.   ∎

As a corollary, or from our previous remarks, we see that a 3-dimensional manifold with a Sasakian 3-structure is of constant curvature $+1$.

Additional curvature properties of 3-Sasakian manifolds include the following. For a vector X orthogonal to $\{\xi_1, \xi_2, \xi_3\}$ $(4n + 3 \geq 7)$, the sum of the three sectional ϕ-sectional curvatures satisfies

$$\sum_{i=1}^{3} K(X, \phi_i X) = 3,$$

Tanno [1971]. If the sectional curvature of plane sections orthogonal to $\{\xi_1, \xi_2, \xi_3\}$ is constant $(4n + 3 \geq 7)$, the 3-Sasakian manifold is of constant curvature $+1$, Konishi and Funabashi [1976].

All complete 3-dimensional 3-Sasakian manifolds were classified by Sasaki [1972] in the following theorem.

Theorem 13.4 *A complete 3-dimensional 3-Sasakian manifold is a quotient S^3/Γ where Γ is one of the following finite subgroups of Clifford translations on S^3:*

1. $\Gamma = \{I\}$,

2. $\Gamma = \{\pm I\}$,

3. Γ is the cyclic group of order $q > 2$ generated by

$$
\begin{pmatrix}
\cos\frac{2\pi}{q} & -\sin\frac{2\pi}{q} & 0 & 0 \\
\sin\frac{2\pi}{q} & \cos\frac{2\pi}{q} & 0 & 0 \\
0 & 0 & \cos\frac{2\pi}{q} & -\sin\frac{2\pi}{q} \\
0 & 0 & \sin\frac{2\pi}{q} & \cos\frac{2\pi}{q}
\end{pmatrix},
$$

4. Γ is a group of Clifford translations corresponding to a binary dihedral group or the binary polyhedral groups of the regular tetrahedron, octahedron or icosahedron.

Boyer, Galicki and Mann [1994] classified all 3-Sasakian homogeneous spaces.

Theorem 13.5 *Let M be a 3-Sasakian homogeneous space. Then M is one of the following:*

$$
\frac{Sp(n+1)}{SP(n)} \simeq S^{4n+3}, \qquad \frac{Sp(n+1)}{SP(n) \times \mathbb{Z}_2} \simeq \mathbb{R}P^{4n+3},
$$

$$
\frac{SU(m)}{S(U(m-2) \times U(1))}, \qquad \frac{SO(k)}{SO(k-4) \times Sp(1)},
$$

$$
\frac{G_2}{Sp(1)}, \quad \frac{F_4}{Sp(3)}, \quad \frac{E_6}{SU(6)}, \quad \frac{E_7}{Spin(12)}, \quad \frac{E_8}{E_7}.
$$

For the first two cases when $n = 0$, $Sp(0)$ is the identity group; $m \geq 3$; and $k \geq 7$. Moreover M fibres over a quaternionic Kähler manifold; the fibre is $Sp(1)$ for S^{4n+3} and $SO(3)$ in the other cases.

There are in addition many inhomogeneous 3-Sasakian spaces, even simply connected ones. Boyer, Galicki and Mann [1994] prove that there are infinitely many homotopically distinct 7-dimensional strongly inhomogeneous compact simply connected 3-Sasakian manifolds.

Let (M, ϕ, ξ, η, g) be a Sasakian manifold with complete Riemannian metric g and denote by $I(M)$ the isometry group of g. Let $A(M)$ be the automorphism group of the Sasakian structure (ϕ, ξ, η, g) i.e., $A(M)$ is the subgroup of isometries which also preserve ϕ, ξ and η. In [1970] Tanno proved the following theorem.

Theorem 13.6 *Let (M, ϕ, ξ, η, g) be a complete Sasakian manifold which is not of constant curvature. Then either*

$$
\dim I(M) = \dim A(M) \text{ or } \dim I(M) = \dim A(M) + 2.
$$

Moreover $\dim I(M) = \dim A(M)$ if and only if M does not admit a Sasakian 3-structure and $\dim I(M) = \dim A(M) + 2$ if and only if M admits a Sasakian 3-structure.

Turning to fibration questions, Tanno [1971, p. 325] proves that of the three dimensional 3-Sasakian manifolds in Theorem 13.4, only $S^3(1)$ and $\mathbb{R}P^3(1)$ are regular with respect to any of the three characteristic vector fields. Consequently we have the following result of Tanno [1971, p. 326].

Lemma 13.2 *Let $(M^{4n+3}, \phi_i, \xi_i, \eta_i, g)$, $i = 1, 2, 3$ be a complete Riemannian manifold with a K-contact 3-structure. If ξ_1 is regular, then so are ξ_2 and ξ_3 and the leaves of the foliation induced by $\{\xi_1, \xi_2, \xi_3\}$ are isometric to $S^3(1)$ and $\mathbb{R}P^3(1)$.*

Moreover M^{4n+3} is a fibre bundle over an almost quaternionic Kähler manifold M^{4n} which is both Einstein and quaternionic Kähler since the structure is 3-Sasakian by Kashiwada's result (Theorem 13.1). These properties of 3-Sasakian manifolds have been proven by various authors at different points in time. Under the assumptions of regularity, Tanno proved in [1971] that the base manifold of the fibration is Einstein. When $n \geq 2$ and the structure is 3-Sasakian, Ishihara [1973] proved that the base manifold is quaternionic Kähler. Conversely if the induced structure on the base manifold is quaternionic Kähler, then the structure is 3-Sasakian, (Konishi [1973]). The reason for the restriction $n \geq 2$ is that the usual definition of a quaternionic Kähler manifold M^{4n} in terms of holonomy breaks down in dimension 4; more precisely, the condition that holonomy group be contained in $Sp(n) \cdot Sp(1)$ $(= Sp(n) \times Sp(1)/\{\pm I\}) \subset SO(4n)$ becomes in dimension 4 simply the orientability of the manifold, since $Sp(1) \cdot Sp(1) = SO(4)$. Thus in dimension 4 a manifold is said to be a quaternionic Kähler manifold if it is Einstein with non-zero scalar curvature and self-dual (see e.g., LeBrun [1991]). With this understanding Tanno [1996] extended Ishihara's result to include $n = 1$. With Kashiwada's recent result in mind, we summarize this discussion in the following theorem.

Theorem 13.7 *Let $(M^{4n+3}, \phi_i, \xi_i, \eta_i, g)$, $i = 1, 2, 3$ be a complete Riemannian manifold with a contact metric 3-structure. If one of the ξ_1's is regular, then M^{4n+3} is an $Sp(1)$ or $SO(3)$ bundle over a quaternionic Kähler manifold M^{4n}.*

The converse question was considered by Konishi in [1975], i.e., starting with a quaternionic Kähler manifold M^{4n} with $n > 1$, she constructs a canonical $SO(3)$ bundle, M^{4n+3}, over M^{4n}. When the scalar curvature of M^{4n} is positive, M^{4n+3} has a Sasakian 3-structure. Tanno [1996] extends this construction to the case $n = 1$ and to the case of negative scalar curvature. We state these results as follows.

Theorem 13.8 *Let M^{4n} be a quaternionic Kähler manifold with non-zero scalar curvature. Then there exists a canonical $SO(3)$ bundle, M^{4n+3}, over M^{4n}. If the scalar curvature of M^{4n} is positive, M^{4n+3} admits a Sasakian 3-*

structure; if the scalar curvature is negative M^{4n+3} admits a pseudo-Sasakian 3-structure, the signature of the metric g being $(3, 4n)$.

In dimensions 7 and 11 the complete principal Riemannian fibrations with Sasakian 3-structure of positive scalar curvature were determined by Boyer, Galicki and Mann [1993, p.253]. They are: S^7, $\mathbb{R}P^7$ and $SU(3)/U(1)$ in dimension 7, and S^{11}, $\mathbb{R}P^{11}$, $SU(4)/S(U(2) \times U(1))$ and $G_2/SU(2)$ in dimension 11.

In general the foliation determined by the vector fields $\{\xi_1, \xi_2, \xi_3\}$ is not a fibration and Boyer, Galicki and Mann [1994] prove the following result.

Theorem 13.9 *Let $(M^{4n+3}, \phi_i, \xi_i, \eta_i, g)$, $i = 1, 2, 3$ be a 3-Sasakian manifold such that the vector fields ξ_1, ξ_2, ξ_3 are complete; then the space of leaves of the foliation determined by $\{\xi_1, \xi_2, \xi_3\}$ is a quaternionic Kähler orbifold of dimension $4n$ with positive scalar curvature $16n(n + 2)$.*

In view of our study of twistor spaces as complex contact manifolds over quaternionic Kähler manifolds, we also consider fibrations of a manifold with a contact metric 3-structure by one of the structure vector fields. The following result of Ishihara and Konishi [1979] is not surprising.

Theorem 13.10 *Let M^{4n+3} be a manifold with a contact metric 3-structure and suppose that one of the structures, say $(\phi_1, \xi_1, \eta_1, g)$, is regular and Sasakian; then the orbit space M^{4n+3}/ξ_1 admits a complex contact metric structure. Moreover if M^{4n+3} is a 3-Sasakian manifold, the complex contact metric structure on the orbit space is a Kähler–Einstein structure of positive scalar curvature.*

With Kashiwada's result the last statement is true without the additional hypothesis. Without the regularity, this question was taken up by Boyer and Galicki [1997] who gave an orbifold version and showed that the space $Z = M^{4n+3}/\xi_1$ is the twistor space of a quaternionic Kähler orbifold.

Some important results on the topology of a compact 3-Sasakian manifold are the following due to Galicki and Salamon [1996]. In Section 6.8 we noted that the odd Betti numbers of a compact Sasakian manifold are even up to middle dimension. Galicki and Salamon prove a stronger result for 3-Sasakian manifolds.

Theorem 13.11 *Let M^{4n+3} be a compact 3-Sasakian manifold. Then the odd Betti numbers b_{2k+1} are all zero for $0 \leq k \leq n$.*

In the regular case Galicki and Salamon [1996] relate the Betti numbers of M^{4n+3} to those of the quaternionic Kähler base space M^{4n} and the intermediate space Z obtained as the Boothby–Wang fibration with respect to one of the characteristic vector fields. In particular the odd Betti numbers of these spaces vanish (cf. Salamon [1982]). The complex structure on Z leads to the fact that $\mathbb{R}P^{4n+3}$ is the only compact regular 3-Sasakian manifold which is not simply-connected. Galicki and Salamon also prove the following remarkable result.

Theorem 13.12 *Up to isometries, there are only finitely many compact regular 3-Sasakian manifolds in each dimension $4n+3$, $n \geq 0$ and the only compact regular 3-Sasakian manifolds with $b_2 > 0$ are the spaces*

$$\frac{U(m)}{U(m-2) \times U(1)}, \quad m \geq 3.$$

Galicki and Salomon also obtain results on certain sums of Betti numbers and the relation $b_{2k}(M^{4n+3}) = b_{2k}(Z) - b_{2k-2}(Z) = b_{2k}(M^{4n}) - b_{2k-4}(M^{4n})$, $k \leq n$. Again an excellent survey of recent work on 3-Sasakian manifolds is Boyer and Galicki [1999].

13.2   Integral submanifolds

When we have one contact structure on a manifold of dimension $2n + 1$ we have seen that the maximum dimension of an integral submanifold is n. In the present context we have three independent contact structures on a manifold of dimension $4n + 3$ and we begin with the following lemma.

Lemma 13.3 *The maximum dimension of a submanifold which is an integral submanifold of all three contact structures is n.*

Proof. If $X_1, \ldots, X_r$ is a local basis tangent to such a submanifold, then the $\phi_i X_a$, $i = 1, 2, 3, a = 1, \ldots, r$ are normal to the submanifold as well as perpendicular to $\xi_i, i = 1, 2, 3$ since $\phi_i \xi_i = \xi_k$. Moreover from $\phi_k = \phi_i \phi_j - \eta_j \otimes \xi_i$ these vectors are independent. Thus the codimension is at least $3r + 3$ and hence $r \leq n$. ∎

Thus in dimension 7 these would be integral (Legendre) curves and that they are plentiful can be seen as follows. The characteristic vector fields of the standard Sasakian 3-structure on S^7 are tangent to the fibres S^3 of S^7 viewed as a principal S^3-bundle over S^4. Thus the horizontal lift of a curve on S^4 gives a curve in S^7 which is a Legendre curve for each of the three contact structures. Curves in S^7 which are Legendre curves for all three contact structures and of constant curvature and unit torsion were classified by Baikoussis and Blair in [1995].

In contrast to this lemma it is possible to have a submanifolds of dimension up to $2n + 1$ which are integral submanifolds of two of the three contact structures. We will discuss this briefly in the case of S^7. Note first that if

M^3 is a 3-dimensional submanifold of a manifold with Sasakian 3-structure which is an integral submanifold of the first Sasakian structure $(\phi_1, \xi_1, \eta_1, g)$ and an invariant submanifold with respect to the third Sasakian structure $(\phi_3, \xi_3, \eta_3, g)$, then it is an integral submanifold of the second Sasakian structure $(\phi_2, \xi_2, \eta_2, g)$. To see this let X be tangent to M^3; then $\eta_2(X) = g(\xi_2, X) = g(\phi_3\xi_1, X) = -g(\xi_1, \phi_3 X) = -\eta_1(\phi_3 X) = 0$.

To give a couple of examples consider Euclidean space E^8 with three complex structures (or quaternionic structure),

$$I = \begin{pmatrix} 0 & -I_4 \\ I_4 & 0 \end{pmatrix}, \quad J = \begin{pmatrix} 0 & 0 & 0 & I_2 \\ 0 & 0 & -I_2 & 0 \\ 0 & I_2 & 0 & 0 \\ -I_2 & 0 & 0 & 0 \end{pmatrix}, \quad K = -IJ$$

where I_n denotes the $n \times n$ identity matrix. (The sign conventions of Baikoussis and Blair [1995] differ from the above.) Let $\mathbf{x}$ denote the position vector of the unit sphere in E^8 and as usual define three vector fields on S^7 by $\xi_1 = -I\mathbf{x}$, $\xi_2 = -J\mathbf{x}$, $\xi_3 = -K\mathbf{x}$. The dual 1-forms η_i are three independent contact structures on S^7. Now consider the linear subspace L of E^8 defined by $x_5 = x_6 = x_7 = x_8 = 0$. The intersection $S^3 = S^7 \cap L$ is then an integral submanifold for η_1 and η_2, but ξ_3 is tangent and this sphere is invariant under the action of K restricted to S^7. Now consider the torus in the 3-sphere we just constructed defined by $x_1^2 + x_3^2 = \frac{1}{2}$, $x_2^2 + x_4^2 = \frac{1}{2}$. This torus is an integral surface for both η_1 and η_2 but ξ_3 is tangent. Note also that the image of the tangent space under ϕ_3 is normal.

We now state the following result from Baikoussis and Blair [1995].

Theorem 13.13 *Let M^3 be a 3-dimensional submanifold of S^7 isometrically immersed as an integral submanifold of the Sasakian structure $(\phi_1, \xi_1, \eta_1, \bar{g})$ (and $(\phi_2, \xi_2, \eta_2, \bar{g})$) and an invariant submanifold of the third Sasakian structure $(\phi_3, \xi_3, \eta_3, \bar{g})$. Then M^3 is a principal circle bundle over a holomorphic Legendre curve in $\mathbb{C}P^3$. Moreover if M^3 is of constant ϕ-sectional curvature, then either M^3 is totally geodesic or a principal circle bundle over the holomorphic (Calabi) curve $\mathbb{C}P^1(\frac{4}{3})$.*

As a corollary we note that a holomorphic Legendre curve of constant Gaussian curvature in $\mathbb{C}P^3$, cannot lie in a totally geodesic subspace $\mathbb{C}P^2$. Similarly if M^3 is a 3-dimensional submanifold of S^7, isometrically immersed as an integral submanifold of one of the Sasakian structures and as an invariant submanifold of one of the other Sasakian structures, and if M^3 is of constant ϕ-sectional curvature and not totally geodesic, then M^3 cannot lie in any totally geodesic Sasakian submanifold (S^5) of S^7.

We now turn to the case of a 2-dimensional submanifold of S^7 which is an integral submanifold of two of the Sasakian structures. Even though it cannot be an integral submanifold for the third structure, we still can have that ϕ_3 maps the tangent bundle into the normal bundle as in the example of a torus. In this regard we have the following result of Baikoussis and Blair [1995].

Theorem 13.14 *Let M^2 be a surface isometrically immersed in S^7 as an integral submanifold of the Sasakian structures $(\phi_1, \xi_1, \eta_1, \bar{g})$ and $(\phi_2, \xi_2, \eta_2, \bar{g})$ and suppose that $\phi_3(TM) \subset T^\perp M$. Then M^2 is flat and ξ_3 is tangent to M^2.*

Bibliography

Abbena, E.,
  [1984] An example of an almost Kähler manifold which is not Kählerian, *Bollettino U.M.I.*, (6) **3**-A, 383–392.

Abe, K.
  [1976] Some examples of non-regular almost contact structures on exotic spheres, *Tôhoku Math. J.*, **28**, 429–435.
  [1977] On a generalization of the Hopf fibration, I, *Tôhoku Math. J.*, **29**, 335–374.

Abe, K. and Erbacher, J.
  [1975] Non-regular contact structures on Brieskorn manifolds, *Bull. Amer. Math. Soc.*, **81**, 407–409.

Alekseevskii, D. V.
  [1968] Riemannian spaces with exceptional holonomy groups, *Funktional Anal. i Priložen*, **2**, 1-10; English translation: *Functional Anal. Appl.*, **2**, 97–105.

Altschuler, S. and Wu, L.
  [2000] On deforming confoliations, *J. Diff. Geom.*, **54**, 75–97.

Anosov, D. V.
  [1967] Geodesic Flows on Closed Riemann Manifolds with Negative Curvature, *Proc. Steklov Inst. Math.*, **90**; English translation: *Amer. Math. Soc.*, (1969).

Apostolov, V. and Draghici, T.
  [1999] Hermitian conformal classes and almost Kähler structures on 4-manifolds, *Differential Geom. and App.*, **11**, 179–195.

Apostolov, V., Draghici, T. and Moroianu, A.
[2001] A splitting theorem for Kähler manifolds whose Ricci tensors have constant eigenvalues, *International J. Math.*, **12**, 769–789.

Armstrong, J.
[1998] Almost Kähler geometry, Ph.D. Thesis, Oxford.

Auslander, L., Green, L. and Hahn, F.
[1963] Flows on Homogeneous Spaces, *Annals of Math. Studies*, **53**, Princeton.

Baikoussis, C. and Blair, D. E.
[1992] Integral surfaces of Sasakian space forms, *J. Geom.*, **43**, 30–40.
[1994] On Legendre curves in contact 3-manifolds, *Geometriae Dedicata*, **49**, 135–142.
[1995] On the geometry of the 7-sphere, *Results in Math.*, **27**, 5–16.

Baikoussis, C., Blair, D. E. and Gouli-Andreou, F.
[1998] Holomorphic Legendre curves in the complex Heisenberg group, *Bull. Inst. Math. Acad. Sinica*, **26**, 179–194.

Baikoussis, C., Blair, D. E. and Koufogiorgos, Th.
[1995] Integral submanifolds of Sasakian space forms $\bar{M}^7(k)$, *Results in Math.*, **27**, 207–226.

Bang, K.
[1994] Riemannian Geometry of Vector Bundles, Thesis, Michigan State University.

Banyaga, A.
[1990] A note on Weinstein's conjecture, *Proc. Amer. Math. Soc.*, **109**, 855–858.

Banyaga, A. and Rukimbira, P.
[1994] Weak stability of almost regular contact foliations, *J. Geom.*, **50**, 16–27.

Bejancu, A.
[1978] CR-submanifolds of a Kaehler manifold I, *Proc. Amer. Math. Soc.*, **69**, 135–142.

Belgun, F. A.
[2000] On the metric structure of non-Kähler complex surfaces, *Math. Ann.* **317**, 1–40.

Bennequin, D.
[1983] Entrelacements et équations de Pfaff, *Astérisque*, **107–108**, 87–161.

Berger, M.
[1966] Remarques sur le groupe d'holonomie des variétés Riemanniennes, *C.R. Acad. Sc. Paris*, **262**, 1316–1318.
[1970] Quelques formules de variation pour une structure riemannienne, *Ann. Sc. École Norm. Sup.*, **3**, 285–294.

Berger, M. and Ebin, D.

[1969] Some decompositions of the space of symmetric tensors on a Riemannian manifold, *J. Diff. Geom.*, **3**, 379–392.

Berndt, J.

[1997] Riemannian geometry of complex two-plane Grassmannians, *Rend. Sem. Mat.*, Univ. Pol. Torino, **55**, 19–83.

Berndt, J. and Vanhecke, L.

[1999] ϕ-symmetric spaces and weak symmetry, *Bollettino U.M.I.*, (8) **2**-B, 389–392.

Besse, A. L.

[1987] *Einstein Manifolds*, Springer-Verlag, Berlin.

Bishop, R. L. and Crittenden, R. J.

[1964] *Geometry of Manifolds*, Academic Press, New York.

Blair, D. E.

[1967] The theory of quasi-Sasakian structures, *J. Diff. Geom.*, **1**, 331–345.

[1971] Almost contact manifolds with Killing structure tensors, *Pacific J. Math.*, **39**, 285–292.

[1976] On the non-existence of flat contact metric structures, *Tôhoku Math. J.*, **28**, 373–379.

[1977] Two remarks on contact metric structures, *Tôhoku Math. J.*, **29**, 319–324.

[1978] On the non-regularity of tangent sphere bundles, *Proc. Royal Soc. Edinburgh* **82A**, 13–17.

[1983] On the set of metrics associated to a symplectic or contact form, *Bull. Inst. Math. Acad. Sinica*, **11**, 297–308.

[1984] Critical associated metrics on contact manifolds, *J. Austral. Math. Soc.*, Series A, **37**, 82–88.

[1989] When is the tangent sphere bundle locally symmetric?, *Geometry and Topology*, World Scientific, Singapore, 15–30.

[1991a] The "total scalar curvature" as a symplectic invariant and related results, *Proc. 3rd Congress of Geometry*, Thessaloniki, 79–83.

[1991b] Critical associated metrics on contact manifolds III, *J. Austral. Math. Soc.*, Series A, **50**, 189–196.

[1996] On the class of contact metric manifolds with a 3-τ-structure, *Note di Mat.*, **16**, 99–104.

[1998] Special directions on contact metric manifolds of negative ξ-sectional curvature, *Ann. Fac. Sc. Toulouse VII*, 365–378.

[2000] Curvature and conformally Anosov flows in contact geometry, Congresso Internazionale in onore di Pasquale Calapso, *Rend. Sem. Mat. Messina*, 41–50.

Blair, D. E., and Carriazo, A.

[toap] The contact Whitney sphere, to appear.

230　　Bibliography

Blair, D. E., and Chen, B.-Y.
  [1979] On CR-submanifolds of Hermitian manifolds, *Israel J. Math.*, **34**, 353–363.

Blair, D. E., and Chen, H.
  [1992] A classification of 3-dimensional contact metric manifolds with $Q\phi = \phi Q$, II, *Bull. Inst. Math. Acad. Sinica*, **20**, 379–383.

Blair, D. E., Dillen, F., Verstraelen, L. and Vrancken, L.
  [1996] Calabi curves as holomorphic Legendre curves and Chen's inequality, *Kyungpook Math. J.*, **35**, 407–416.

Blair, D. E. and Goldberg, S. I.
  [1967] Topology of almost contact manifolds, *J. Diff. Geom.*, **1**, 347–354.

Blair, D. E. and Ianus, S.
  [1986] Critical associated metrics on symplectic manifolds, *Contemp. Math.*, **51**, 23–29.

Blair, D. E., Ishihara, S. and Ludden, G. D.
  [1978] Projectable almost complex contact structures, *Kōdai Math. J.*, **1**, 75–84.

Blair, D. E., Korkmaz, B. and Vrancken, L.
  [2000] The Calabi (Veronese) imbeddings as integral submanifolds of $\mathbb{C}P^{2n+1}$, *Glasgow Math. J.*, **42**, 183–193.

Blair, D. E. and Koufogiorgos, Th.
  [1994] When is the tangent sphere bundle conformally flat?, *J. Geom.* **49**, 55–66.

Blair, D. E., Koufogiorgos, Th. and Papantoniou, B. J.
  [1995] Contact metric manifolds satisfying a nullity condition, *Israel J. Math.*, **91**, 189–214.

Blair, D. E., Koufogiorgos, Th. and Sharma, R.
  [1990] A classification of 3-dimensional contact metric manifolds with $Q\phi = \phi Q$, *Kōdai Math. J.*, **13**, 391–401.

Blair, D. E. and Ledger, A. J.
  [1986] Critical associated metrics on contact manifolds II, *J. Austral. Math. Soc.*, Series A, **41**, 404–410.

Blair, D. E. and Oubina, J. A.
  [1990] Conformal and related changes of metric on the product of two almost contact metric manifolds, *Publicacions Matemàtiques*, **34**, 199–207.

Blair, D. E. and Perrone, D.
  [1992] A variational characterization of contact metric manifolds with vanishing torsion, *Canad. Math. Bull.*, **35**, 455–462.
  [1995] Second variation of the "total scalar curvature" on contact manifolds, *Canad. Math. Bull.*, **38**, 16–22.

[1998] Conformally Anosov flows in contact metric geometry, *Balkan J. Geom. and App.*, **3**, 33–46.

Blair, D. E. and Sharma, R.
   [1990a] Generalization of Myers' theorem on a contact manifold, *Ill. J. Math.*, **34**, 837–844.
   [1990b] Three dimensional locally symmetric contact metric manifolds, *Bollettino U.M.I.*, (7) 4-A, 385–390.

Blair, D. E., Showers, D. K. and Yano, K.
   [1976] Nearly Sasakian structures, *Kōdai Math. Sem. Rep.*, **27**, 175–180.

Blair, D. E. and Vanhecke, L.
   [1987a] New characterizations of ϕ-symmetric spaces, *Kōdai Math. J.*, **10**, 102–107.
   [1987b] Symmetries and ϕ-symmetric spaces, *Tôhoku Math. J.*, **39**, 373–383.
   [1987c] Volume-preserving ϕ-geodesic symmetries, *C. R. Math. Rep. Acad. Sci. Canada*, IX, 31–36.
   [1987d] Geodesic spheres and Jacobi vector fields on Sasakian space forms, *Proc. Royal Soc. Edinburgh*, **105A**, 17–22.

Boeckx, E.
   [1999] A class of locally ϕ-symmetric contact metric spaces, *Arch. Math.*, **72**, 466–472.
   [2000] A full classification of contact metric (k, μ)-spaces, *Ill. J. Math.*, **44**, 212–219.

Boeckx, E., Bueken, P. and Vanhecke, L.
   [1999] ϕ-symmetric contact metric spaces, *Glasgow Math. J.*, **41**, 409–416.

Boeckx, E. and Vanhecke, L.
   [1997] Characteristic reflections on unit tangent sphere bundles, *Houston J. Math.* **23** 427–448.

Boothby, W. M.
   [1961] Homogeneous complex contact manifolds, Proc. Symp. Pure Math. III, *Amer. Math. Soc.*, 144–154.
   [1962] A note on homogeneous complex contact manifolds, *Proc. Amer. Math. Soc.*, **13**, 276–280.
   [1969] Transitivity of the automorphisms of certain geometric structures, *Trans. Amer. Math. Soc.*, **137**, 93–100.

Boothby, W. M. and Wang, H. C.
   [1958] On contact manifolds, *Ann. of Math.*, **68** (1958), 721–734.

Borisenko, A. A. and Yampol'skii, A. L.
   [1987a] The sectional curvature of the Sasaki metric of $T_1 M^n$, *Ukr. Geom. Sb.*, **30**, 10–17; English translation: *J. Sov. Math.*, **51** (1990), 2503–2508.

[1987b] On the Sasaki metric of the normal bundle of a submanifold in a Riemannian space, *Mat. Sb.* **134** (176); English translation: *Math. USSR Sb.*, **62** (1989), 157–175.

Borrelli, V. Chen, B.-Y., and Morvan, J.-M.
[1995] Une caractérisation géométrique de la sphère de Whitney, *C. R. Acad. Sci. Paris*, **321**, 1485–1490.

Bouche, T.
[1990] La cohomologie coeffective d'une variété symplectique, *Bull. Sc. Math.*, **114**, 115–122.

Boyer, C. P. and Galicki, K.
[1997] The twistor space of a 3-Sasakian manifold, *Internat. J. Math.*, **8**, 31–60.
[1999] *3-Sasakian Manifolds, Essays on Einstein Manifolds*, International Press.
[2000] On Sasakian–Einstein geometry, *Internat. J. Math.*, **11**, 873–909.
[2001] Einstein manifolds and contact geometry, *Proc. Amer. Math. Soc.*, **129**, 2419–2430.

Boyer, C. P., Galicki, K. and Mann, B. M.
[1993] Quaternionic reduction and Einstein manifolds, *Communications in Analysis and Geometry*, **1**, 229–279.
[1994] The geometry and topology of 3-Sasakian manifolds, *J. reine angew. math.*, **455**, 183–220.

Brieskorn, E. and Van de Ven, A.
[1968] Some complex structures on products of homotopy spheres, *Topology*, **7**, 389–393.

Bryant, R. L.
[1982] Conformal and minimal immersions of compact surfaces into the 4-sphere, *J. Diff. Geom.*, **17**, 455–473.

Bueken, P. and Vanhecke, L.
[1988] Geometry and symmetry on Sasakian manifolds, *Tsukuba J. Math.*, **12**, 403–422.
[1989] Reflections in K-contact geometry, *Math. Rep. Toyama Univ.*, **12**, 41–49.
[1993] Reflections with respect to submanifolds in contact geometry, *Archivum Math.*, **29**, 43–57.

Calabi, E.
[1953a] *Metric Riemann Surfaces, Contributions to the Theory of Riemann Surfaces*, Princeton University Press, Princeton, 77–85.
[1953b] Isometric imbeddings of complex manifolds, *Ann. of Math.*, **58**, 1–23.

Calabi, E. and Eckmann, B.
[1953] A class of compact complex manifolds which are not algebraic, *Ann. of Math.* **58**, 494–500.

Calvaruso, G. and Perrone, D.
[2000] Torsion and homogeneity on contact metric three-manifolds, *Ann. Mat. Pura Appl.*, (4) **178**, 271–285.

Calvaruso, G., Perrone, D. and Vanhecke, L.
[1999] Homogeneity on three-dimensional contact metric manifolds, *Israel J. Math.*, **114**, 301–321.

Capursi, M.
[1984] Some remarks on the product of two almost contact manifolds, *An. Sti. Univ. "Al. I. Cuza"*, XXX, 75–79.

Cartan, E.
[1983] *Geometry of Riemannian Spaces*, Math. Sci. Press, Brookline, MA.

Chen, B.-Y.
[1997a] Interaction of Legendre curves and Lagrangian submanifolds, *Israel J. Math.*, **99**, 69–108.
[1997b] Complex extensors and Lagrangian submanifolds in complex Euclidean spaces, *Tôhoku Math. J.*, **49**, 277–297.
[1998] Intrinsic and extrinsic structures of Lagrangian surfaces in complex space forms, *Tsukuba J. Math.*, **22**, 657–680.

Chen, B.-Y. and Houh, C.-S.
[1979] Totally real submanifolds of a quaternion projective space, *Ann. Mat. Pura Appl.*, **120**, 185–199.

Chen, B.-Y. and Ogiue, K.
[1974a] On totally real submanifolds, *Trans. Amer. Math. Soc.*, **193**, 257–266.
[1974b] Two theorems on Kaehler manifolds, *Michigan Math. J.*, **21**, 225–229.

Chen, B.-Y. and Vanhecke, L.
[1989] Isometric, holomorphic and symplectic reflections, *Geometria Dedicata*, **29**, 259–277.

Chern, S. S.
[1953] Pseudo-groupes continus infinis, colloques Internationaux du C. N. R. S., Strasbourg, 119–136.

Chern, S. S., Cowen, M. J. and Vitter III, A. L.
[1974] Frenet Frames Along Holomorphic Curves, *Value Distribution Theory*, part A, Dekker, New York, 191–203.

Chern, S. S., do Carmo, M. P. and Kobayashi, S.
[1970] Minimal submanifolds of a sphere with second fundamental form of constant length, *Functional Analysis and Related Fields*, Springer, Berlin, 59–75.

Chern, S. S. and Hamilton, R. S.
[1985] On Riemannian metrics adapted to three-dimensional contact manifolds, *Lecture Notes in Mathematics*, Vol. 1111, Springer, Berlin, 279–308.

Chevalley, C.
    [1946] *Theory of Lie Groups*, Princeton University Press, Princeton.

Chinea, D.
    [1985] Invariant submanifolds of a quasi-K-Sasakian manifold, *Riv. Mat. Univ. Parma*, **11**, 25–29.

Chinea, D., de Leon, M. and Marrero, J. C.
    [1993] Topology of cosymplectic manifolds, *J. Math. Pures Appl.*, **72**, 567–591.
    [1995] Coeffective cohomology on almost cosymplectic manifolds, *Bull. Sci. Math.*, **119**, 3–20.
    [1997] Spectral sequences on Sasakian and cosymplectic manifolds, *Houston J. Math.*, **23**, 631–649.

Cho, J.
    [1999] A new class of contact Riemannian manifolds, *Israel J. Math.*, **109**, 299–318.

Cordero, L.A., Fernández, M. and Gray, A.
    [1986] Symplectic manifolds with no Kähler structure, *Topology*, **25**, 375–380.

Cordero, L.A., Fernández, M. and de Leon, M.
    [1985] Examples of compact non-Kähler almost Kähler manifolds, *Proc. Amer. Math. Soc.*, **95**, 280–286.

Davidov, J. and Muskarov, O.
    [1990] Twistor spaces with Hermitian Ricci tensor, *Proc. Amer. Math. Soc.*, **109**, 1115–1120.

Deng, S.
    [1991] The second variation of the Dirichlet energy on contact manifolds, *Kōdai Math. J.*, **14**, 470–476.

Dillen, F. and Vrancken, L.
    [1989] C-totally real submanifolds of $S^7(1)$ with non-negative sectional curvature, *Math. J. Okayama Univ.*, **31**, 227–242.
    [1990] C-totally real submanifolds of Sasakian space forms, *J. Math. Pures Appl.* **69**, 85–93.

Dolbeault, S.
    [1977] Moving frames in Hermitian geometry, Proc. Symp. Pure Math. XXX, *Amer. Math. Soc.*, 103–106.

Dombrowski, P.
    [1962] On the geometry of the tangent bundle, *J. Reine und Angew. Math.* **210**, 73–88.

Douady, A.
[1982/83] Noeuds et structure de contact en dimension 3,
Astérisque **105-106**, 129–148.

Draghici, T.
[1994] On the almost Kähler manifolds with Hermitian Ricci tensor,
Houston J. Math. **20**, 293–298.
[1999] Almost Kähler 4-manifolds with J-invariant Ricci tensor, *Houston J. Math.*, **25**, 133–145.

Dragomir, S.
[1995] On pseudohermitian immersions between strictly pseudoconvex CR manifolds, *Amer. J. Math.* **117**, 169–202.

Ebin, D.
[1970] The manifold of Riemannian metrics, Proc. Symp. Pure Math. XV, *Amer. Math. Soc.*, 11–40.

Ejiri, N.
[1982] Totally real minimal immersions of n-dimensional real space forms into n-dimensional complex space forms, *Proc. Amer. Math. Soc.*, **84**, 243–246.

Eliashberg, Y.
[1989] Classification of overtwisted contact structures on 3-manifolds, *Invent. Math.* **98**, 623–637.

Elisahberg, Y. and Thurston, W.
[1998] Confoliations, University Lecture Series **13**, *Amer. Math. Soc.*

Endo, H.
[1985] Invariant submanifolds in a contact Riemannian manifold, *Tensor* (N.S.), **42**, 86–89.
[1986] Invariant submanifolds in contact metric manifolds, *Tensor* (N.S.), **43**, 83–86.

Fernández M. and Gray, A.
[1986] The Iwasawa manifold, *Lecture Notes in Mathematics*, Vol. 1209, Springer, Berlin, 157–159.

Fernández M., Ibánez, R. and de Leon, M.
[1997] A Nomizu's theorem for the coeffective cohomology, *Math. Z.*, **226**, 11–23.
[1998] Coeffective and de Rham cohomologies on almost contact manifolds, *Differential Geom. Appl.*, **8**, 285–303.

Foreman, B.
[1996] Variational Problems on Complex Contact Manifolds with

Applications to Twistor Space Theory, Thesis, Michigan State University.

[1999] Three-dimensional complex homogeneous complex contact manifolds, *Balkan, J. Geom. and Appl.*, **4**, 53–67.

[2000a] Boothby–Wang fibrations on complex contact manifolds, *Differential Geom. Appl.*, **13**, 179–196.

[2000b] Complex contact manifolds and hyperkähler geometry, *Kōdai Math. J.*, **23**, 12–26.

Fujitani, T.

[1966] Complex-valued differential forms on normal contact Riemannian manifolds, *Tôhoku Math. J.*, **18**, 349–361.

Galicki, K. and Salamon, S.

[1996] Betti numbers of 3-Sasakian manifolds, *Geometriae Dedicata*, **63**, 45–68.

Geiges, H.

[1997a] Constructions of contact manifolds, *Proc. Camb. Phil. Soc.*, **121**, 455–464.

[1997b] Normal contact structures on 3-manifolds, *Tôhoku Math. J.*, **49**, 415–422.

Geiges, H. and Gonzalo, J.

[1995] Contact geometry and complex surfaces, *Invent. Math.*, **121**, 147–209.

Ghosh, A., Koufogiorgos, Th. and Sharma, R.

[2001] Conformally flat contact metric manifolds, *J. Geom.*, **70**, 66–76.

Ghys E.

[1987] Flots d'Anosov dont les feuilletages stables sont différentiables, *Ann. Sc. École Norm. Sup.*, **20**, 251–270.

Ginzburg, V.

[1995] An embedding $S^{2n-1} \to \mathbb{R}^{2n}$, $2n - 1 \geq 7$, whose Hamiltonian flow has no periodic trajectories, *International Math. Res. Notices*, 83–98.

Goldberg, S. I.

[1962] *Curvature and Homology*, Academic Press, New York.

[1967] Rigidity of positively curved contact manifolds, *J. London Math. Soc.*, **42**, 257–263.

[1968a] Totally geodesic hypersurfaces of Kaehler manifolds, *Pacific J. Math.*, **27**, 275–281.

[1968b] On the topology of compact contact manifolds, *Tôhoku Math. J.*, **20**, 106–110.

[1969] Integrability of almost Kähler manifolds, *Proc. Amer. Math. Soc.*, **21**, 96–100.

[1986] Nonnegatively curved contact manifolds, *Proc. Amer. Math. Soc.*, **96**, 651–656.

Gompf, R.

[1994] Some new symplectic 4-manifolds, *Turkish J. Math.*, **18**, 7–15.

González-Dávila, J. C., González-Dávila, M. C. and Vanhecke, L.

[1995] Reflections and isometric flows, *Kyungpook Math. J.*, **35**, 113–144.

Gonzalo, J.

[1987] Branched covers and contact structures, *Proc. Amer. Math. Soc.*, **101**, 347–352.

Gouli-Andreou, F. and Xenos, Ph. J.

[1998] On 3-dimensional contact metric manifolds with $\nabla_\xi \tau = 0$, *J. Geom.*, **62**, 154–165.

[1999] Two classes of conformally flat contact metric 3-manifolds, *J. Geom.*, **64**, 80–88.

Gray, A. and Hervella, L. M.

[1980] The sixteen classes of almost Hermitian manifolds and their linear invariants, *Ann. Mat. pura appl.*, **123**, 35–58.

Gray, J.

[1959] Some global properties of contact structures, *Ann. of Math.*, **69**, 421–450.

Greenfield, S.

[1968] Cauchy-Riemann equations in several variable, *Ann. Sc. Norm. Sup. Pisa*, **22**, 275–314.

Gromov, M.

[1971] A topological technique for the construction of solutions of differential equations and inequalities, Internat. Congr. Math. (Nice 1970) **2**, Gauthier-Villars, Paris, 221–225.

[1985] Pseudo holomorphic curves in symplectic manifolds, *Invent. Math.*, **82**, 307–347.

Hadjar, A.

[1998] Sur les structures de contact du tore T^5, *L'Enseign Math.*, **44**, 91–93.

Haefliger, A.

[1971] Lectures on the theorem of Gromov, *Lecture Notes in Mathematics*, Vol. 209 Springer, Berlin, 128–141.

Harada, M.
[1973a] On Sasakian submanifolds, *Tôhoku Math. J.*, **25**, 103–109.
[1973b] On Sasakian submanifolds II, *Bull. Tokyo Gakugei Univ.*, **25**, 19–23.

Hasegawa, I. and Seino, M.
[1981] Some remarks on Sasakian geometry, *J. Hokkaido Univ. Education*, **32**, 1–7.

Hatakeyama, Y.
[1962] On the existence of Riemann metrics associated with a 2-form of rank $2r$, *Tôhoku Math. J.*, **14**, 162–166.
[1963] Some notes on differentiable manifolds with almost contact structures, *Tôhoku Math. J.*, **15**, 176–181.
[1966] Some notes on the group of automorphisms of contact and symplectic structures, *Tôhoku Math. J.*, **18**, 338–347.

Hatakeyama, Y., Ogawa, Y. and Tanno, S.
[1963] Some properties of manifolds with contact metric structures, *Tôhoku Math. J.*, **15**, 42–48.

Hilbert, D.
[1915] Die grundlagen der Physik, *Nachr. Ges. Wiss. Göttingen*, 395–407.

Hitchin, N. J.
[1987] The self-duality equations on a Riemann surface, *Proc. London Math. Soc.*, **55**, 59–126.

Hofer, H.
[1993] Pseudoholomorphic curves in symplectizations with applications to the Weinstein conjecture in dimension three, *Invent. Math.*, **114**, 515–563.

Hofer, H. and Viterbo, C.
[1988] The Weinstein conjecture in cotangent bundles and related results, *Ann. Sc. Norm. Sup. Pisa*, **15**, 411–445.

Hofer, H. and Zehnder, E.
[1987] Periodic solutions on hypersurfaces and a result by C. Viterbo, *Invent. Math.* **90**, 1–9.

Holubowicz, R. and Mozgawa, W.
[1998] On compact non-Kählerian manifolds admitting an almost Kähler structure, *Rend. Cir. Mat. Palermo*, Serie II, Suppl., **54**, 53–57.

Houh, C.-S.
  [1973] Some totally real minimal surfaces in $\mathbb{C}P^2$, *Proc. Amer. Math. Soc.*, **40**, 240–244.
  [1976] On the holonomy group of a normal complex almost contact manifold, *Kōdai Math. Sem. Rep.*, **28**, 72–77.

Hsu, C. J.
  [1960] On some structures which are similar to the quaterion structure, *Tôhoku Math. J.*, **12**, 403–428.

Huygens, C.
  [1690] *Traité de la Lumiere*, Vander Aa, Leiden.

Ianus, S.
  [1972] Sulle varietà di Cauchy-Rieman, *Rend. dell'Accademia di Scienze Fisiche e Matematiche*, Napoli, XXXIX, 191–195.

Ishihara, S.
  [1973] Quaternion Kähler manifolds and fibred Riemannian spaces with Sasakian 3-structure, *Kōdai Math. Sem. Rep.*, **25**, 321–329.
  [1974] Quaternion Kählerian manifolds, *J. Diff. Geom.*, **9**, 483–500.

Ishihara, S and Konishi M.
  [1979] Real contact 3-structure and complex contact structure, *Southeast Asian Bulletin of Math.*, **3**, 151–161.
  [1980] Complex almost contact manifolds, *Kōdai Math. J.*, **3**, 385–396.
  [1982] Complex almost contact structures in a complex contact manifold, *Kōdai Math. J.*, **5**, 30–37.

Itoh, M.
  [1997] Odd dimensional tori and contact structure, *Proc. Japan Acad.*, **72**, 58–59.

Janssens, D. and Vanhecke, L.
  [1981] Almost contact structures and curvature tensors, *Kōdai Math. J.*, **4**, 1–27.

Jayne, N.
  [1992] Legendre Foliations on Contact Metric Manifolds, Thesis, Massey University.

Jelonek, W.
  [1996] Some simple examples of almost Kähler non-Kähler structures, *Math. Ann.*, **305**, 639–649.

Jerison, D. and Lee, J. M.
  [1984] A subelliptic, nonlinear eigenvalue problem and scalar curvature on CR manifolds, *Contemp. Math.*, **27**, 57–63.

240   Bibliography

Jiménez, J. A. and Kowalski, O.
  [1993] The classification of ϕ-symmetric Sasakian manifolds, *Mh. Math.*, **115**, 83–98.

Kashiwada, T.
  [1971] A note on a Riemannian space with Sasakian 3-structure, *Natural Sci. Rep. Ochanomizu Univ.*, **22**, 1–2.
  [1998] A note on Hitchin's lemma, *Tensor* (N.S.) **60**, 323–326.
  [toap] On a contact 3-structure, to appear.

Kenmotsu, K.
  [1969] Invariant submanifolds in a Sasakian manifold, *Tôhoku Math. J.*, **21**, 495–500.
  [1972] A class of almost contact Riemannian manifolds, *Tôhoku Math. J.*, **24**, 93–103.

Kobayashi, S.
  [1956] Principal fibre bundles with the 1-dimensional toroidal group, *Tôhoku Math. J.*, **8**, 29–45.
  [1959] Remarks on complex contact manifolds, *Proc. Amer. Math. Soc.*, **10**, 164–167.
  [1963] Topology of positively pinched Kaehler manifolds, *Tôhoku Math. J.* **15**, 121–139.

Kobayashi, S. and Nomizu, K.
  [1963-69] *Foundations of Differential Geometry*, I, II, Wiley-Interscience, New York.

Kobayashi, S. and Wu H.
  [1983] *Complex Differential Geometry*, Birkhäuser, Basel.

Kodaira, K.
  [1964] On the structure of compact complex analytic surfaces, I, *Amer. J. Math.*, **86**, 751–798.

Kon, M.
  [1976] Invariant submanifolds in Sasakian manifolds, *Math. Ann.*, **219**, 277–290.

Konishi, M.
  [1973] Note on regular K-contact 3-structures, *Natural Sci. Rep. Ochanomizu Univ.*, **22**, 1–5.
  [1975] On manifolds with Sasakian 3-structure over quaternion Kaehler manifolds, *Kōdai Math. Sem. Rep.*, **26**, 194–200.

Konishi, M. and Funabashi, S.
  [1976] On Riemannian manifolds with Sasakian 3-structure of constant

horizontal sectional curvature, *Kōdai Math. Sem. Rep.*, **27**, 362–366.

Korkmaz, B.
[1998] A curvature property of complex contact metric structures, *Kyung-pook Math. J.* **38**, 473–488.
[2000] Normality of complex contact manifolds, *Rocky Mountain J. Math.*, **30**, 1343–1380.
[toap] A nullity condition for complex contact metric manifolds, to appear.

Koufogiorgos, Th.
[1997a] Contact Riemannian manifolds with constant ϕ-sectional curvature, *Tokyo J. Math.*, **20**, 13–22.
[1997b] Contact strongly pseudo-convex integrable CR metrics as critical points, *J. Geom.*, **59**, 94–102.

Koufogiorgos, Th. and Tsichlias, C.
[2000] On the existence of a new class of contact metric manifolds, *Canad. Math. Bull.*, **43**, 440–447.
[toap] On a class of generalized (κ, μ)-contact metric manifolds, to appear.

Kowalski, O.
[1971] Curvature of the induced Riemannian metric on the tangent bundle of a Riemannian manifold, *J. reine und angew. Math.*, **250**, 124–129.

Kowalski, O. and Vanhecke, L.
[1984] A generalization of a theorem on naturally reductive homogeneous spaces, *Proc. Amer. Math. Soc.*, **91**, 433–435.

Kowalski, O. and Wegrzynowski, S.
[1987] A classification of five-dimensional ϕ-symmetric spaces, *Tensor* (N.S.) **46**, 379–386.

Kriegl, A. and Michor P. W.
[1997] *The Convenient Setting of Global Analysis*, Math. Surveys and Monog. **53**, Amer. Math. Soc.

Kuo, Y.-Y.
[1970] On almost contact 3-structure, *Tôhoku Math. J.*, **22**, 325–332.

Lawson, H. B.
[1970] The Riemannian geometry of holomorphic curves, Proc. Carolina Conf. on Holomorphic Mappings and Minimal Surfaces, 86–107.

LeBrun, C.

[1991] On complete quaternionic-Kähler manifolds, *Duke Math. J.*, **63**, 723–743.

[1995] Fano manifolds, contact structures and quaternionic geometry, *International J. Math.*, **6**, 419–437.

Libermann, P.

[1959] Sur les automorphismes infinitésimaux des structures symplectiques et de structures de contact, Coll. Géom. Diff. Globale (Bruxelles 1958), Gauthier-Villars, Paris, 37–59.

Libermann, P. and Marle, C.-M.

[1987] *Symplectic Geometry and Analytical Mechanics*, D. Reidel, Dordrecht.

Lie, S.

[1890] *Theorie der Transformationgruppen*, Vol. 2, Leipzig, Teubner.

Ludden, G. D., Okumura, M. and Yano, K.

[1975] A totally real surface in $\mathbb{C}P^2$ that is not totally geodesic, *Proc. Amer. Math. Soc.*, **53**, 186–190.

Lutz, R.

[1979] Sur la géométrie des structures de contact invariantes, *Ann. Inst. Fourier*, **29**, 283–306.

Lutz, R. and Meckert, C.

[1976] Structures de contact sur certaines sphères exotiques, *C. R. Acad. Sc. Paris*, **282**, 591–593.

Lychagin, V.

[1977] On sufficient orbits of a group of contact diffeomorphisms, *Math. USSR Sb.*, **33**, 223–242.

MacLane, S.

[1968] Geometrical Mechanics II, Lecture notes, University of Chicago.

Marrero, J. C.

[1992] The local structure of trans-Sasakian manifolds, *Ann. Mat. Pura Appl.*, **162**, 77–86.

Marrero, J. C. and Padron, E.

[1998] New examples of compact cosymplectic solvmanifolds, *Arch. Math.*, **34**, 337–345.

Martinet, J.

[1970] Sur les singularitiés des formes différentielles, *Ann. Inst. Fourier*, **20**, 95–178.

[1971] Formes de contact sur les variétiés de dimension 3, *Lecture Notes in Mathematics*, Vol. 209, Springer, Berlin, 142–163.

McCarthy, J. D. and Wolfson, J. G.
[1994] Symplectic normal connect sum, *Topology*, **33**, 729–764.

McDuff, D.
[1984] Examples of simply-connected symplectic non-Kählerian manifolds, *J. Diff. Geom.*, **20**, 267–277.

Milnor, J.
[1976] Curvature of left invariant metrics on Lie groups, *Advances in Math.*, **21**, 293–329.

Mitric, G.
[1991] C. R. -structures on the unit sphere bundle in the tangent bundle of a Riemannian manifold, *Semin. Mec. Univ. Timis. Fac. Mat.*, **32**.

Mitsumatsu, Y.
[1995] Anosov flows and non-Stein symplectic manifolds, *Ann. Inst. Fourier*, **45**, 1407–1421.

Monna, G.
[1983] ƅ-platitude des structures de contact, *Mh. Math.*, **95**, 221–227.

Morimoto, A.
[1963] On normal almost contact structures, *J. Math. Soc. Japan*, **15**, 420–436.
[1964] On normal almost contact structure with a regularity, *Tôhoku Math. J.*, **16**, 90–104.

Morimoto, S.
[1992] Almost complex foliations and its application to contact geometry, *Natural Sci. Rep. Ochanomizu Univ.*, **43**, 11–23.

Moroianu, A. and Semmelmann U.
[1994] Kählerian Killing spinors, complex contact structures and twistor spaces, Quaternionic Structures in Mathematics and Physics, SISSA, Trieste, 225–231.

Morrow, J. and Kodaira, K.
[1971] *Complex Manifolds*, Holt, Rinehart and Winston, New York.

Morvan, J.-M.
[1983] On Lagrangian immersions, *Res. Notes in Math.*, Vol. 80, Pitman, Boston, 167–171.

Moser, J.

[1965] On the volume element on a manifold, *Trans. Amer. Math. Soc.*, **120**, 286–294.

Moskal, E.

[1966] Contact Manifolds of Positive Curvature, Thesis, University of Illinois.

[1977] On the tridegree of forms on f-manifolds, *Kōdai Math. Sem. Rep.*, **28**, 115–128.

Musso, E. and Tricerri, F.

[1988] Riemannian metrics on tangent bundles, *Ann. Mat. Pura Appl.*, **150**, 1–20.

Muto, Y.

[1974a] On Einstein metrics, *J. Diff. Geom.*, **9**, 521–530.

[1974b] Curvature and critical Riemannian metric, *J. Math. Soc. Japan*, **26**, 686–697.

[1975] Critical Riemannian metrics, *Tensor* (N. S.), **29**, 125–133.

Myers, S. B.

[1941] Riemannian manifolds with positive mean curvature, *Duke Math. J.*, **8**, 401–404.

Nagano, T.

[1967] A problem on the existence of an Einstein metric, *J. Math. Soc. Japan*, **19**, 30–31.

Nakagawa, H. and Ogiue, K.

[1976] Complex space forms immersed in complex space forms, *Trans. Amer. Math. Soc.*, **219**, 289–297.

Newlander, A. and Nirenberg, L.

[1957] Complex analytic coordinates in almost complex manifolds, *Ann. of Math.*, **65**, 391–404.

Nurowski, P. and Przanowski, M.

[1999] A four dimensional example of Ricci flat metric admitting almost-Kähler non-Kähler structure, *Class. Quantum Grav.*, **16**, L9–L13.

Obata, M.

[1965] Riemannian manifolds admitting a solution of a certain system of differential equations, Proc. U.S.-Japan Seminar in Differential Geometry, 101–114.

Ogiue, K.

[1964] On almost contact manifolds admitting axiom of planes or axiom of free mobility, *Kōdai Math. Sem. Rep.*, **16**, 115–128.

[1965] On fiberings of almost contact manifolds, *Kōdai Math. Sem. Rep.*, **17**, 53–62.

[1972] Positively curved complex submanifolds immersed in a complex projective space, *J. Diff. Geom.*, **7**, 603–606.

[1974] Differential geometry of Kaehler submanifolds, *Advances in Math.*, **13**, 73–114.

[1976a] Positively curved complex submanifolds immersed in a complex projective space III, *J. Diff. Geom.*, **11**, 613–615.

[1976b] Positively curved totally real minimal submanifolds immersed in a complex projective space, *Proc. Amer. Math. Soc.*, **56**, 264–266.

Okumura, M.

[1962a] Some remarks on space with a certain contact structure, *Tôhoku Math. J.*, **14**, 135–145.

[1962b] On infinitesimal conformal and projective transformations of normal contact spaces, *Tôhoku Math. J.*, **14**, 398–412.

[1964a] Certain almost contact hypersurfaces in Euclidean spaces, *Kōdai Math. Sem. Rep.*, **16**, 44–54.

[1964b] Certain almost contact hypersurfaces in Kaehlerian manifolds of constant holomorphic sectional curvatures, *Tôhoku Math. J.*, **16**, 270–284.

[1965] Cosymplectic hypersurfaces in Kaehlerian manifold of constant holomorphic sectional curvature, *Kōdai Math. Sem. Rep.*, **17**, 63–73.

[1966] Contact hypersurfaces in certain Kaehlerian manifolds, *Tôhoku Math. J.*, **18**, 74–102.

Olszak, Z.

[1979] On contact metric manifolds, *Tôhoku Math. J.*, **31**, 247–253.

O'Neill, B.

[1966] The fundamental equations of a submersion, *Michigan Math. J.*, **13**, 459–469.

Oubina, J. A.

[1985] New classes of almost contact metric structures, *Publicationes Mathematicae Debrecen*, **32**, 187–193.

Palais, R. S.

[1957] A Global Formulation of the Lie Theory of Transformation Groups, *Mem. Amer. Math. Soc.* #22.

Park, J.-S. and Oh, W. T.

[1996] The Abbena-Thurston manifold as a critical point, *Canad. Math. Bull.*, **39**, 352–359.

Pastore, A. M.
  [1998] Classification of locally symmetric contact metric manifolds, *Balkan J. Geom. and App.*, **3**, 89–96.

Perrone, D.
  [1989] 5-dimensional contact manifolds with second Betti number $b_2 = 0$, *Tôhoku Math. J.*, **41**, 163–170.
  [1990] Torsion and critical metrics on contact three-manifolds, *Kōdai Math. J.*, **13**, 88–100.
  [1992a] Contact Riemannian manifolds satisfying $R(X, \xi) \cdot R = 0$, *Yokohama Math. J.*, **39**, 141–149.
  [1992b] Torsion tensor and critical metrics on contact (2n+1)-manifolds, *Mh. Math.*, **114**, 245–259.
  [1994] Tangent sphere bundles satisfying $\nabla_\xi \tau = 0$, *J. Geom.*, **49**, 178–188.
  [1998] Homogeneous contact Riemannian three-manifolds, *Ill. J. Math.*, **42**, 243–256.
  [2000] Special directions on contact metric three-manifolds, *J. Geom.*, **69**, 180–191.

Perrone, D. and Vanhecke, L.
  [1991] Five-dimensional homogeneous contact and related problems, *Tôhoku Math. J.*, **43**, 243–248.

Petkov, V. and Popov, G.
  [1995] On the Lebesgue measure of the periodic points of a contact manifold, *Math. Z.*, **218**, 91–102.

Priest, E. R.
  [1982] *Solar Magnetohydrodynamics*, D. Reidel, Dordrecht.

Rabinowitz, P. H.
  [1978] Periodic solutions of Hamiltonian systems, *Comm. Pure Appl. Math.*, **XXXI**, 157–184.

Reckziegel, H.
  [1979] On the eigenvalues of the shape operator of an isometric immersion into a space of constant curvature, *Math. Ann.*, **243**, 71–82.

Reeb, G.
  [1952] Sur certaines propriétés topologiques des trajectoires des systèmes dynamiques, Mémoires de l'Acad. Roy. de Beligique, Sci. Ser. 2, **27**, 1–62.

Ros, A.
  [1985a] Positively curved Kaehler submanifolds, *Proc. Amer. Math. Soc.*, **93**, 329–331.

[1985b] A characterization of seven compact Kaehler submanifolds by holomorphic pinching, *Ann. of Math.*, **121**, 377–382.

Ros, A. and Verstraelen, L.
[1984] On a conjecture of K. Ogiue, *J. Diff. Geom.*, **19**, 561–566.

Ros, A. and Urbano, F.
[1998] Lagrangian submanifolds of $\mathbb{C}^n$ with conformal Maslov form and the Whitney sphere, *J. Math. Soc. Japan*, **50**, 203–226.

Rukimbira, P.
[1993] Some remarks on R-contact flows, *Ann. Global Anal. Geom.*, **11**, 165–171.
[1994] The dimension of leaf closures of K-contact flows, *Ann. Global Anal. Geom.*, **12**, 103–108.
[1995a] Topology and closed characteristics of K-contact manifolds, *Bull. Belg. Math. Soc.*, 349–356.
[1995b] Vertical sectional curvature and K-contactness, *J. Geom.*, **53**, 163–166.
[1995c] Chern-Hamilton's conjecture and K-contactness, *Houston J. Math.*, **21**, 709–718.
[1998] A characterization of flat contact metric geometry, *Houston J. Math.*, **24**, 409–414.
[1999] On K-contact manifolds with minimal number of closed characteristics, *Proc. Amer. Math. Soc.*, **127**, 3345–3351.

Salamon, S.
[1982] Quaternionic Kähler manifolds, *Invent. Math.*, **67**, 143–171.

Salingaros, N. A.
[1986], On solutions of the equation $\nabla \times \mathbf{a} = k\mathbf{a}$, *J. Phys. A: Math. Gen.*, **19**, L101-L104.
[1990] Lorentz force and magnetic stress in force-free configurations, *Appl. Phys. Lett.*, **56**, 617–619.

Sasaki, S.
[1958] On the differential geometry of tangent bundles of Riemannian manifolds, *Tôhoku Math. J.*, **10**, 338–354.
[1962] On the differential geometry of tangent bundles of Riemannian manifolds II, *Tôhoku Math. J.*, **14**, 146–155.
[1964] A characterization of contact transformations, *Tôhoku Math. J.*, **16**, 285–290.
[1965] Almost Contact Manifolds, Lecture Notes, Mathematical Institute, Tôhoku University, Vol. 1.
[1972] Spherical space forms with normal contact metric 3-structure, *J.*

Diff. Geom., **6**, 307–315.

[1985] Contact structures on Brieskorn manifolds (lecture Japan Mathematical Society 1975), *Shigeo Sasaki Selected Papers*, Kinokuniya, Tokyo, 349–363.

Sasaki, S. and Hatakeyama, Y.

[1961] On differentiable manifolds with certain structures which are closely related to almost contact structures II, *Tôhoku Math. J.*, **13**, 281–294.

[1962] On differentiable manifolds with contact metric structures, *J. Math. Soc. Japan*, **14**, 249–271.

Sasaki, S. and Hsu, C. J.

[1976] On a property of Brieskorn manifolds, *Tôhoku Math. J.*, **28**, 67–78.

Sasaki, S. and Takahashi, T.

[1976] Almost contact structures on Brieskorn manifolds, *Tôhoku Math. J.*, **28**, 619–624.

Sato, H.

[1977] Remarks concerning contact manifolds, *Tôhoku Math. J.*, **29**, 577–584.

Sekigawa, K.

[1987] On some compact Einstein almost Kähler manifolds, *J. Math. Soc. Japan*, **39**, 677–684.

Sharma, R. and Koufogiorgos, T.

[1991] Locally symmetric and Ricci-symmetric contact metric manifolds, *Ann. Global Anal. Geom.*, **9**, 177–182.

Shibuya, Y.

[1978] On the existence of a complex almost contact structure, *Kōdai Math. J.*, **1**, 197–204.

[1982] Some isospectral problems, *Kōdai Math. J.*, **5**, 1–12.

Sikorav, J.-C.

[1986] Non-existence de sous-variété lagrangienne exacte dans $\mathbb{C}^n$ (d'après Gromov), Aspects Dynamiques et Topologiques des Groupes Infinis de Transformation de la Méchanique, Travaux en Cours **25**, Hermann, Paris, 95–110.

Simons, J.

[1968] Minimal varieties in Riemannian manifolds, *Ann. of Math.*, **88**, 62–105.

Smolentsev, N. K.

[1995] Critical associated metrics on a symplectic manifold, *Sibirsk. Mat.*

Zh., **36**, 409–418; English translation: *Siberian Math. J.*, **36**, 359–367.

Steenrod, N.

[1951] *Topology of fibre Bundles*, Princeton University Press, Princeton.

Stong, R. E.

[1974] Contact manifolds, *J. Diff. Geom.*, **9** (1974), 219–238.

Tachibana, S.

[1965] On harmonic tensors in compact Sasakian spaces, *Tôhoku Math. J.*, **17**, 271–284.

Tachibana, S. and Ogawa, Y.

[1966] On the second Betti number of a compact Sasakian space, *Natural Sci. Rep. Ochanomizu Univ.*, **17**, 27–32.

Tachibana, S. and Okumura, M.

[1962] On the almost complex structure of tangent bundles of Riemannian spaces, *Tôhoku Math. J.*, **14**, 156–161.

Tachibana, S. and Yu, W. N.

[1970] On a Riemannian space admitting more than one Sasakian structure, *Tôhoku Math. J.*, **22**, 536–540.

Takahashi, T.

[1977] Sasakian ϕ-symmetric spaces, *Tôhoku Math. J.*, **29**, 91–113.

Tanaka N.

[1976] On non-degenerate real hypersurfaces, graded Lie algebras and Cartan connections, *Japan J. Math.*, **2**, 131–190.

Tanno S.

[1963] Some transformations on manifolds with almost contact and contact metric structures II, *Tôhoku Math. J.*, **15**, 322–331.

[1965] A theorem on regular vector fields and its applications to almost contact structures, *Tôhoku Math. J.*, **17**, 235–238.

[1967a] Harmonic forms and Betti numbers of certain contact Riemannian manifolds, *J. Math. Soc. Japan*, **19**, 308–316.

[1967b] Locally symmetric K-contact Riemannian manifolds, *Proc. Japan Acad.*, **43**, 581–583.

[1968] The topology of contact Riemannian manifolds, *Ill. J. Math.*, **12**, 700–717.

[1969] Sasakian manifolds with constant φ-holomorphic sectional curvature, *Tôhoku Math. J.*, **21**, 501–507.

[1970] On the isometry group of Sasakian manifolds, *J. Math. Soc. Japan*, **22**, 579–590.

[1971] Killing vectors on contact Riemannian manifolds and fiberings related to the Hopf fibrations, *Tôhoku Math. J.*, **23**, 313–333.

[1978a] Some differential equations on Riemannian manifolds, *J. Math. Soc. Japan*, **30**, 509–531.

[1978b] Geodesic flows on C_L-manifolds and Einstein metrics on $S^3 \times S^2$, *Minimal Submanifolds and Geodesics*, Kaigai, Tokyo, 283–292.

[1988] Ricci curvatures of contact Riemannian manifolds, *Tôhoku Math. J.*, **40**, 441–448.

[1989] Variational problems on contact Riemannian manifolds, *Trans. Amer. Math. Soc.*, **314**, 349–379.

[1991] Pseudo-conformal invariants of type (1,3) of CR manifolds, *Hokkaido Math. J.*, **20**, 195–204.

[1992] The standard CR structure on the unit tangent bundle, *Tôhoku Math. J.*, **44**, 535–543.

[1996] Remarks on a triple of K-contact structures, *Tôhoku Math. J.*, **48**, 519–531.

Tashiro, Y.

[1963] On contact structures of hypersurfaces in complex manifolds I, *Tôhoku Math. J.*, **15**, 62–78.

[1969] On contact structures of tangent sphere bundles, *Tôhoku Math. J.*, **21**, 117–143.

Thomas, C. B.

[1976] Almost regular contact structures, *J. Diff. Geom.*, **11**, 521–533.

Thurston, W.

[1976] Some simple examples of symplectic manifolds, *Proc. Amer. Math. Soc.*, **55**, 467–468.

Tischler, D.

[1970] On fibering certain foliated manifolds over S^1, *Topology*, **9**, 153–154.

Udriste, C.

[1969] Structures presque coquaternioniennes, *Bull. Math. Soc. Sci. Math. R. S. Roumanie*, **13**, 487–507.

Urbano, F.

[1985] Totally real minimal submanifolds of a complex projective space, *Proc. Amer. Math. Soc.*, **93**, 332–334.

Vaisman, I.

[1978] On the Sasaki-Hsu contact structure of the Brieskorn manifolds, *Tôhoku Math. J.*, **30**, 553–560.

Vanhecke, L.
  [1988] Geometry in normal and tubular neighborhoods, *Rend. Sem. Fac. Sci. Univ. Cagliari*, LVIII (Supplement), 73–176.

Van Lindt D., Verheyen, P. and Verstraelen, L.
  [1986] Minimal submanifolds in Sasakian space forms, *J. Geom.*, **27**, 180–187.

Vernon, M.
  [1987] Contact hypersurfaces of a complex hyperbolic space, *Tôhoku Math. J.*, **39**, 215–222.

Verstraelen, L. and Vrancken, L.
  [1988] Pinching theorems for C-totally real submanifolds of Sasakian space forms, *J. Geom.*, **33**, 172–184.

Viterbo, C.
  [1987] A proof of Weinstein's conjecture in $\mathbf{R}^{2n}$, *Ann. Inst. Henri Poincaré*, 4, 337–356.

Walters, P.
  [1975] Ergodic Theory-Introductory Lectures, *Lecture Notes in Mathematics*, Vol. 458, Springer, Berlin.

Watanabe, Y.
  [1980] Geodesic symmetrices in Sasakian locally ϕ-symmetric spaces, *Kōdai Math. J.*, **3**, 48–55.

Watanabe, Y. and Fujita, H.
  [1988] A family of homogeneous Sasakian structures on $S^2 \times S^3$, *C. R. Math. Rep. Acad. Sci. Canada*, **X**, 57–61.

Watson, B.
  [1983] New examples of strictly almost Kähler manifolds, *Proc. Amer. Math. Soc.*, **88**, 541–544.

Weinstein, A.
  [1971] Symplectic manifolds and their Lagrangian submanifolds, *Advances in Math.*, **6**, 329–346.
  [1977] Lectures on Symplectic Manifolds, CBMS regional conference series in Mathematics, **29**, Amer. Math. Soc., Providence.
  [1978] Periodic orbits for convex hamiltonian systems, *Ann. of Math.*, **108**, 507–518.
  [1979] On the hypothesis of Rabinowitz's periodic orbit theorems. *J. Diff. Eq.*, **33**, 353–358.

Weyl H.
  [1939] *The Classical Groups*, Princeton Univ. Press, Princeton.

Wolf, J. A.
   [1965] Complex homogeneous contact manifolds and quaternionic sym-
       metric spaces, *J. Math. and Mech.*, **14**, 1033–1047.
   [1968] A contact structure for odd dimensional spherical space forms,
       Proc. Amer. Math. Soc., **19**, 196.

Wood, C. M.
   [1995] Harmonic almost-complex structures, *Compositio Math.*, **99**,
       183–212.

Woodhouse, N.
   [1980] *Geometric Quantization*, Oxford Mathematical Monographs, Claren-
       don Press, Oxford.

Yamaguchi, S. and Chūman, G.
   [1983] Critical Riemannian metrics on Sasakian manifolds, *Kōdai Math.
       J.*, **6**, 1–13.

Yamaguchi, S., Kon, M. and Ikawa, T.
   [1976] C-totally real submanifolds, *J. Diff. Geom.*, **11**, 59–64.

Yamaguchi, S., Kon, M. and Miyahara, Y.
   [1976] A theorem on C-totally real minimal surface, *Proc. Amer. Math.
       Soc.*, **54**, 276–280.

Yampol'skii, A. L.
   [1985] The curvature of the Sasaki metric of tangent sphere bundles, *Ukr.
       Geom. Sb.*, **28**, 132–145; English translation: *J. Sov. Math.*, **48**
       (1990), 108–117.

Yano, K.
   [1965] *Differential Geometry on Complex and Almost Complex
       Spaces*, Pergamon, New York.

Yano, K., Ishihara, S. and Konishi, M.
   [1973] Normality of almost contact 3-structure, *Tôhoku Math. J.*, **25**,
       167–175.

Yano, K. and Kon, M.
   [1983] *CR-submanifolds of Kaelerian and Sasakian Manifolds*, Birkhäuser,
       Boston.

Yau, S.-T.
   [1974] Submanifolds with constant mean curvature I., *Amer. J. Math.*,
       96, 346–366.

Ye, Y.-G.
   [1994] A note on complex projective threefolds admitting holomorphic
       contact structures, *Invent. Math.*, **115**, 311–314.

Zeghib, A.
   [1995] Subsystems of Anosov systems, *Amer. J. Math.*, **117**, 1431–1448.

Subject Index

Author Index